CW00459550

* Atlas 1265.3.1911
Accessions
Shelf No.
SIGILLUM CIVITATIS BOSTONIAE
BOSTONIA CONDITA A.D. 1630
JONATHAN PHILLIPS, DIED, 1860.
BOSTON PUBLIC LIBRARY
FROM THE
Phillips Fund.
Added

STANFORD'S

INDEXED ATLAS OF

THE

COUNTY OF LONDON

WITH PARTS OF

THE ADJACENT BOROUGHS

AND

URBAN DISTRICTS

WITH A PREFACE BY

SIR LAURENCE GOMME

LONDON

EDWARD STANFORD, CARTOGRAPHER TO HIS MAJESTY

12, 13, AND 14, LONG ACRE, W.C.

1911

PREFACE

LONDON has developed from a city of one square mile in area to a county of one hundred and twenty square miles with numerous populous centres all round beyond this huge area. It is the largest city in the world, with a population of close upon five millions, or, reckoning the outlying suburban fringes, of nearly seven millions. Its network of streets, as indicated by this map, extends to a total length of two thousand one hundred and fifty-one miles and includes some of the most beautiful as also some of the most miserable streets that a great capital city can possess. There are practically no straight streets in London, unless Regent Street be so considered, but the irregularities constitute a great charm which wins upon the lovers of London. And even such monotony as exists in our irregular streets is relieved first of all by a series of admirably placed parks and open spaces (coloured green on the map) and secondly, by the almost unique feature of "squares" (the principal ones also coloured green). London squares came into vogue about the end of the eighteenth century. Houses are built round a common garden of a more or less square shape, sometimes with their backs to the common garden, more generally with frontages to a roadway which skirts the common garden in the centre.

The river Thames flows through the centre of London and is spanned by a series of bridges. It is the origin and cause of London's greatness, and helps to make it a beautiful city, a city from which Turner has drawn inspiration for his pictures and Wordsworth for his poetry.

The name of London is Celtic in origin, and was originally applied to a Celtic stronghold constructed on the site upon part of which St. Paul's Cathedral now stands. The Roman city was the foundation of the city and the city was, until the middle of the eighteenth century, enclosed within the mediæval walls constructed on the site planned by the Romans.

The name of London is now properly and legally applied to two well defined areas, the original city and the county constituted by the Local Government Act, 1888. The city is the centre of the business world of London. It contains the Bank of England, the head offices of the great banking and insurance companies, and the offices and warehouses of the principal merchants. The city churches, nearly one hundred in number, are a conspicuous feature, while the great cathedral of St. Paul's, Wren's masterpiece, dominates the city as the principal feature from whatever point the city is seen. The county completely surrounds the city, extending from Bow on the east to Hammersmith on the west, from Hampstead on the north to Tooting on the south, and taking in Hackney on the north-east, Woolwich on the south-east, and extending to Putney on the south-west. It contains the Houses of Parliament and the whole of the Government offices, the Mint, Trinity House, the Law Courts, the Inns of Court, the Tower of London, the cathedral church of Southwark, the Abbey Church of Westminster, the Roman Catholic cathedral church, the arsenal of Woolwich, the King's Palace, the town residence of the Prince of Wales, the town houses of most of the nobility and aristocracy, and the palaces of the Archbishop of Canterbury and the Bishop of London. It is the seat of two bishoprics, namely London and Southwark. It contains Eltham Palace, one of the favourite residences of the Kings of England, particularly of Henry VIII. and including Charles I.; Greenwich Palace, which Henry VIII., Queen Mary and Queen Elizabeth occupied; Kensington Palace, where William III., Mary II., Anne, and George II. died and Queen Victoria was born; and many beautiful buildings of historic importance. All the principal railway terminus stations are in the county. It has also very large manufacturing and industrial centres which make London the largest manufacturing city in the world. And it is the centre of all institutions for amusements, the Albert Hall, and all the theatres, music halls, hotels. The University of London and its schools and colleges are also situated in the county, as are the British Museum, the Natural History Museum, South Kensington Museum, the National Gallery, the National Portrait Gallery, the Tate Gallery, the Wallace Collection, and other galleries and museums. It is the seat also of all the principal learned societies, located chiefly at Burlington House, and of Greenwich Observatory upon which practically the time of the world is based.

The county is administered for municipal purposes by the County Council, the City of London Corporation and twenty-eight metropolitan borough councils. For Poor Law purposes it is administered by the Metropolitan Asylums Board and thirty-one Boards of Guardians. And it will be gathered that the jurisdiction of these local bodies, covering though it does the whole County of London, is not conterminous in its local organization. Indeed the complexity and confusion of the London administrative areas are considerable drawbacks to its efficiency and economy. The Guildhall of the City of London is the principal municipal building at present in London; the new County Hall now being erected on the banks of the Thames at the eastern end of Westminster Bridge will not be completed until about 1916.

The term "metropolis" was formerly applied to what is now the county as well as to other administrative areas. This term was first used in connection with the commissions of sewers which were instituted in the reign of Henry VIII., the several commissions being finally amalgamated into the Metropolitan Commission in 1847. It now only applies to those areas which, sometimes including both the city and county of London, sometimes excluding the city and including the county only in a larger area, are subject to the special jurisdiction of various authorities. These jurisdictions are of a most conflicting and puzzling character, and have arisen during the long period before the year 1888 when London was growing to its greatness without being endowed with any municipal government. These jurisdictions may be shortly described.

The area of jurisdiction of the Central Criminal Court includes the City of London, the counties of London and Middlesex, together with parts of the counties of Essex, Surrey and Kent, and the high seas.

The Metropolitan Police area extends to parishes coming within a radius of fifteen miles from Charing Cross, and thus extends from Hatfield on the north to Epsom on the south, from Staines on the west to Dartford on the east.

The Metropolitan Police Court District extends over the county of London with the exception of a part of Hampstead.

The Metropolitan Water Board administers an area different from all of these, being composed of the areas formerly within the supplies of each of the eight water companies, not so extensive on the whole as the Metropolitan Police area though extending beyond that area in various directions, particularly into Ware, Romford, and Sevenoaks.

The London Parliamentary area is practically the County of London except that certain adjustments have been made in the county boundary which have not been carried out in the Parliamentary boundary.

The Port of London Authority has the administration of all the Docks and of the river from about two miles below the Nore to Teddington.

Besides these there are some special jurisdictions in London which should be mentioned. The Thames Conservancy has charge of the river above Teddington and therefore of some of the sources of London's water supply. The Main Drainage system of the County Council includes not only the whole county, but West Ham, a small part of East Ham, Tottenham, Wood Green, parts of Hornsey and Willesden, Acton, and small portions of Chiswick and Ealing outside the county on the north side of the Thames, and Penge, with parts of Croydon and Beckenham and a small portion of Mitcham on the south side. The Trinity House has charge of the port of London for purposes of lighthouses and other lights, pilotage and other services. The Lord Chamberlain has licensing jurisdiction over London Theatres other than patent theatres in respect of plays to be performed within an area comprising practically the whole of the eastern and central portions of the county north of the Thames, and the north part of the central portion of the county south of the river.

Gas supply is in the hands of private companies and extends beyond the county area on the west, in the south-west and on the east. Electricity is confined to the smaller areas within the county and is partly administered by the Borough Councils and partly by private companies.

To these administrative areas constituted by statutory authority must be added the Metropolitan Postal District, which is created by H.M. Post Office for the convenience of telegraph and postal purposes. It extends far beyond the London area (except at Woolwich) and has no relationship to other areas, but depends simply upon the facilities for collecting and distributing the postal service.

All these different jurisdictions constitute areas for the purpose only of special services which have resulted for the most part from the past state of chaos in London Government. They make it almost impossible for resident Londoners to understand London administration, and quite impossible for strangers, unless they devote some time to its study.

Looking over this map, one is conscious of the vast area which it embraces. Workers in the centre no longer live at or near their place of work—neither the employer nor the employed. They travel to and fro morning and night. The principal roadways which form the main arteries from and into London are gradually being supplied with tramways by the County Council, though there are still considerable districts, principally in the north-west and west, which are without this necessary means of popular locomotion. Tube underground railways and motor omnibuses supply to some extent this deficiency, but London more than any other city in the world needs a perfect system of locomotion between all its parts, for workers cannot walk the distances.

Changes will come over this great city as time produces its inevitable changes in modes of life, but there is a solidity about London which, with its wonderful capacity to attract even the most superficial people, tells for practical continuity of existing conditions under improved and improving circumstances. London is the capital of the British Empire in no formal sense, neither by statute nor appointment. It is so by right of its pre-eminence among the cities of the whole world. It is the capital because it has no rival.

LAURENCE GOMME.

MARLBOROUGH PLACE, N.W.
December 1910.

NOTE.

The maps in the Atlas numbered 1 to 84, are coloured to show :—

In red, the boundaries of the Metropolitan Boroughs, and of the adjacent Municipal Boroughs, Urban Districts, and Rural Districts ;

In green, the Parks and Open Spaces ;

In blue, the principal waters (rivers, lakes, canals) ;

In brown, the roads on which Tramway rails are laid.

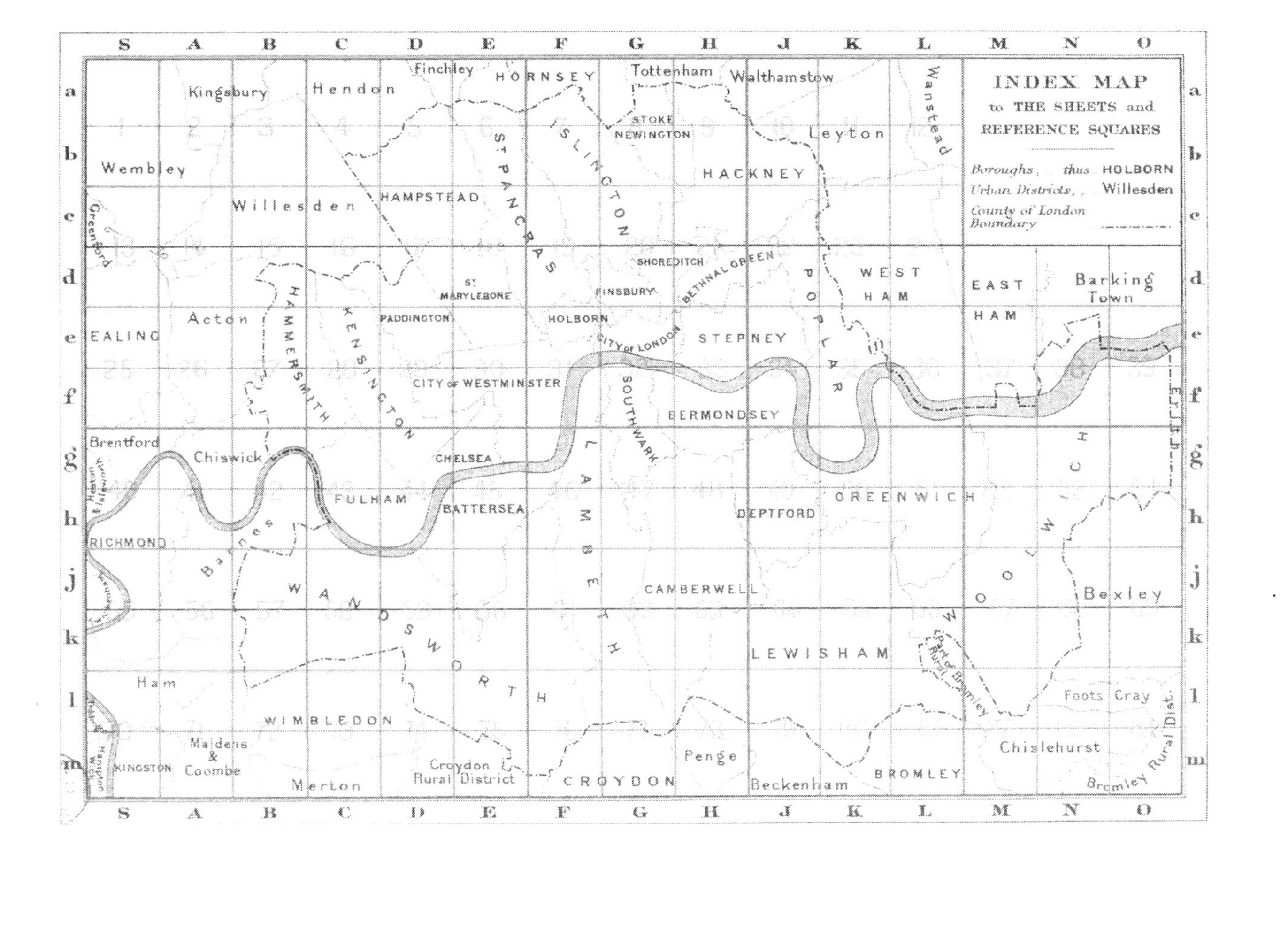

INDEX MAP
to THE SHEETS and
REFERENCE SQUARES
Boroughs, thus HOLBORN
Urban Districts, , Willesden
County of London Boundary
Kingsbury
Hendon
Finchley
HORNSEY
Tottenham
Walthamstow
Wanstead
STOKE NEWINGTON
Leyton
Wembley
HACKNEY
Willesden
HAMPSTEAD
ST PANCRAS
ISLINGTON
Greenford
SHOREDITCH
BETHNAL GREEN
POPLAR
WEST HAM
EAST HAM
Barking Town
ST. MARYLEBONE
FINSBURY
HOLBORN
PADDINGTON
EALING
Acton
HAMMERSMITH
KENSINGTON
CITY of LONDON
STEPNEY
CITY of WESTMINSTER
SOUTHWARK
BERMONDSEY
Brentford
Chiswick
CHELSEA
LAMBETH
FULHAM
BATTERSEA
DEPTFORD
GREENWICH
WOOLWICH
RICHMOND
Barnes
WANDSWORTH
CAMBERWELL
Bexley
LEWISHAM
Part of Rural Bromley
Ham
Foots Cray
WIMBLEDON
Maidens & Coombe
KINGSTON
Merton
Croydon Rural District
CROYDON
Penge
Beckenham
BROMLEY
Chislehurst
Bromley Rural Dist.
Hampton Wick

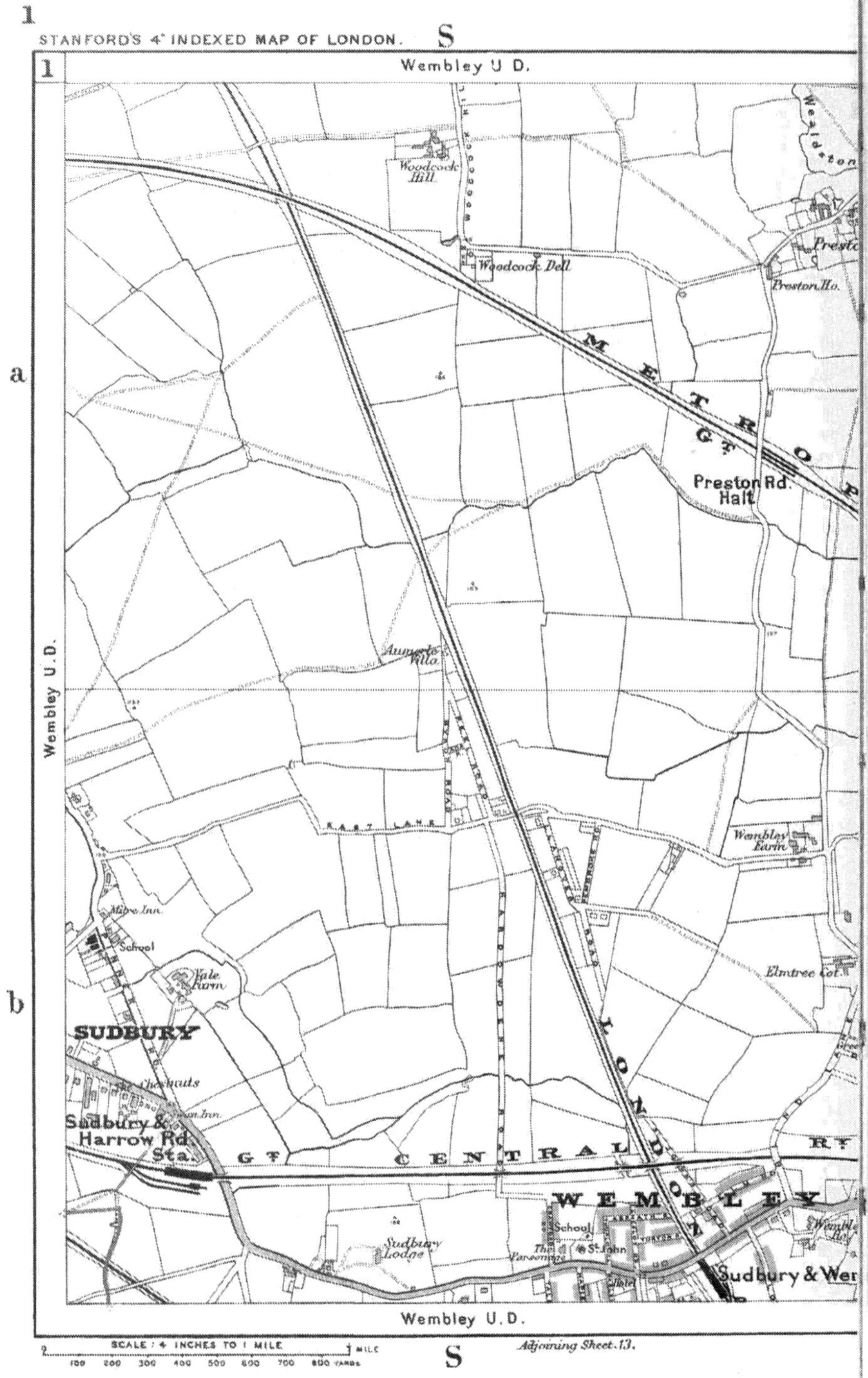
1
STANFORD'S 4" INDEXED MAP OF LONDON.
S
1
Wembley U D.
Wembley U.D.
a
b
Woodcock Hill
Woodcock Dell
Preston Ho.
Wealdston
M E T R O P
G.T.
Preston Rd. Halt
Aumeto Villa
EAST LANE
Wembley Farm
Mitre Inn
School
Vale Farm
Elmtree Cot.
SUDBURY
The Chestnuts
Sudbury & Harrow Rd. Sta.
L O N D
G.T. C E N T R A L R Y
W E M B L E Y
School
The Parsonage
St. John
Sudbury Lodge
Sudbury & Wer
Wembley U.D.
SCALE 4 INCHES TO 1 MILE
100 200 300 400 500 600 700 800 YARDS
1 MILE
S
Adjoining Sheet 13.

A

Wembley U.D.
Kingsbury U.D.
2
Bush Farm
Lit. Bush F.
Preston Farm
Uxendon Farm
Barn Hill
Hillhouse
Kingsbury U.D.
a
Wembley Paddocks
North Forty Farm
Forty Farm
Chalkhill Ho.
Blackpothill F.
KINGSBURY
Wembley Park Sta.
The Lake
Refreshment Room
Wembley Park
Variety Hall
Fountain
Band Stand
Pavilion
WEMBLEY
PARK
Tea Ho.
Wembley Cot.
Willesden U.D.
b
Wembley Hill Sta.
Wakington Farm
Brent
Wembley U.D.
Willesden U.D.
Adjoining Sheet 3.
Adjoining Sheet 14

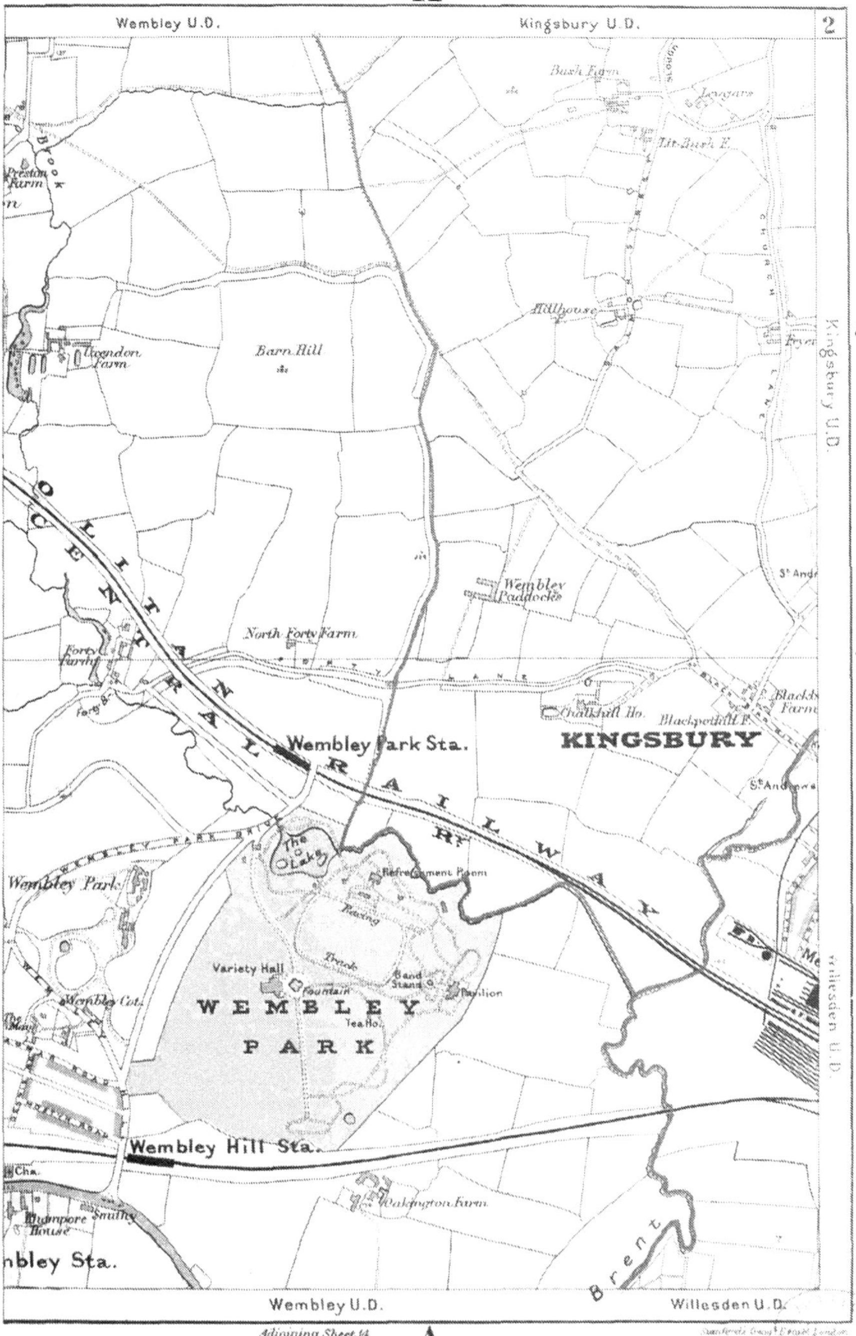

A

STANFORD'S 4" INDEXED MAP OF LONDON. B

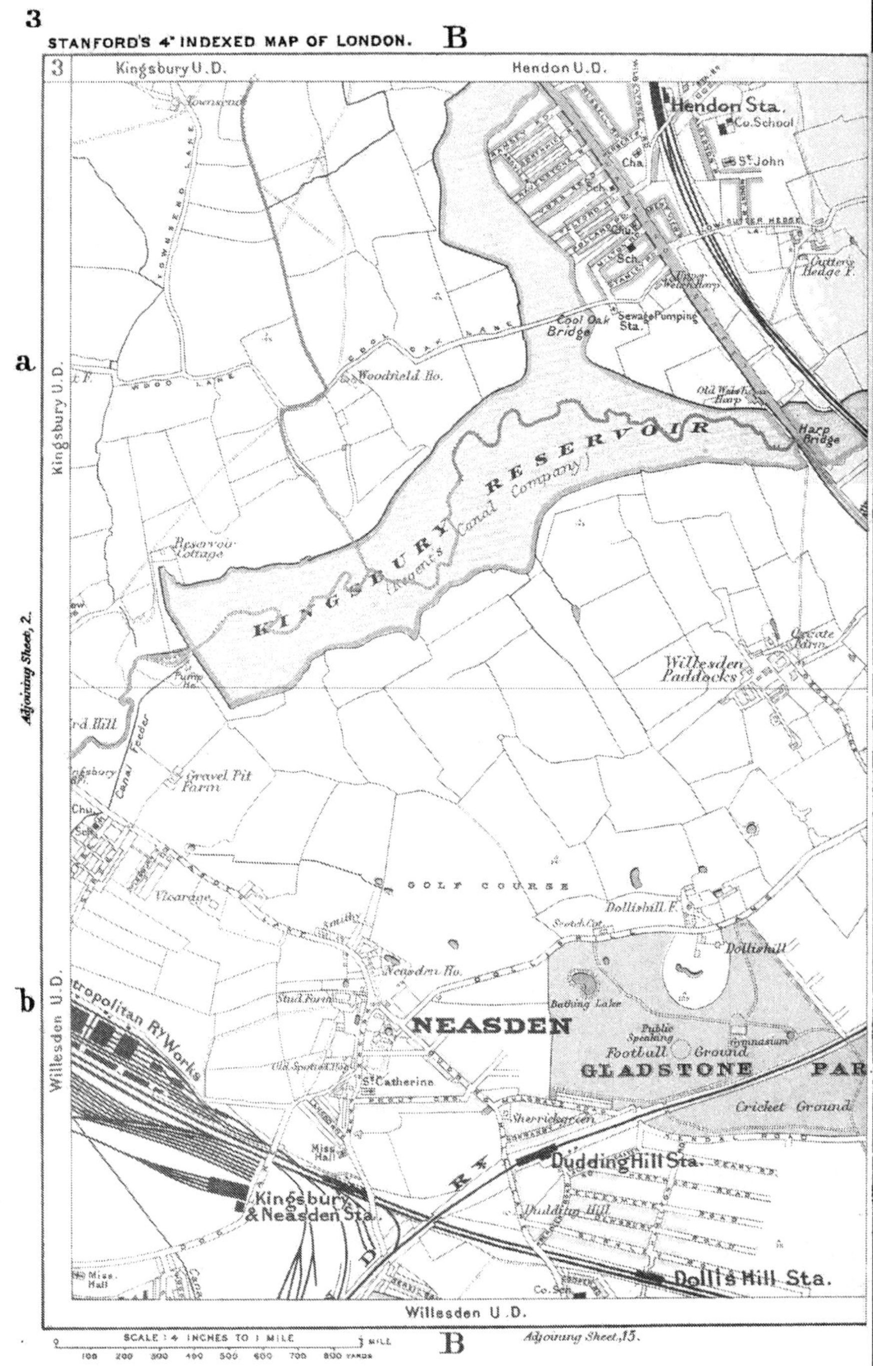

SCALE: 4 INCHES TO 1 MILE

0 100 200 300 400 500 600 700 800 YARDS ½ MILE

B

Adjoining Sheet, 15.

4
C
Hendon U.D.
4
Gooseberry Gardens
Shire Hall
River Brent
Renters Farm
The Homestead
Highfield
Elmer Lodge
The Elms
GOLDERS GREEN
Portsdown Lodge
Sewage Works
(Hendon U.D.)
Fever Hosp.l
a
Brent Gas Works
Hendon U.D.
Clitterhouse Farm
MIDLAND
Adjoining Sheet, 5.
Hendon Allotments
Night Light Factory
Bd. Sch.
School
Lower Oxgate Farm
Cricklewood Sta.
b
CRICKLEWOOD
Askes' Haberdashers' School
HAMPSTEAD
CRICKET GROUND
Willesden U.D.
HAMPSTEAD
Adjoining Sheet, 16.
C

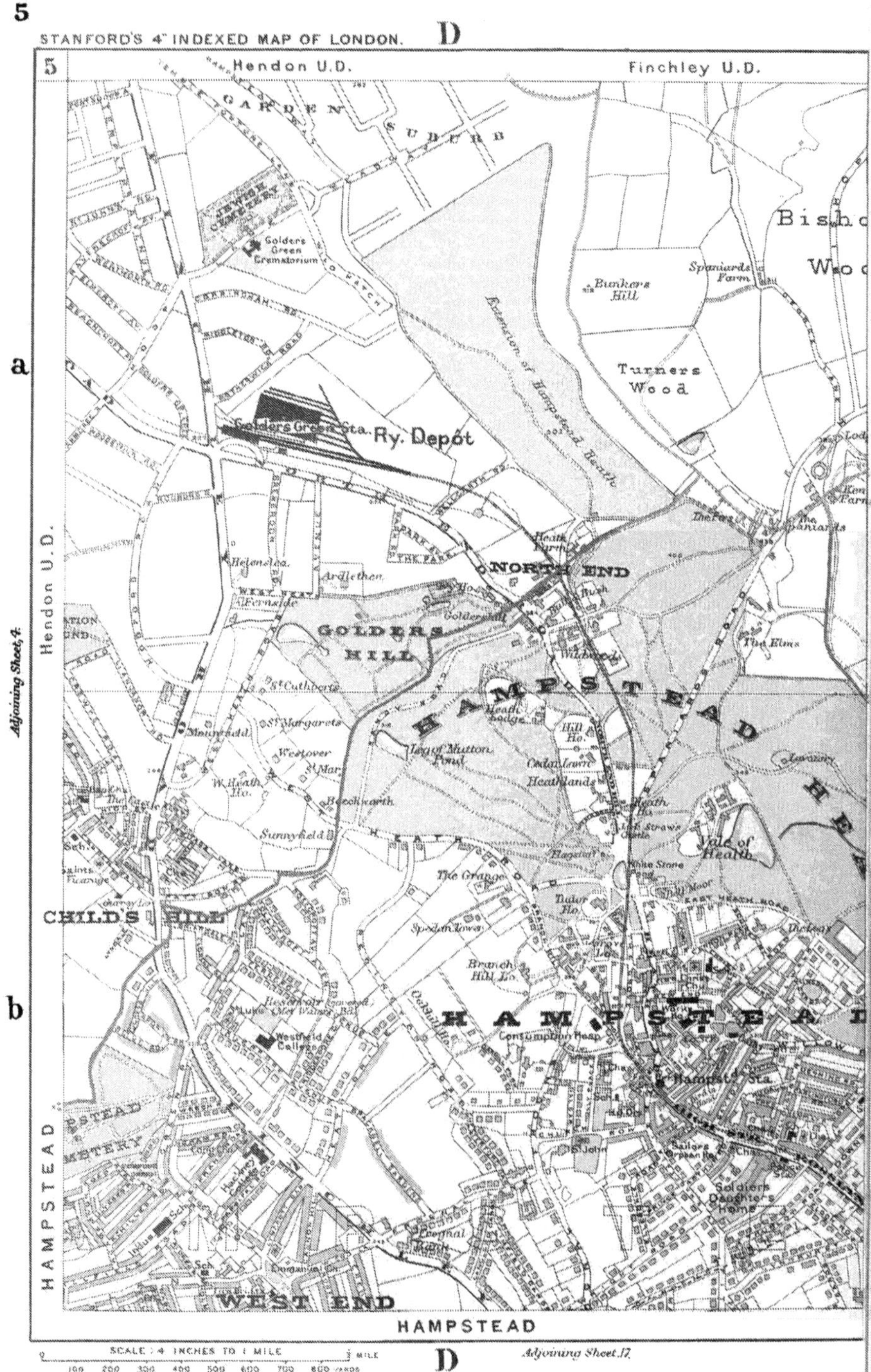
STANFORD'S 4" INDEXED MAP OF LONDON.
D
5
Hendon U.D.
Finchley U.D.
GARDEN SUBURB
Jewish Cemetery
Golders Green Crematorium
Bishops Wood
Spaniards Farm
Bunkers Hill
Turners Wood
a
Golders Green Sta.
Ry. Depot
Extension of Hampstead Heath
The Firs
The Spaniards
Heath Farm
NORTH END
Helenslea
Ardlethen
Fernside
GOLDERS HILL
Hendon U.D.
Adjoining Sheet 4.
The Elms
St Cuthberts
St Margarets
Westover
HAMPSTEAD
Heath Lodge
Hill Ho.
Leg of Mutton Pond
Cedar Lawn
Heathlands
W. Heath Ho.
Beechworth
Sunnyfield
Jack Straws Castle
Vale of Health
Flagstaff
Whitestone
The Grange
Tudor Ho.
East Heath Road
CHILD'S HILL
Branch Hill Lo.
b
HAMPSTEAD
Consumption Hosp.
Westfield College
Hampstead Sta.
Sailors Orphan Ho.
Soldiers Daughters Home
HAMPSTEAD
Frognal
WEST END
HAMPSTEAD
SCALE: 4 INCHES TO 1 MILE
D
Adjoining Sheet 17.

E

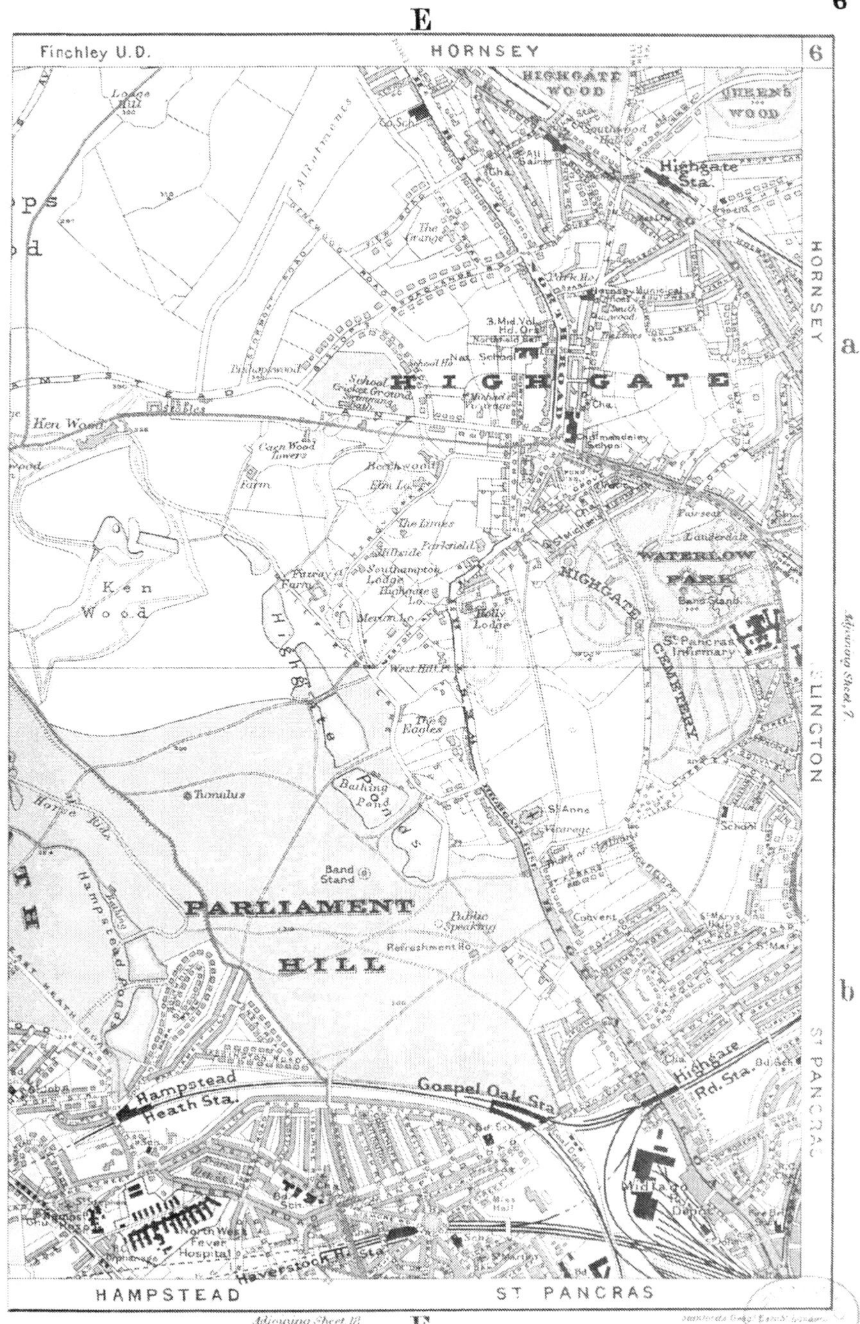

Finchley U.D.
HORNSEY
6
HIGHGATE WOOD
QUEENS WOOD
Highgate Sta.
HORNSEY
a
HIGHGATE
Ken Wood
Caen Wood Towers
Farm
Beechwood
The Limes
Parkfield
Hillside
Southampton Lodge
Highgate Lo.
Fitzroy Farm
Merton Lo.
Holly Lodge
West Hill Pl.
Ken Wood
HIGHGATE CEMETERY
WATERLOW PARK
Band Stand
St. Pancras Infirmary
Adjoining Sheet 7.
ISLINGTON
The Eagles
Bathing Pond
Tumulus
St. Anne
Vicarage
School
Band Stand
PARLIAMENT HILL
Public Speaking
Refreshment Ho.
Convent
b
Hampstead Ponds
Hampstead Heath Sta.
Gospel Oak Sta.
Highgate Rd. Sta.
ST. PANCRAS
Midland Ry. Depot
North West Fever Hospital
Haverstock H. Sta.
HAMPSTEAD
ST. PANCRAS
Adjoining Sheet 18.

E

STANFORD'S 4" INDEXED MAP OF LONDON. F

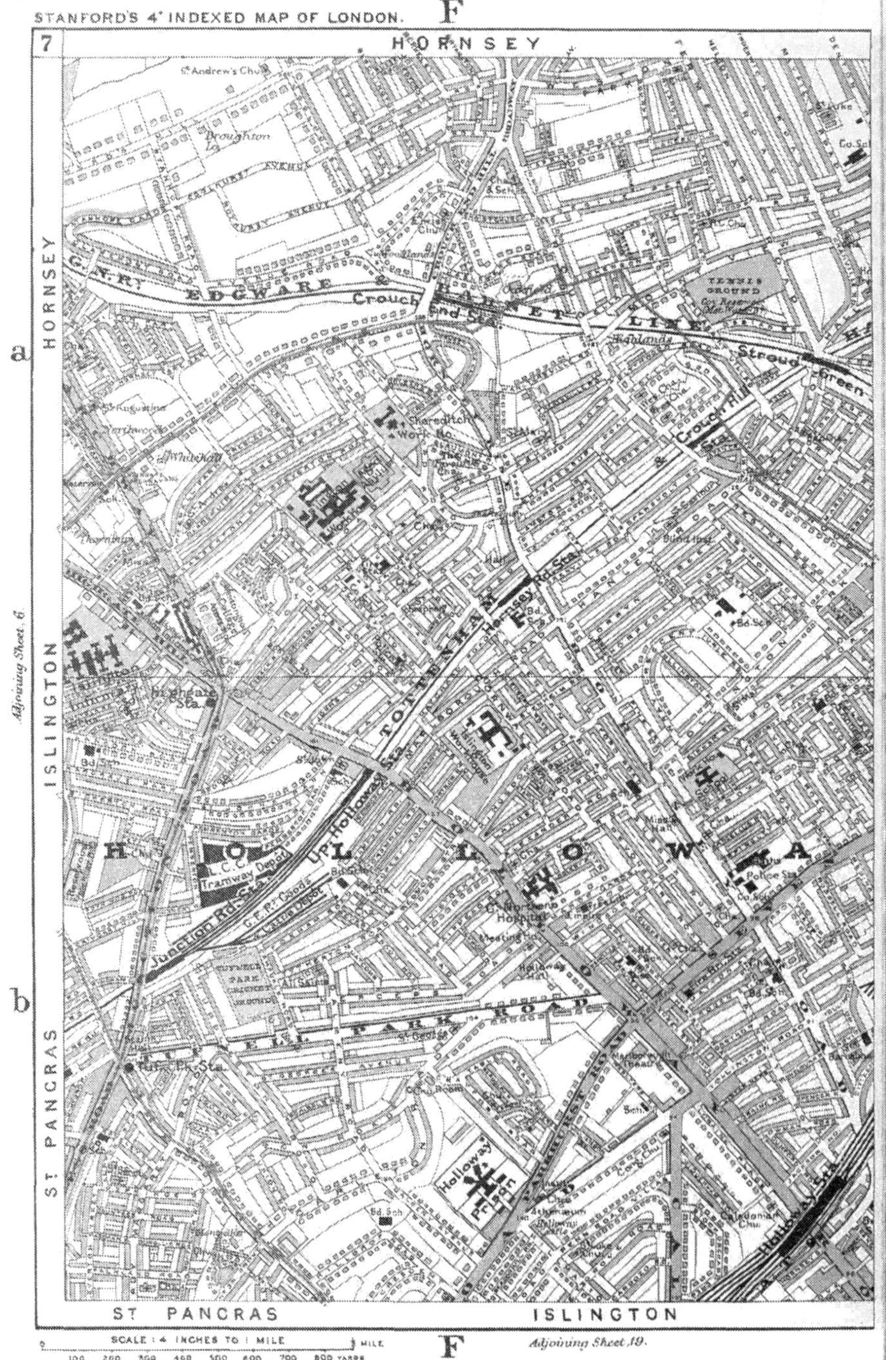

G

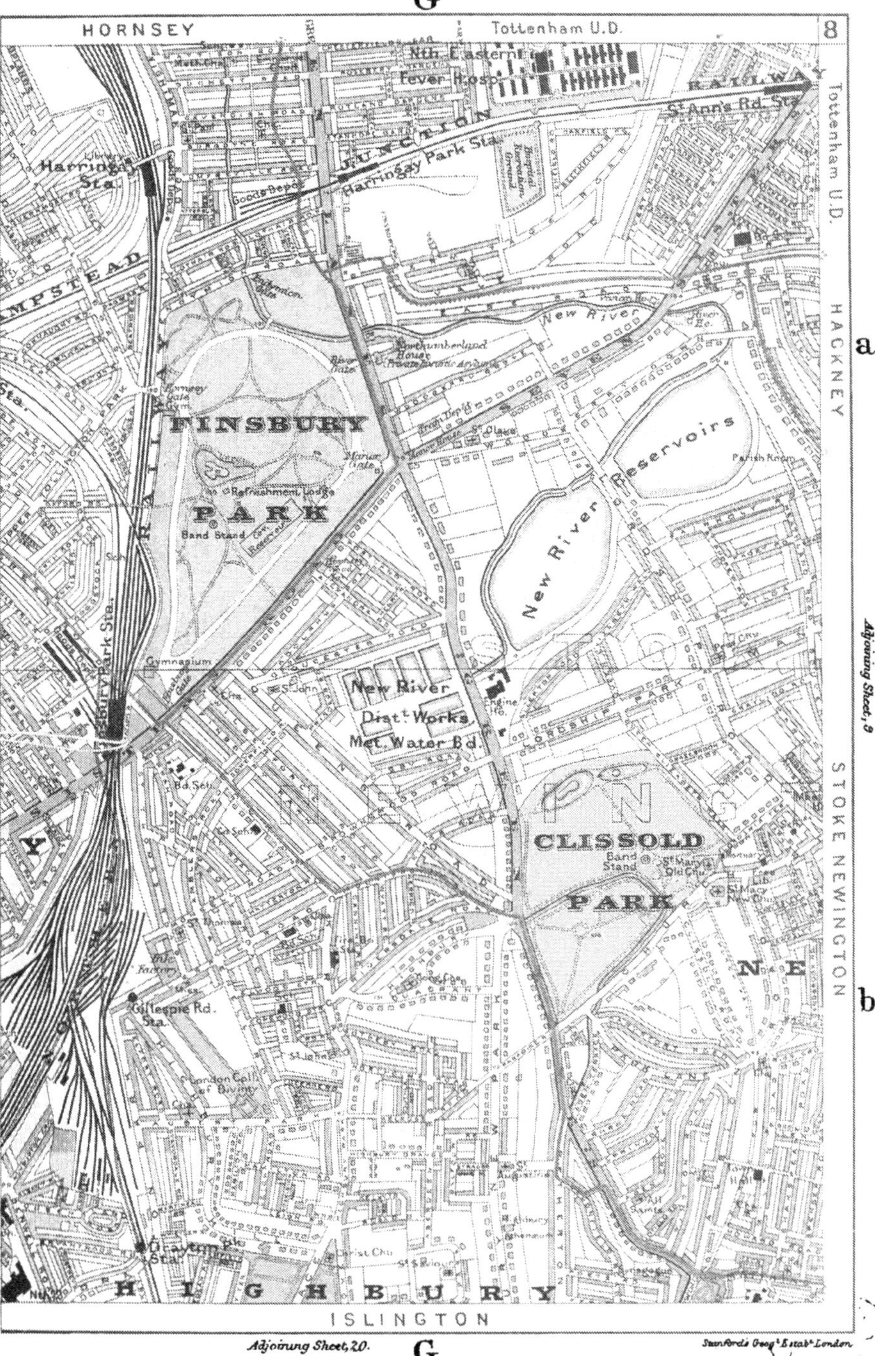
HORNSEY
Tottenham U.D.
8
Nth. Eastern Fever Hosp.
RAILWAY
JUNCTION
St. Ann's Rd. Sta.
Harringay Park Sta.
Harringay Sta.
Goods Depot
HAMPSTEAD
Tottenham U.D.
HACKNEY
New River
Northumberland House
FINSBURY
PARK
Refreshment Lodge
Band Stand
Gymnasium
Finsbury Park Sta.
New River Reservoirs
Parish Knoll
New River
New River Dist.t Works, Met. Water Bd.
Engine Ho.
Adjoining Sheet, 9
STOKE NEWINGTON
CLISSOLD
PARK
Band Stand
St. Mary's Old Chu.
St. Mary New Chu.
NE
Gillespie Rd. Sta.
London Coll. of Divinity
Drayton Pk. Sta.
Christ Chu.
HIGHBURY
ISLINGTON
Adjoining Sheet, 20.
Stanford's Geog.l Estab.t London
a
b

G

STANFORD'S 4" INDEXED MAP OF LONDON. **H**

a

b

Adjoining Sheet, 8.

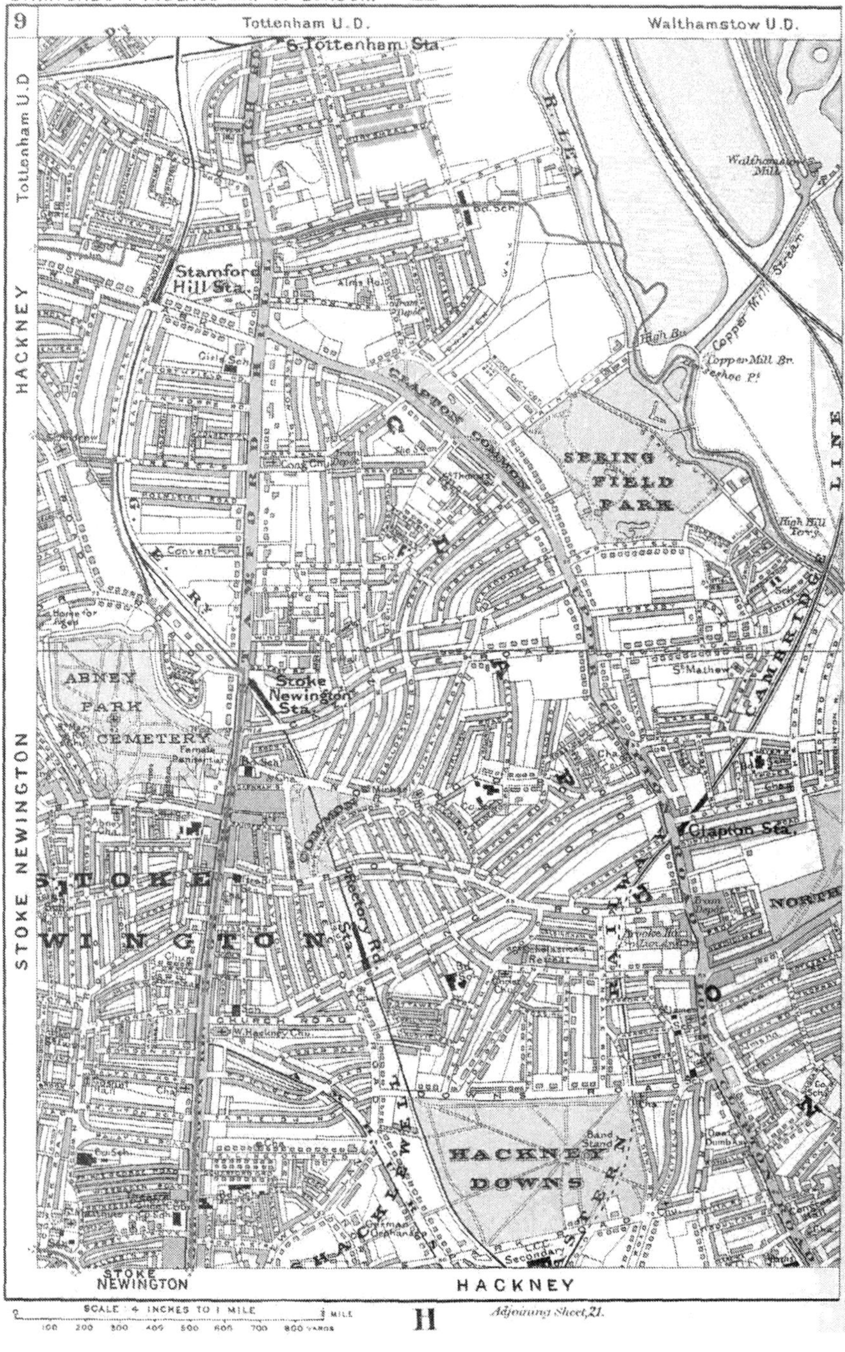

SCALE: 4 INCHES TO 1 MILE — 0 100 200 300 400 500 600 700 800 YARDS — ½ MILE

H

Adjoining Sheet, 21.

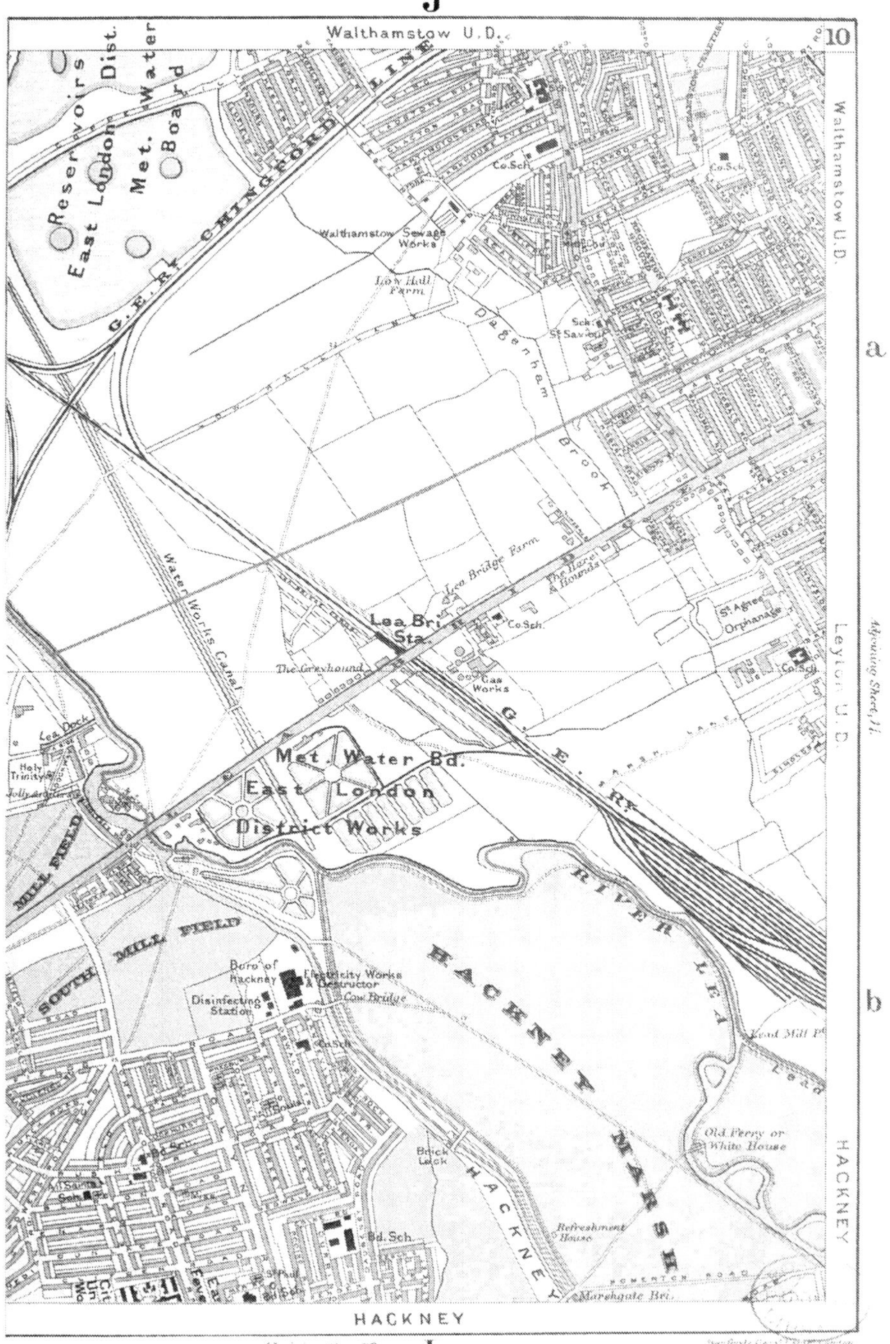
J
Walthamstow U.D.
10
Reservoirs
East London Dist.
Met. Water Board
G. E. Ry. CHINGFORD LINE
Walthamstow Sewage Works
Low Hall Farm
Co.Sch.
Sch.
St. Saviour
Dagenham Brook
Walthamstow U.D.
a
Water Works Canal
Lea Bridge Farm
The Hare & Hounds
Lea Bri. Sta.
Co.Sch.
St. Agnes Orphanage
Co.Sch.
The Greyhound
Gas Works
G. E. Ry.
Leyton U.D.
Adjoining Sheet, 11.
Lea Dock
Holy Trinity
Jolly Anglers
Met. Water Bd.
East London
District Works
MILL FIELD
SOUTH MILL FIELD
Boro' of Hackney
Electricity Works & Destructor
Disinfecting Station
Cow Bridge
Co.Sch.
HACKNEY MARSH
RIVER LEA
b
Lead Mill Pt.
Old Ferry or White House
Brick Lock
HACKNEY
Bd. Sch.
Bd. Sch.
Refreshment House
Marshgate Bri.
HOMERTON ROAD
HACKNEY
HACKNEY
Adjoining Sheet 32.
J

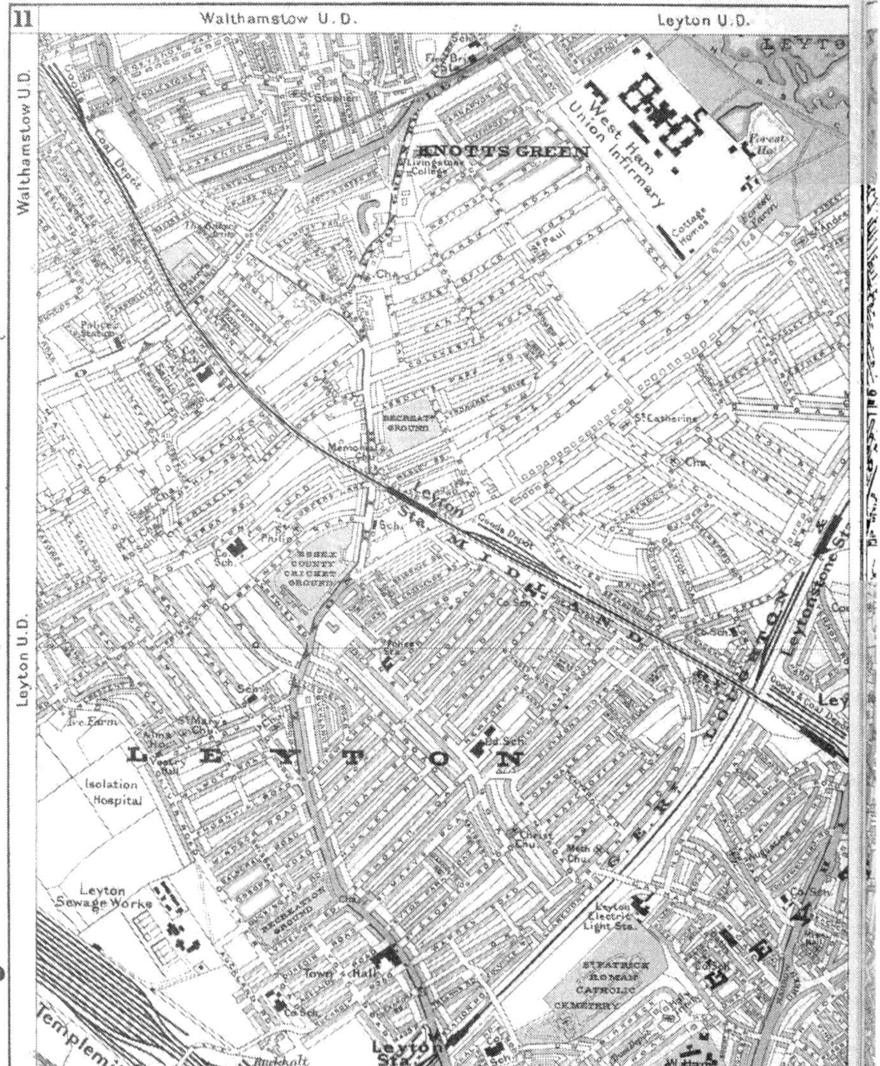

11 Walthamstow U.D. Leyton U.D.

Walthamstow U.D.

a

Adjoining Sheet, 10.

Leyton U.D.

b

HACKNEY

KNOTT'S GREEN

West Ham Union Infirmary

Cottage Homes

Forest Hall

Coal Depot

Police Station

Recreatn Ground

Memorial Hall

Leyton Sta.

MIDLAND RY

Goods Depot

St. Catherine

Essex County Cricket Ground

Ive Farm

St. Mary's Chu.

LEYTON

Isolation Hospital

Leyton Sewage Works

Christ Chu.

Meth. Chu.

Leyton Electric Light Sta.

St. Patrick Roman Catholic Cemetery

Town Hall

Co. Sch.

Leyton Sta.

Ruckholt Farm

Templemills Sidings

West Ham Union Work Ho.

St. Alban

Leytonstone Sta.

Bd. Sch.

HACKNEY WEST HAM Leyton U.D. WEST HAM

SCALE: 4 INCHES TO 1 MILE
0 100 200 300 400 500 600 700 800 YARDS ¼ MILE

Adjoining Sheet, 23.

L

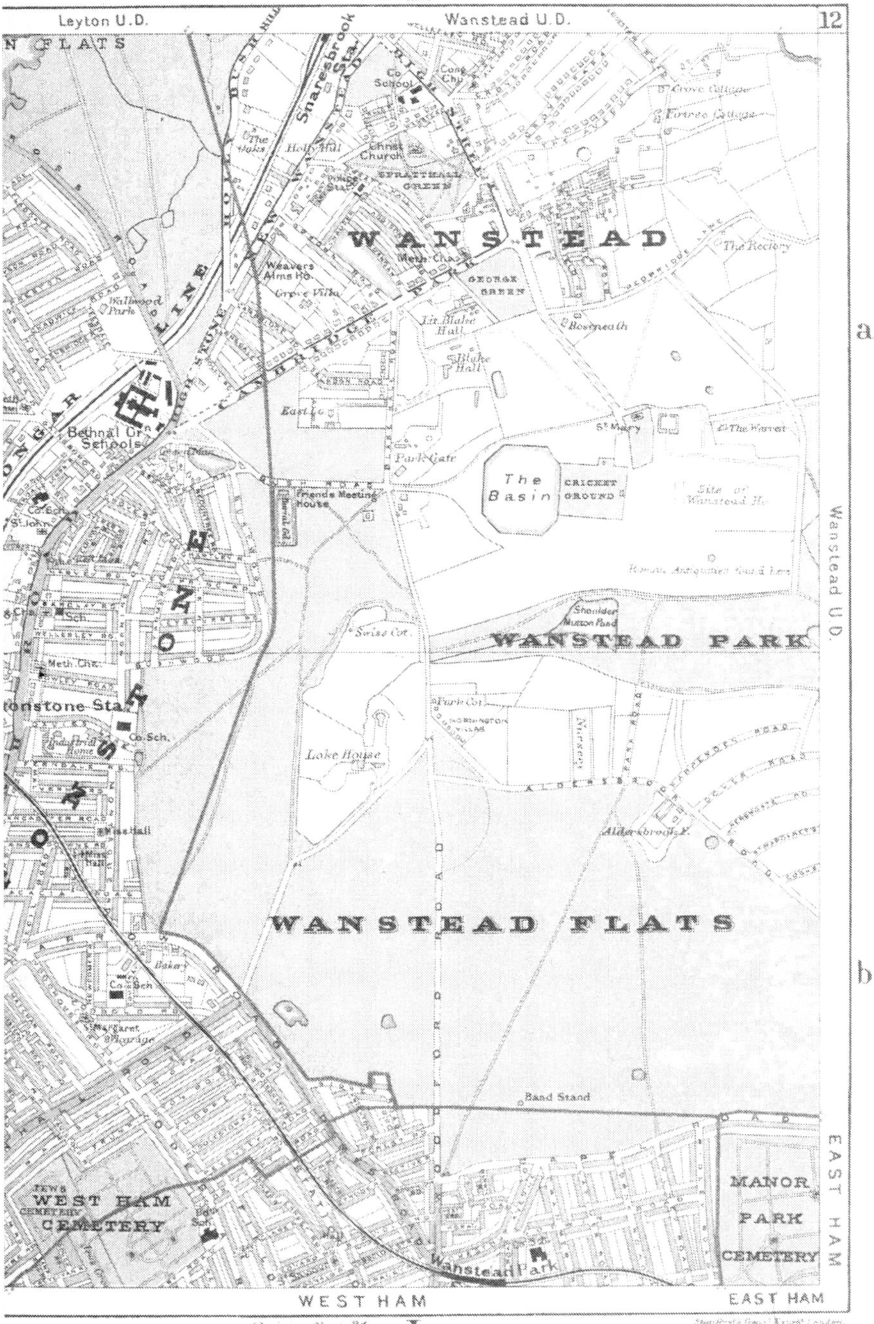

Leyton U.D.

Wanstead U.D.

12

Snaresbrook Sta.

Co. School

Cong. Chu.

Grove Cottage

Fortree Cottage

The Oaks

Holly Hall

Christ Church

Strathall Green

WANSTEAD

The Rectory

Weavers Alms Ho.

Grove Villa

Meth. Cha.

George Green

Wallwood Park

Lit. Blake Hall

Blake Hall

Bosgneath

Bethnal Gr. Schools

East Lo.

Park Gate

St. Mary

The Warren

The Basin

Cricket Ground

Friends Meeting House

Site of Wanstead Ho.

Co. Sch.

St. John

Roman Antiquities found here

Shoulder Mutton Pond

Swiss Cot.

WANSTEAD PARK

Wanstead U.D.

Meth. Ch.

Park Cot.

Leytonstone Sta.

Co. Sch.

Lake House

Nursery

Alders Brook

Aldersbrook F.

a

WANSTEAD FLATS

b

Bakery

Co. Sch.

Band Stand

West Ham Cemetery

Jews Cemetery

Manor Park Cemetery

EAST HAM

Wanstead Park

WEST HAM

EAST HAM

Adjoining Sheet 24.

L

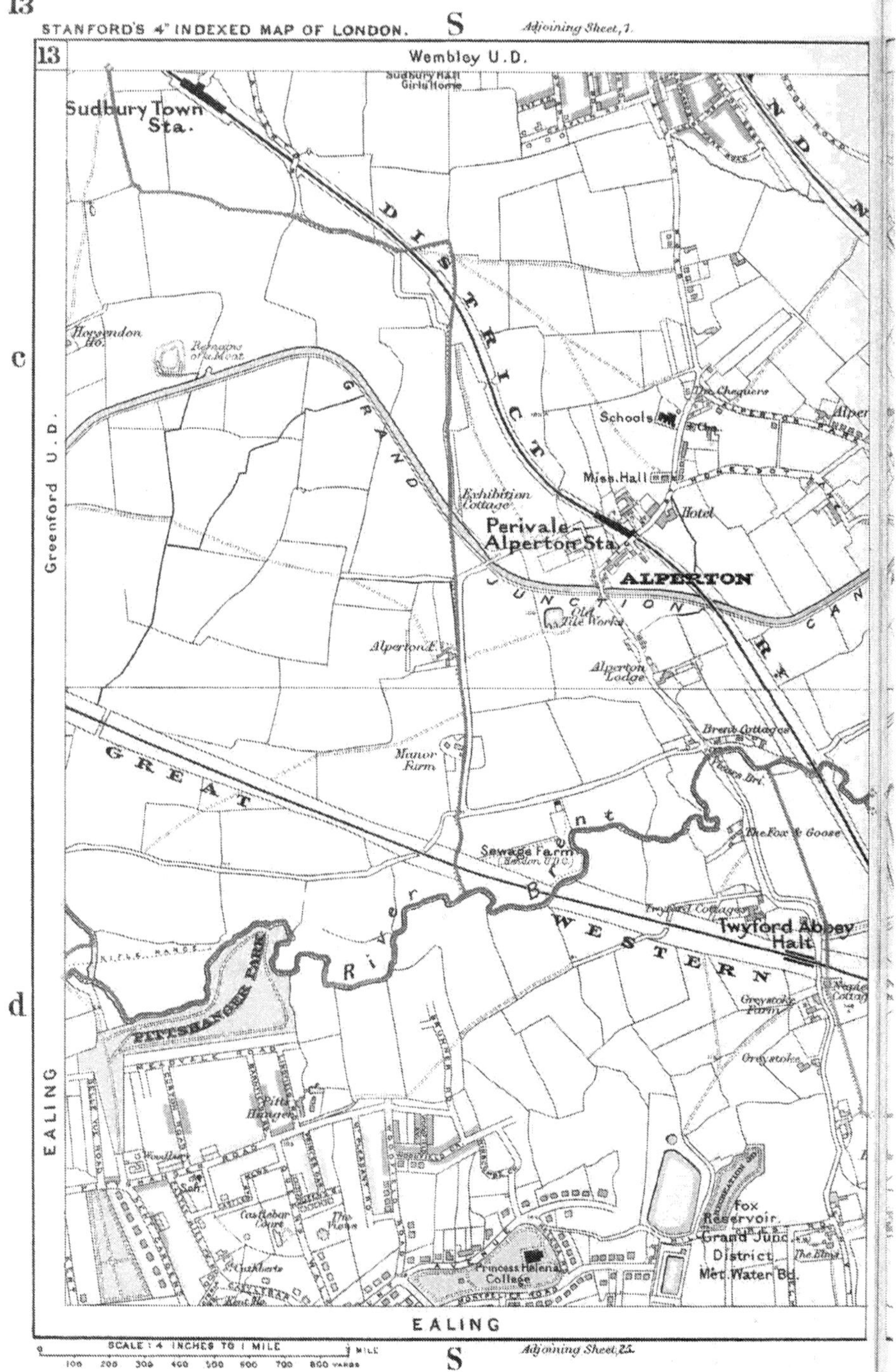
13
Wembley U.D.
Sudbury Hall Girls Home
Sudbury Town Sta.
DISTRICT
Horsendon Ho.
Remains of a Moat
GRAND JUNCTION CANAL
Greenford U.D.
The Chequers
Schools
Miss. Hall
Hotel
Exhibition Cottage
Perivale-Alperton Sta.
ALPERTON
Old Tile Works
Alperton F.
Alperton Lodge
Brent Cottages
Manor Farm
GREAT WESTERN
The Fox & Goose
Sewage Farm
River Brent
Twyford Cottages
Twyford Abbey Halt
Greystoke Farm
Greystoke
PITSHANGER PARK
Pitts Hanger
EALING
Castlebar Court
St Cuthberts
Fox Reservoir
Grand Junc. District Met. Water Bd.
The Elms
Princess Helena College
EALING

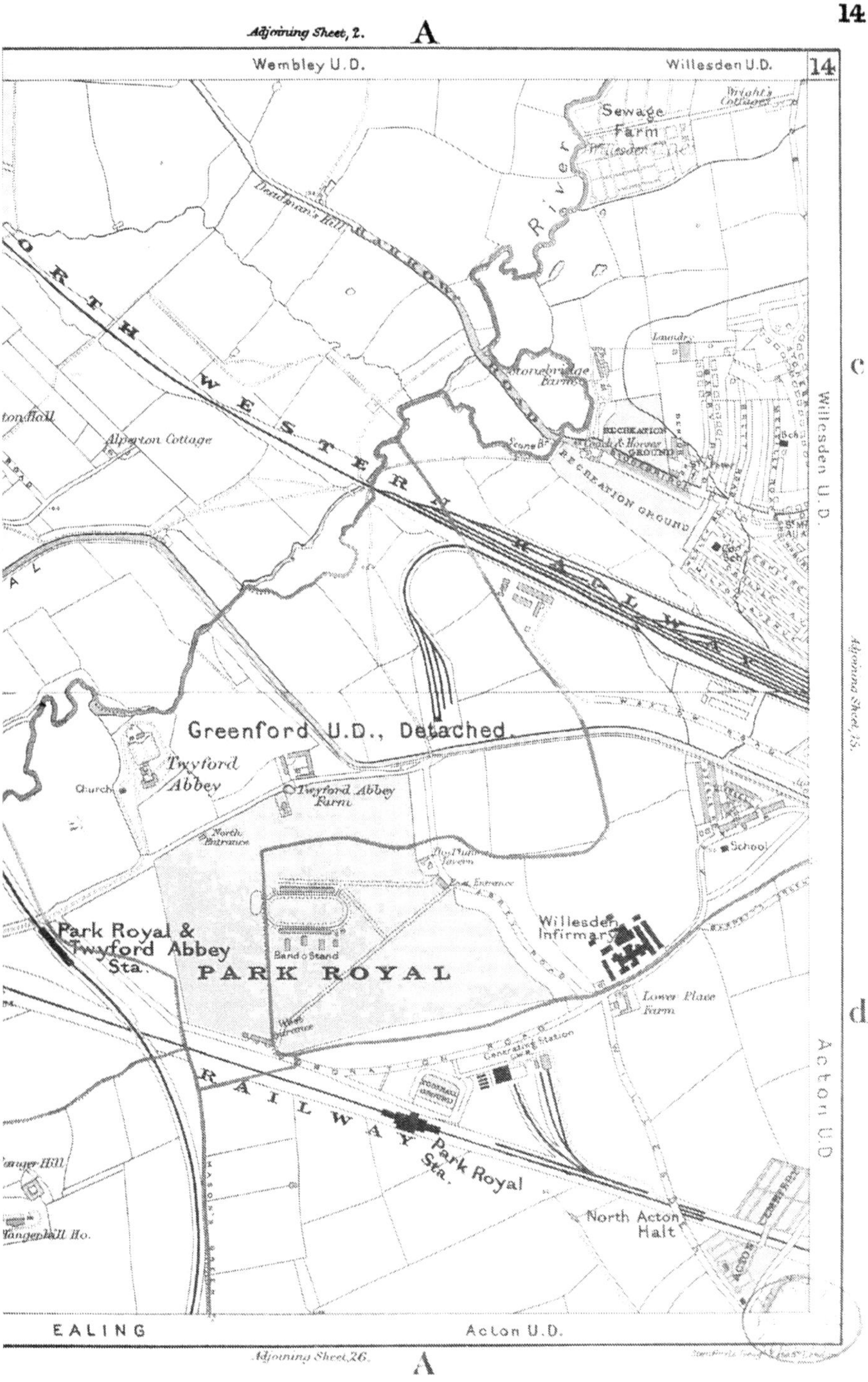

Adjoining Sheet, 2.

A

Wembley U.D.

Willesden U.D.

14

Sewage Farm

River

Wright's Cottages

Deadman's Hill

HARROW ROAD

NORTH WESTERN RAILWAY

Stonebridge Farm

Laundry

RECREATION GROUND

Alperton Cottage

Greenford U.D., Detached.

Twyford Abbey

Church

Twyford Abbey Farm

North Entrance

East Entrance

School

Willesden Infirmary

Park Royal & Twyford Abbey Sta.

Band Stand

PARK ROYAL

Lower Place Farm

Generating Station

RAILWAY

Park Royal Sta.

North Acton Halt

Willesden U.D.

c

d

Acton U.D.

Adjoining Sheet, 15.

EALING

Acton U.D.

Adjoining Sheet, 26.

A

STANFORD'S 4" INDEXED MAP OF LONDON. B *Adjoining Sheet, 3.*

15

Willesden U.D.

C

d

Willesden U.D.

Acton U.D.

Adjoining Sheet, 14.

Willesden Isolation Hospital

Co. Sch.

Willesden Council Elect. Wks.

WILLESDEN

WILLESDEN CEMETERY

JEWS CEMETERY

CRICKET GROUND

Dudding Hill

Church End

St Mary's

Harlesden Sta.

ROUNDWOOD PARK

Band Stand

Allotment Gardens

W. Midd. Dist. Reservoir (Met. Water Bd.)

Willesden Cott. Hosp.

Knowles Tower

Roundwood Ho.

Haycroft F.

Convent

HARLESDEN GREEN

Met Electric Supply Co.

Carriage Shed

Engine Shed

Willesden Junction

Atlas Brick & Tile Works

Paper Works

Chandos Metal Works

ST MARY'S KENSAL ROMAN CATH.

CANAL

Engine Shed

Engineering Works

Gas Works

Carriage Shed

Non Military portion for Public at all times

Wells Ho.

WORMWOOD SCRUBS

Acton U.D. HAMMERSMITH

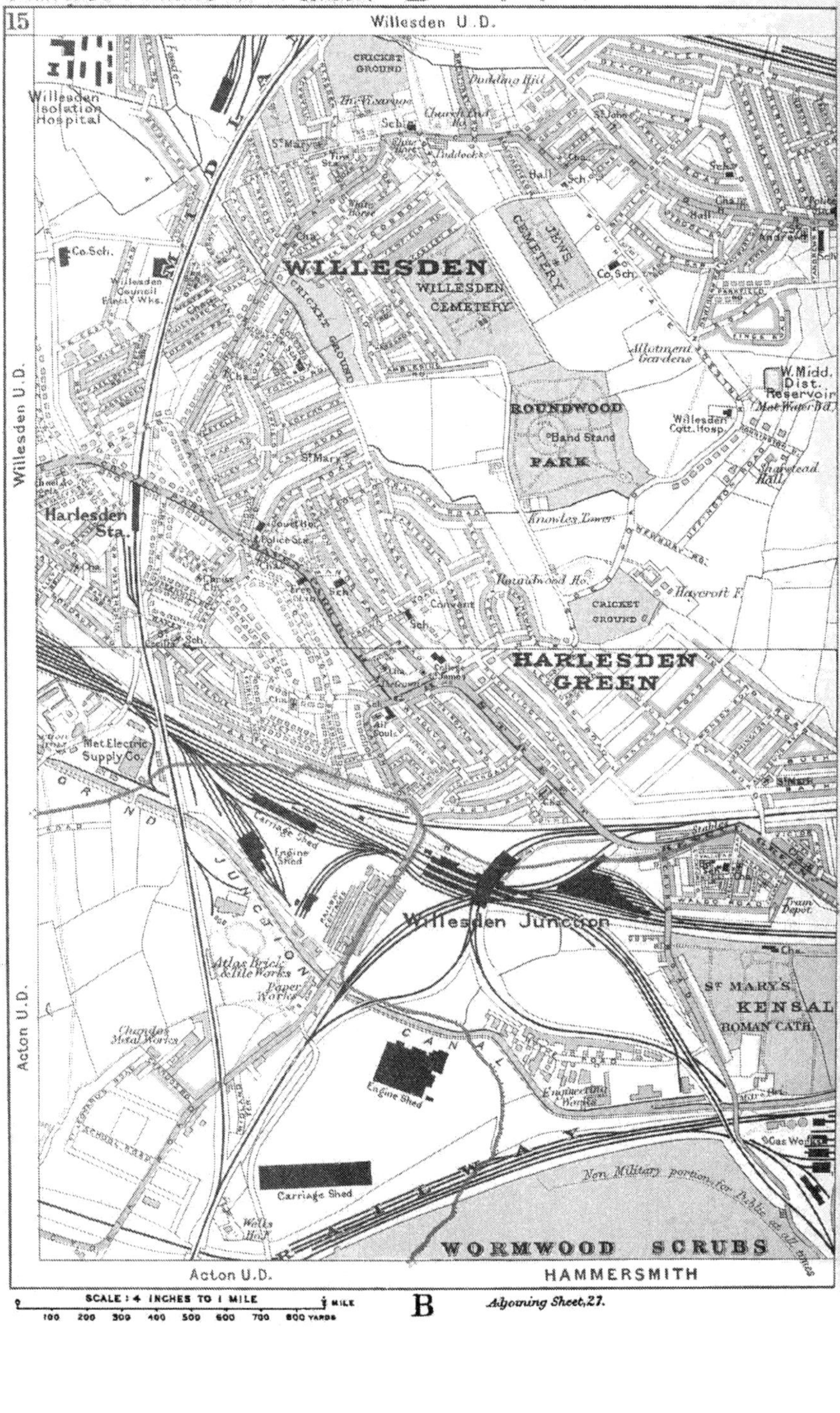

SCALE: 4 INCHES TO 1 MILE ¼ MILE

100 200 300 400 500 600 700 800 YARDS

B *Adjoining Sheet, 27.*

Adjoining Sheet, 4. C

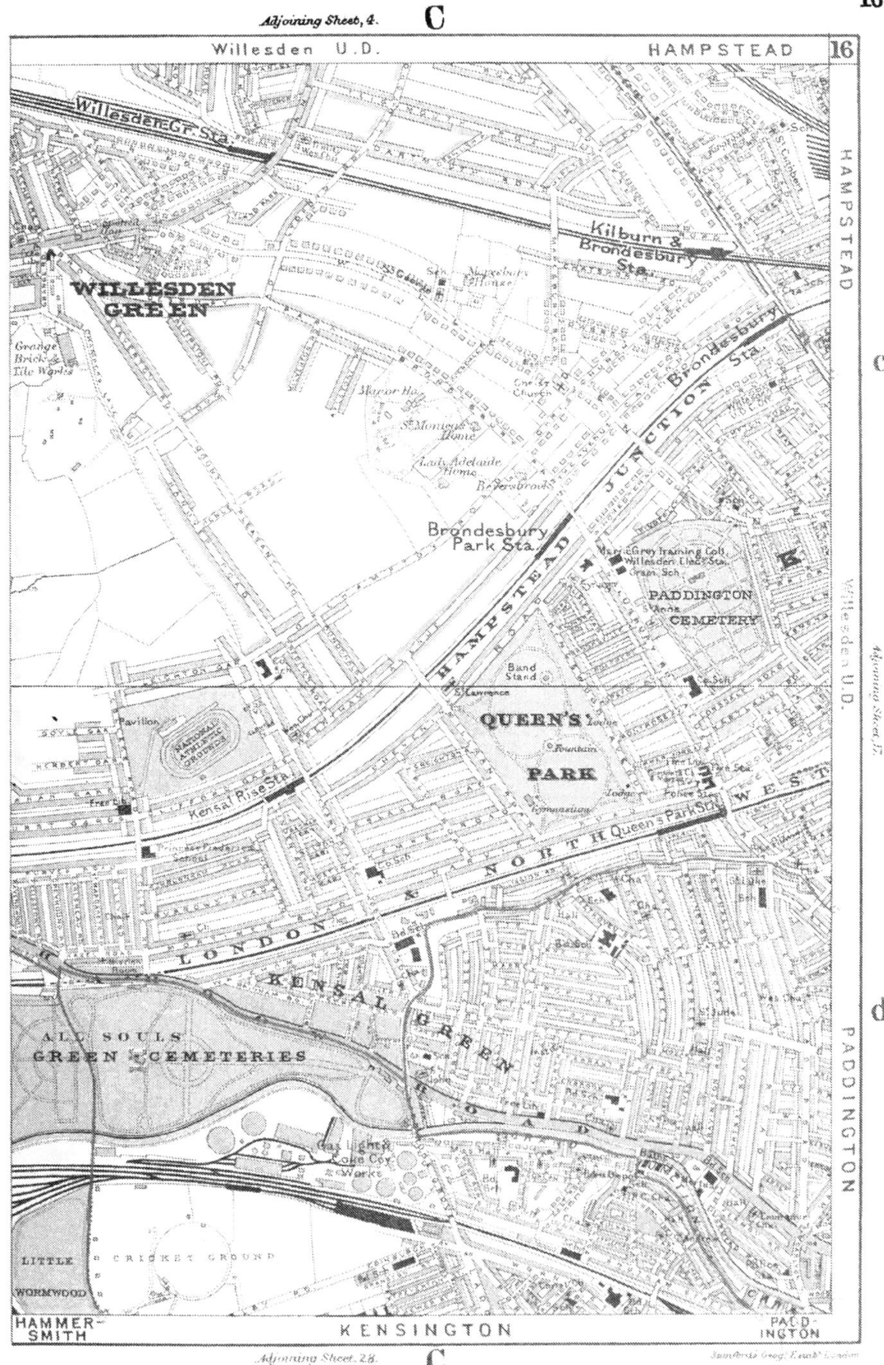

Adjoining Sheet, 28. C

Adjoining Sheet, 17.

c

d

STANFORD'S 4" INDEXED MAP OF LONDON. D *Adjoining Sheet 5.*

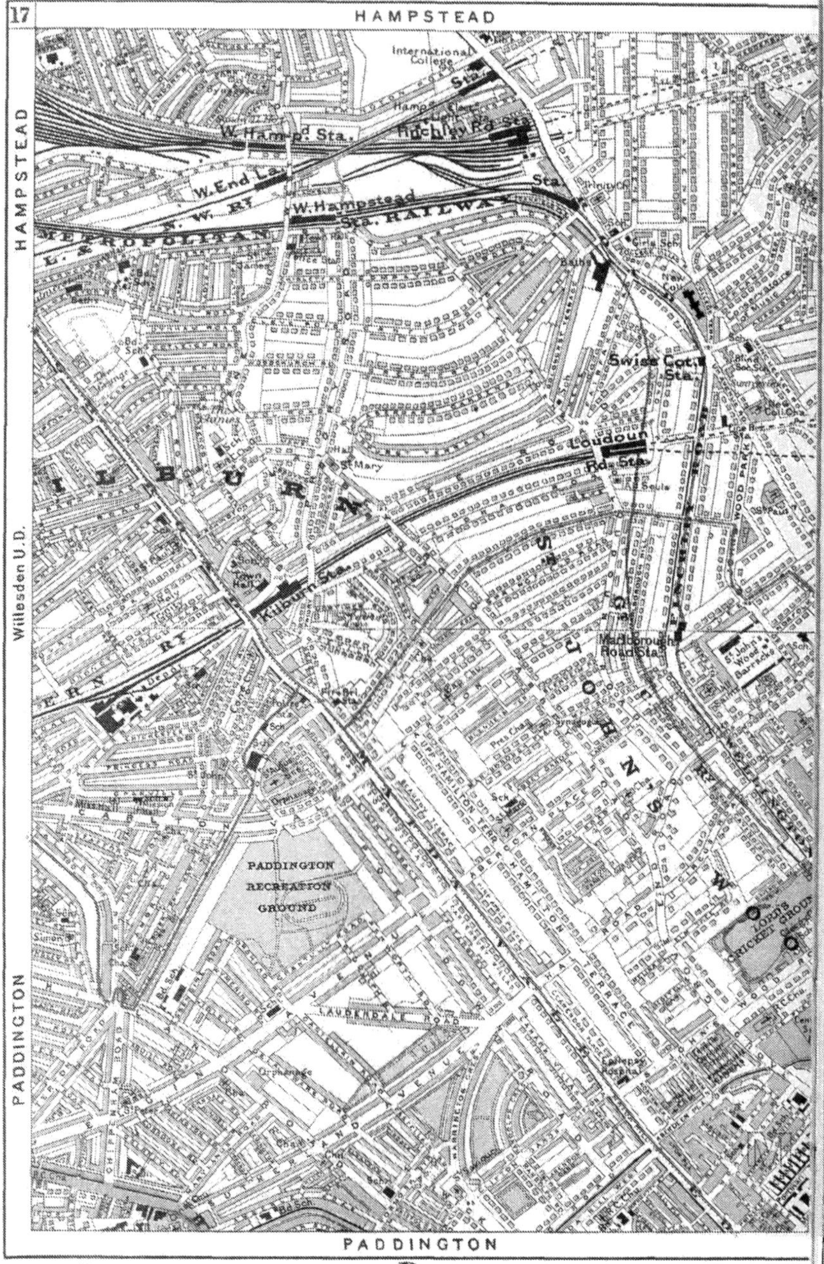

17 HAMPSTEAD

HAMPSTEAD

C

Willesden U.D. *Adjoining Sheet, 16.*

d

PADDINGTON

PADDINGTON

SCALE: 4 INCHES TO 1 MILE

D *Adjoining Sheet, 23.*

Adjoining Sheet 5 E

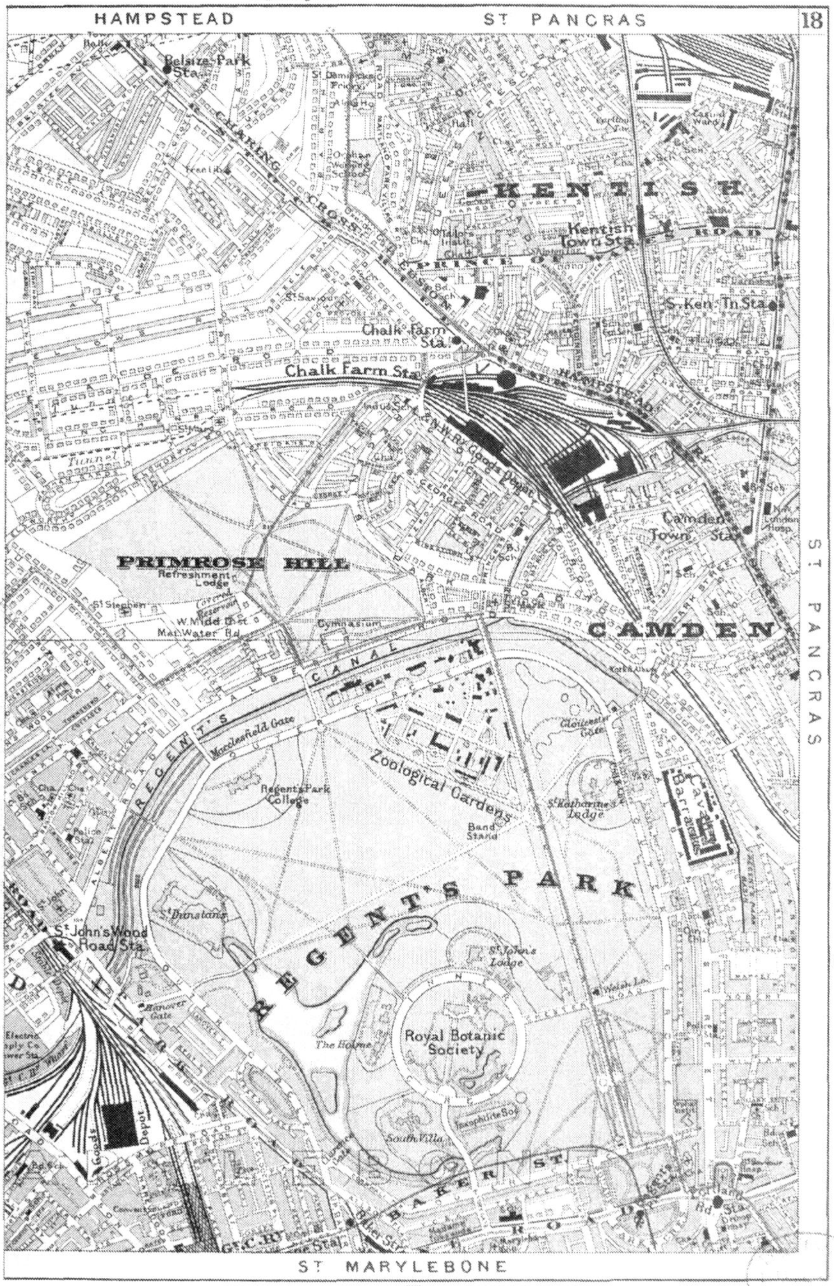

C

d

Adjoining Sheet 19

Adjoining Sheet 30 E

Stanford's Geog.l Estab.t London

STANFORD'S 4" INDEXED MAP OF LONDON. F *Adjoining Sheet. 7.*

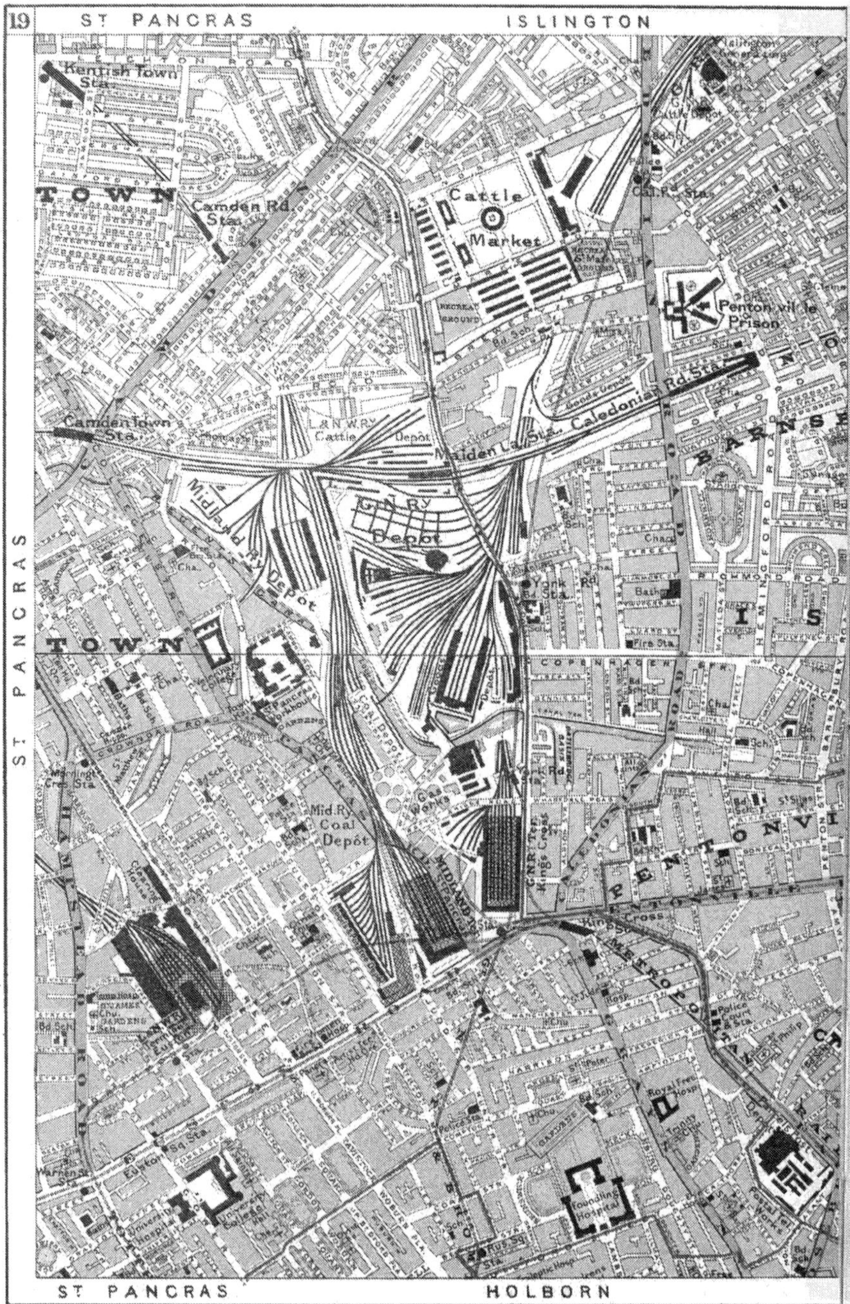

19 ST PANCRAS ISLINGTON

C

Adjoining Sheet. 18. ST PANCRAS

d

ST PANCRAS HOLBORN

SCALE : 4 INCHES TO 1 MILE ½ MILE

0 100 200 300 400 500 600 700 800 YARDS

F *Adjoining Sheet. 31.*

Adjoining Sheet, 8. G

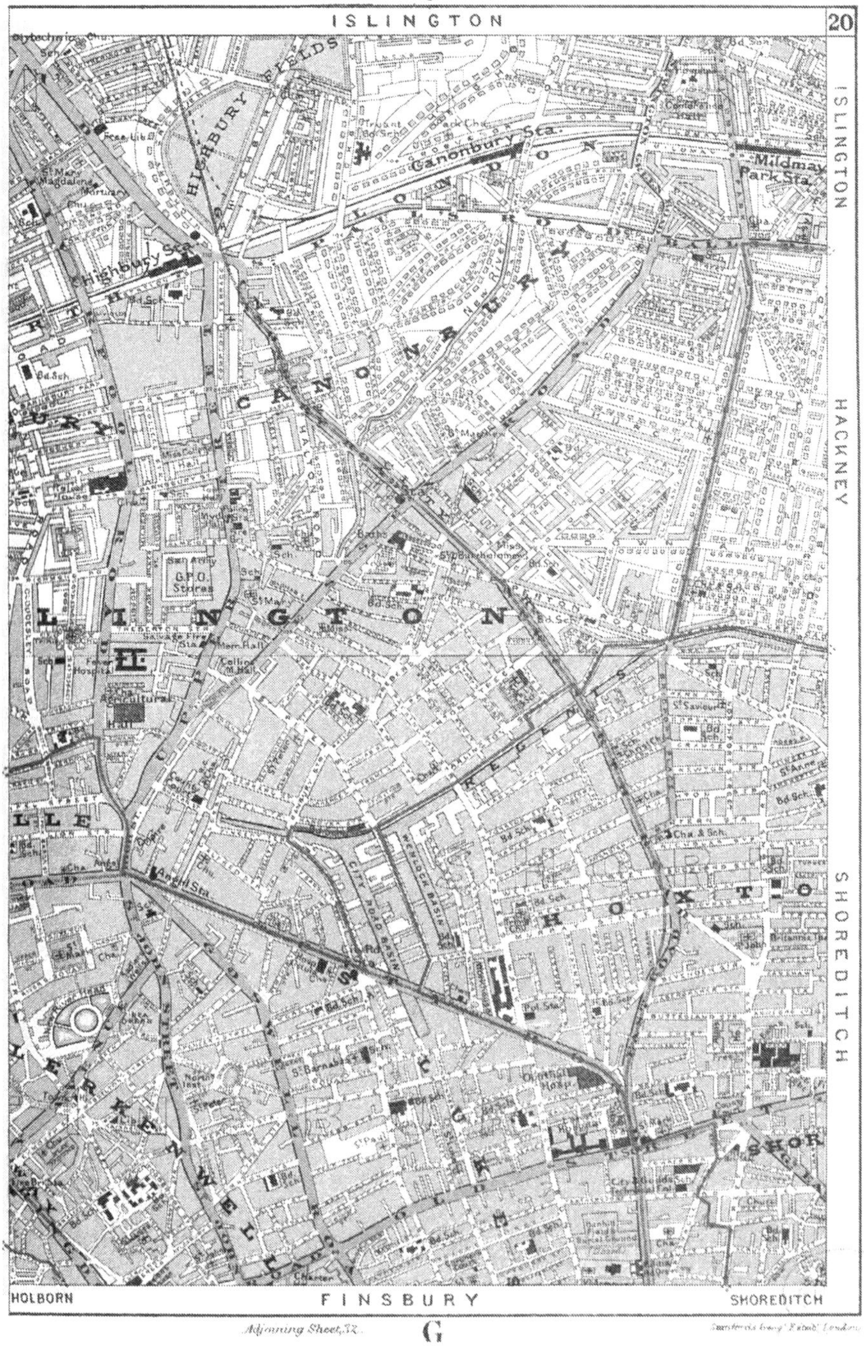
ISLINGTON
20
ISLINGTON
HACKNEY
SHOREDITCH
HOLBORN
FINSBURY
SHOREDITCH
HIGHBURY FIELDS
Highbury Sta.
Canonbury Sta.
Mildmay Park Sta.
CANONBURY
ISLINGTON
G.P.O. Stores
Agricultural Hall
Angel Sta.
City Road Basin
HOXTON
CLERKENWELL

C

Adjoining Sheet, 21.

d

Adjoining Sheet, 32. G

Sanders's Engr Estab. London

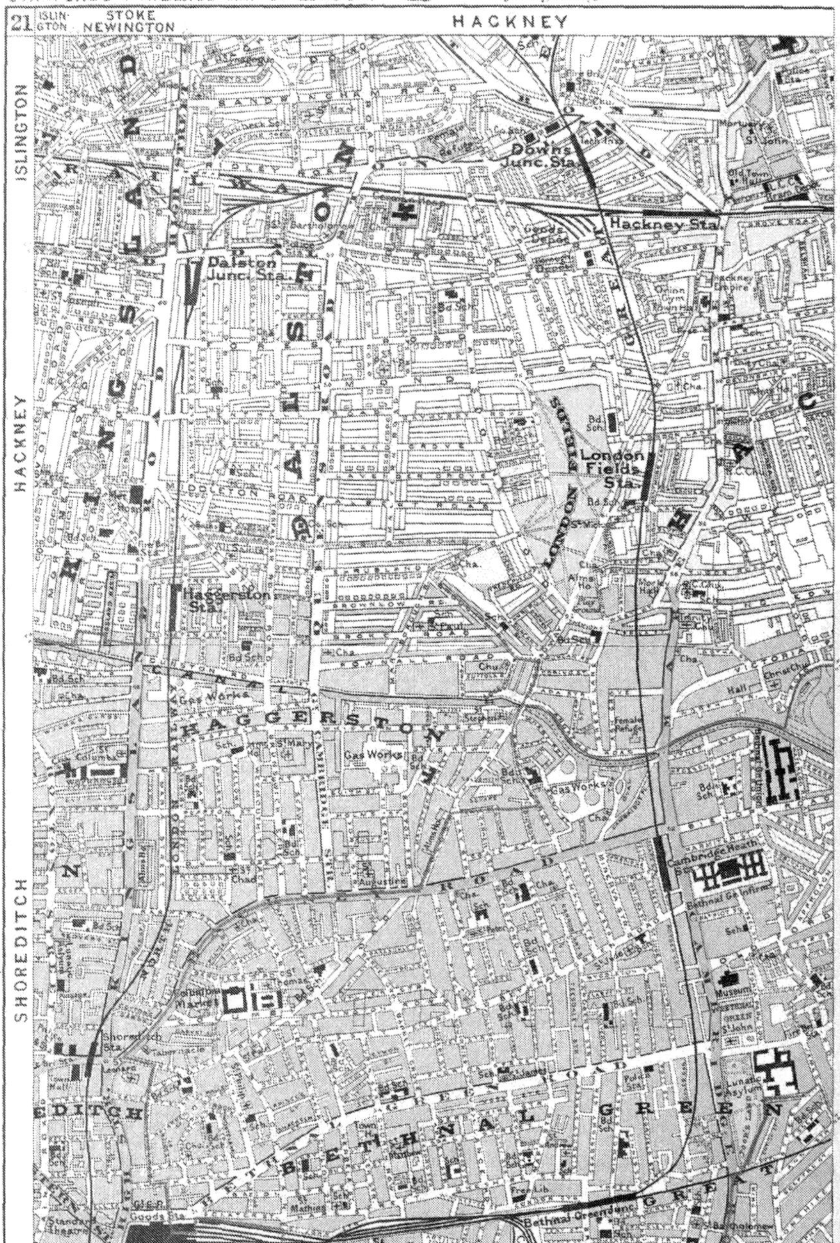

21 ISLINGTON STOKE NEWINGTON HACKNEY

ISLINGTON

HACKNEY

SHOREDITCH

c

d

Adjoining Sheet, 20.

Dalston Junc. Sta. Downs Junc. Sta. Hackney Sta. London Fields Sta. Haggerston Sta. Shoreditch Sta.

KINGSLAND DALSTON HAGGERSTON BETHNAL GREEN SHOREDITCH

SHOREDITCH BETHNAL GREEN STEPNEY

SCALE: 4 INCHES TO 1 MILE ¼ MILE
0 100 200 300 400 500 600 700 800 YARDS

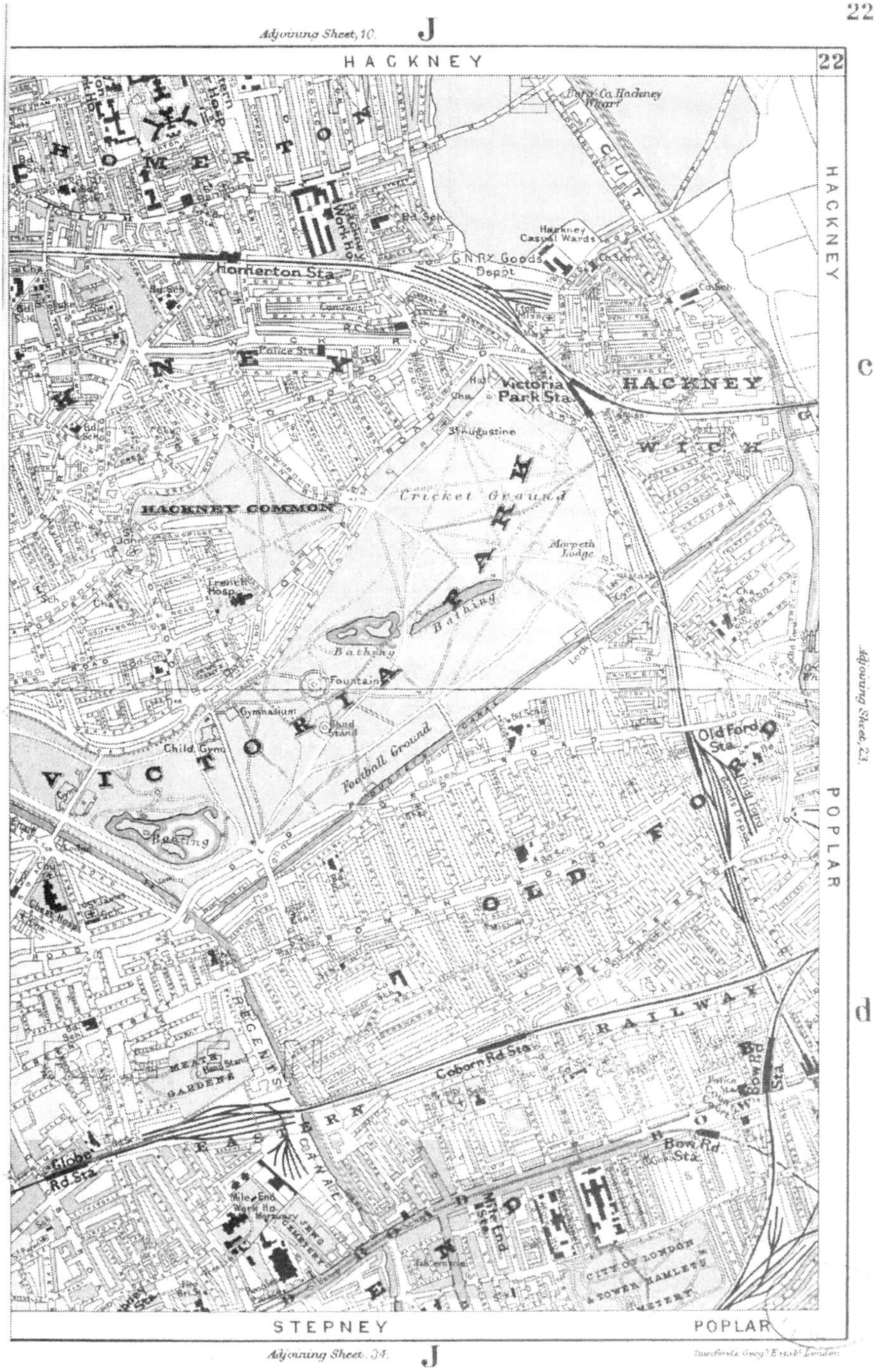
Adjoining Sheet, 10.
J
HACKNEY
22
HACKNEY
C
Adjoining Sheet, 23.
POPLAR
d
STEPNEY
POPLAR
Adjoining Sheet, 34.
J
Stanford's Geog.l Estab.t London
HOMERTON
Homerton Sta.
G.N.R.Y. Goods Depot
Hackney Casual Wards
HACKNEY WICK
Victoria Park Sta.
St. Augustine
Cricket Ground
HACKNEY COMMON
Morpeth Lodge
French Hosp.
Bathing
Fountain
Gymnasium
Band Stand
Football Ground
Child. Gym.
VICTORIA PARK
Old Ford Sta.
OLD FORD
REGENTS CANAL
Coborn Rd. Sta.
RAILWAY
EASTERN
MEATH GARDENS
Globe Rd. Sta.
Mile End Work Ho.
Bow Rd. Sta.
CITY OF LONDON & TOWER HAMLETS CEMETERY
BOW
CUT

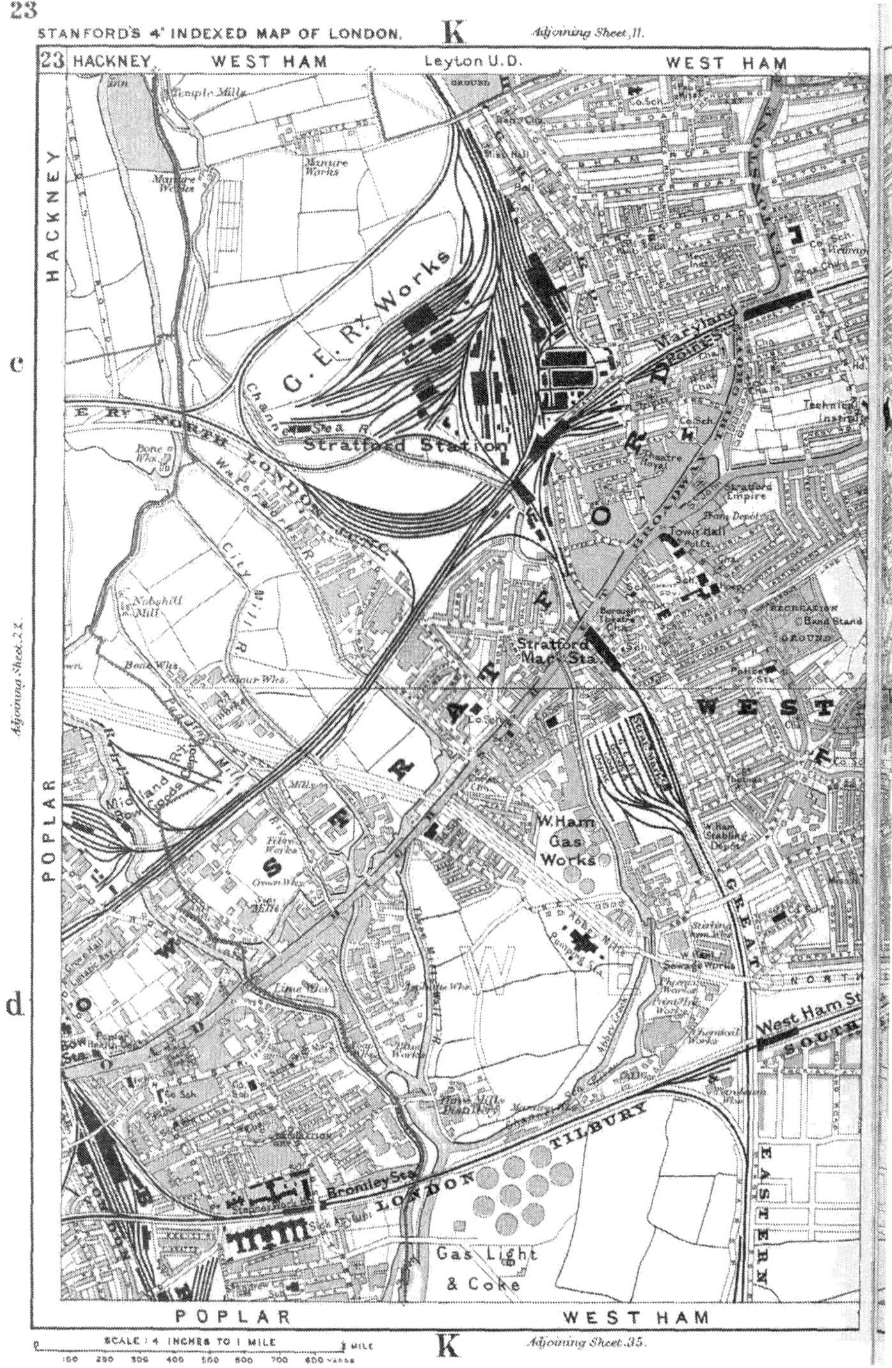
STANFORD'S 4" INDEXED MAP OF LONDON.
K
Adjoining Sheet. 11.
23 HACKNEY
WEST HAM
Leyton U.D.
WEST HAM
HACKNEY
POPLAR
c
d
Adjoining Sheet. 22.
G. E. R^{y} Works
Stratford Station
Stratford Market Sta.
Maryland Point
Theatre Royal
Stratford Empire
Town Hall
Borough Theatre
RECREATION GROUND
Band Stand
W. Ham Gas Works
West Ham Sta.
Bromley Sta.
Gas Light & Coke
TILBURY
LONDON
GREAT EASTERN
NORTH
SOUTH
STRATFORD
WEST
Manure Works
Temple Mills
Bone Whs.
City Mill R.
Bow Sta.
Midland Bow Goods Depot
POPLAR
WEST HAM
SCALE: 4 INCHES TO 1 MILE
K
Adjoining Sheet. 35.

Adjoining Sheet 12. L

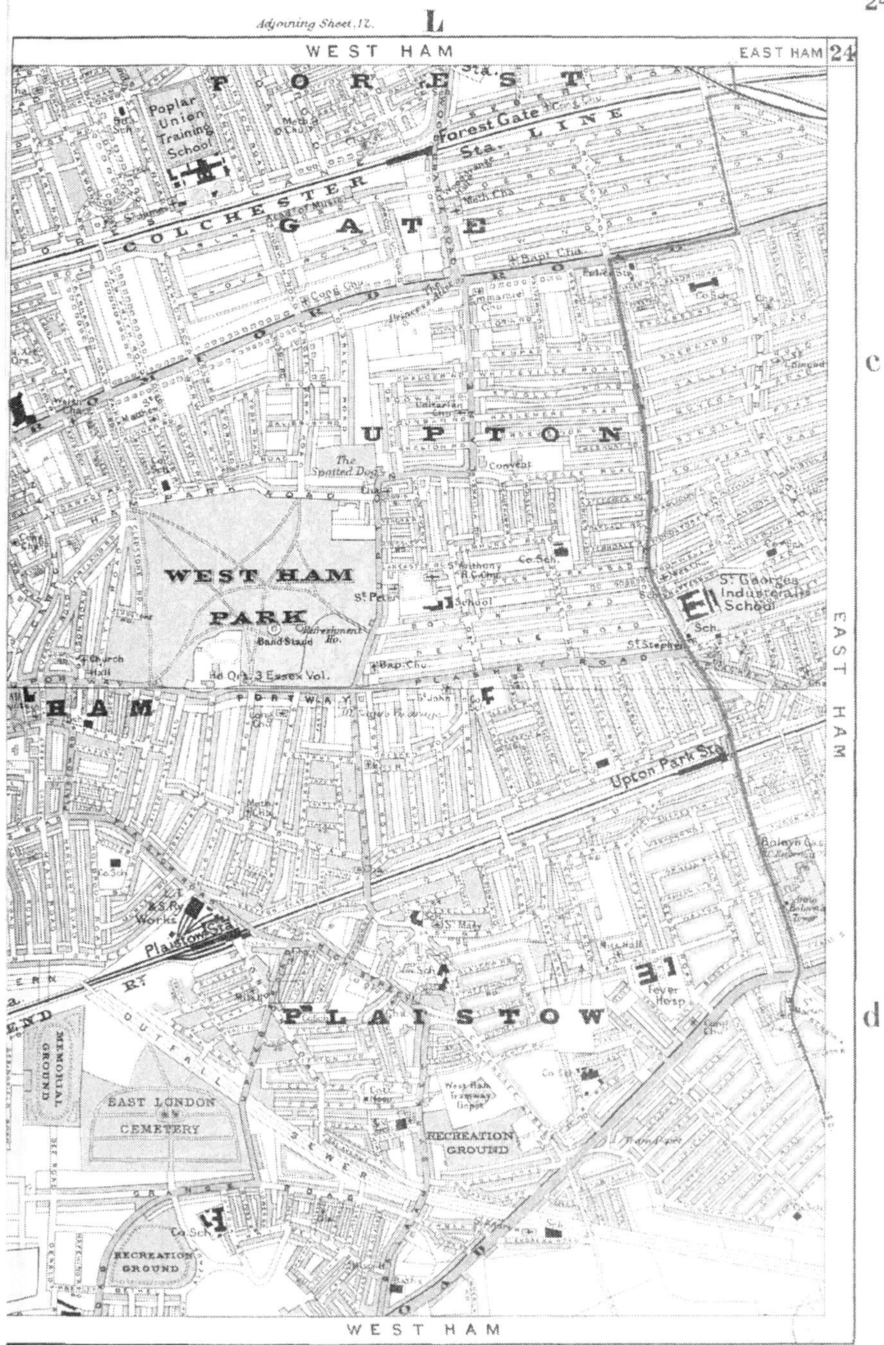

Adjoining Sheet 36. L

Stanford's Geog.l Estab.t London

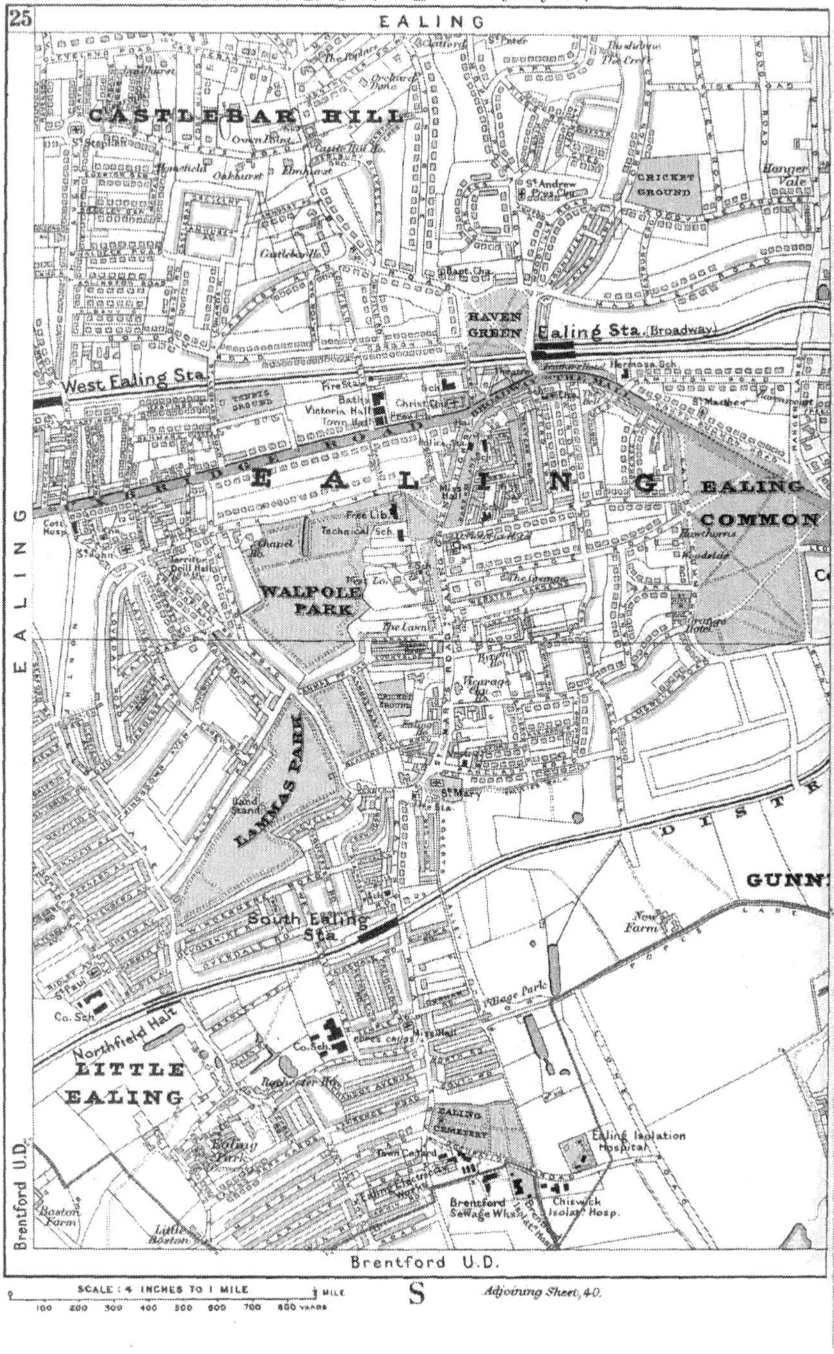

25 EALING

CASTLEBAR HILL

St Stephen

Castlebar Hill

St Andrew

CRICKET GROUND

Hanger Vale

Bapt. Cha.

RAVEN GREEN

Ealing Sta. (Broadway)

West Ealing Sta.

TENNIS GROUND

Fire Sta.

Baths

Victoria Hall

Town Hall

Christ Church

Hermosa Sch.

St Matthew

UXBRIDGE ROAD

E A L I N G

EALING COMMON

Cott. Hosp.

Free Lib.

Technical Sch.

Chapel Ho.

St John

Drill Hall

WALPOLE PARK

The Lawn

The Grange

Grange Hotel

Vicarage

St Mary

LAMMAS PARK

Band Stand

D I S T R

GUNN

South Ealing Sta.

New Farm

Village Park

St Paul

Co. Sch.

Northfield Halt

LITTLE EALING

Co. Sch.

EALING CEMETERY

Ealing Isolation Hospital

Ealing Park

Town Co. Yard

Brentford Sewage Wks.

Chiswick Isolatn Hosp.

Boston Farm

Little Boston

e

f

EALING

Brentford U.D.

Brentford U.D.

SCALE: 4 INCHES TO 1 MILE

0 100 200 300 400 500 600 700 800 YARDS ¼ MILE

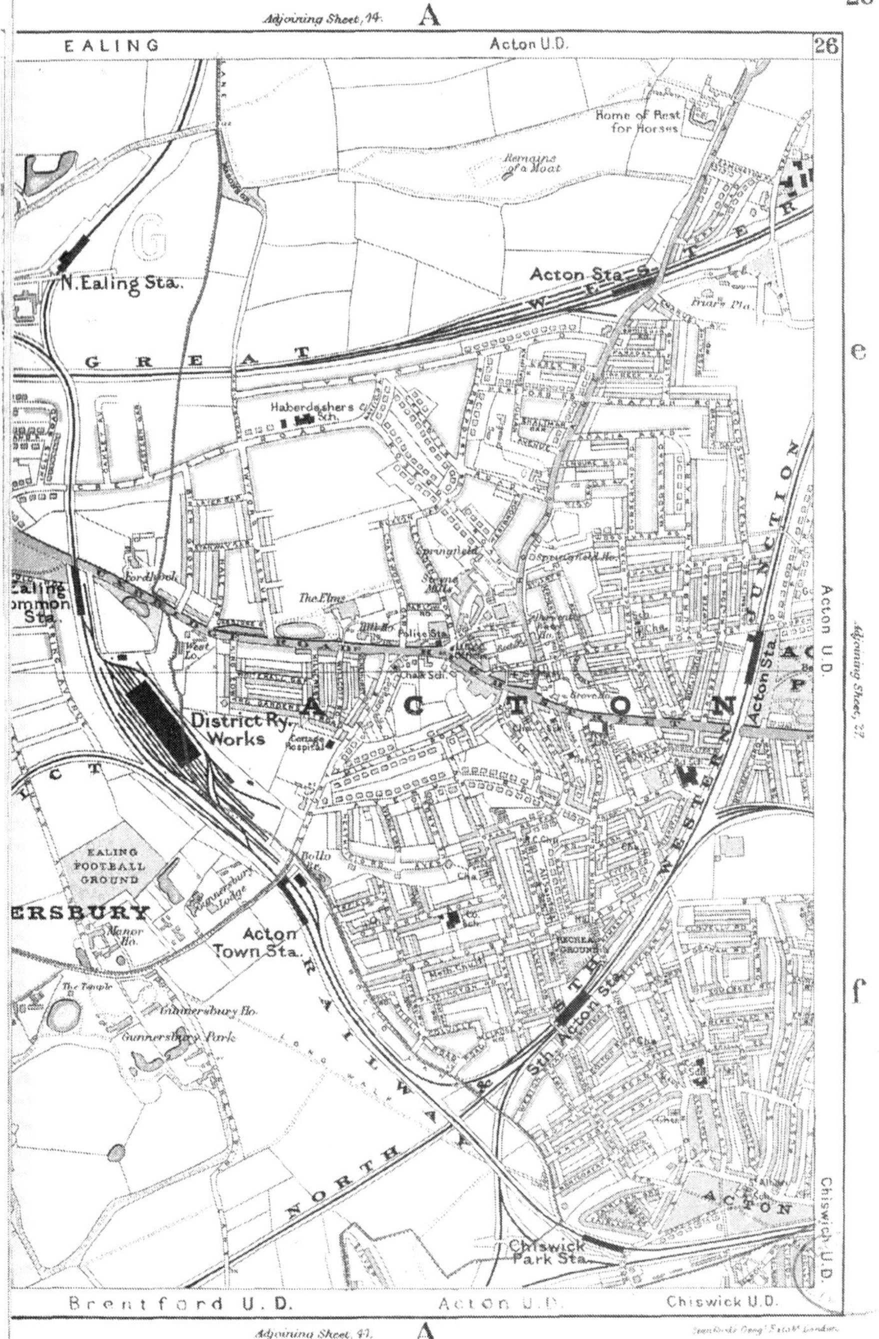
Adjoining Sheet, 14.
A
EALING
Acton U.D.
26
Home of Rest for Horses
Remains of a Moat
N. Ealing Sta.
Acton Sta.
G R E A T W E S T E R N
Friar's Pla.
Haberdashers Sch.
Springfield
Springfield Ho.
Fordhook
The Elms
Ealing Common Sta.
West Lo.
Police Sta.
Chalk Sch.
A C T O N
District Ry. Works
Cottage Hospital
Acton Sta.
Acton U.D.
Adjoining Sheet, 27.
e
EALING FOOTBALL GROUND
ERSBURY
Gunnersbury Lodge
Manor Ho.
Acton Town Sta.
RECREATION GROUND
Sth. Acton Sta.
N. & S. W. JUNCTION
The Temple
Gunnersbury Ho.
Gunnersbury Park
f
NORTH RAILWAY
ACTON
Chiswick Park Sta.
Chiswick U.D.
Brentford U.D.
Acton U.D.
Chiswick U.D.
Adjoining Sheet, 41.
A

STANFORD'S 4" INDEXED MAP OF LONDON. B *Adjoining Sheet, 15.*

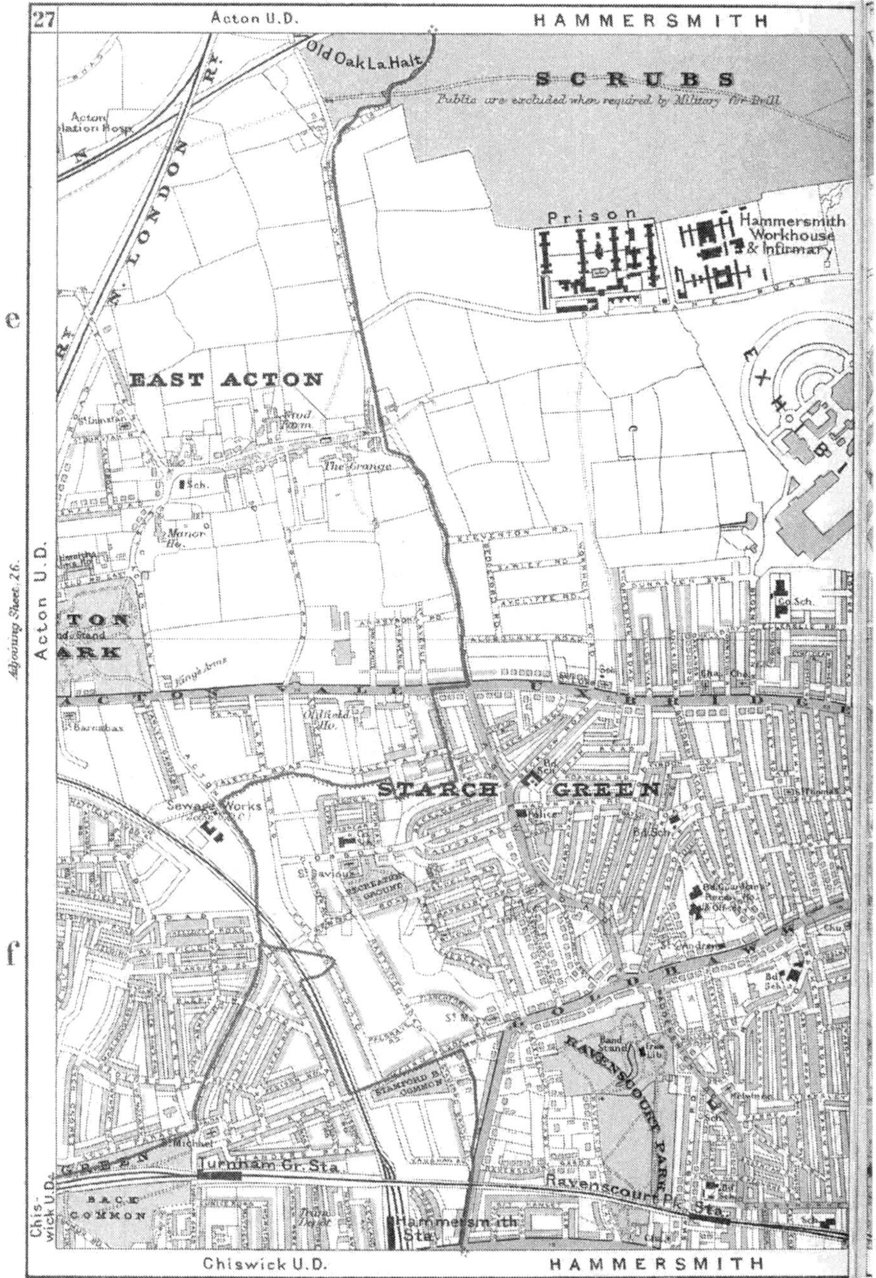
27
Acton U.D.
HAMMERSMITH
Old Oak La. Halt
SCRUBS
Public are excluded when required by Military for Drill
Prison
Hammersmith Workhouse & Infirmary
EAST ACTON
The Grange
Manor Ho.
Sch.
Adjoining Sheet, 26.
Acton U.D.
Kings Arms
Oldfield Ho.
Sewage Works
STARCH GREEN
St. Saviours
RECREATION GROUND
St. Andrews
RAVENSCOURT PARK
Band Stand
STAMFORD B. COMMON
St. Michael
Turnham Gr. Sta.
BACK COMMON
Hammersmith Sta.
Ravenscourt Pk. Sta.
Chiswick U.D.
Chiswick U.D.
HAMMERSMITH
e
f

SCALE: 4 INCHES TO 1 MILE ¼ MILE

100 200 300 400 500 600 700 800 YARDS

B *Adjoining Sheet, 42.*

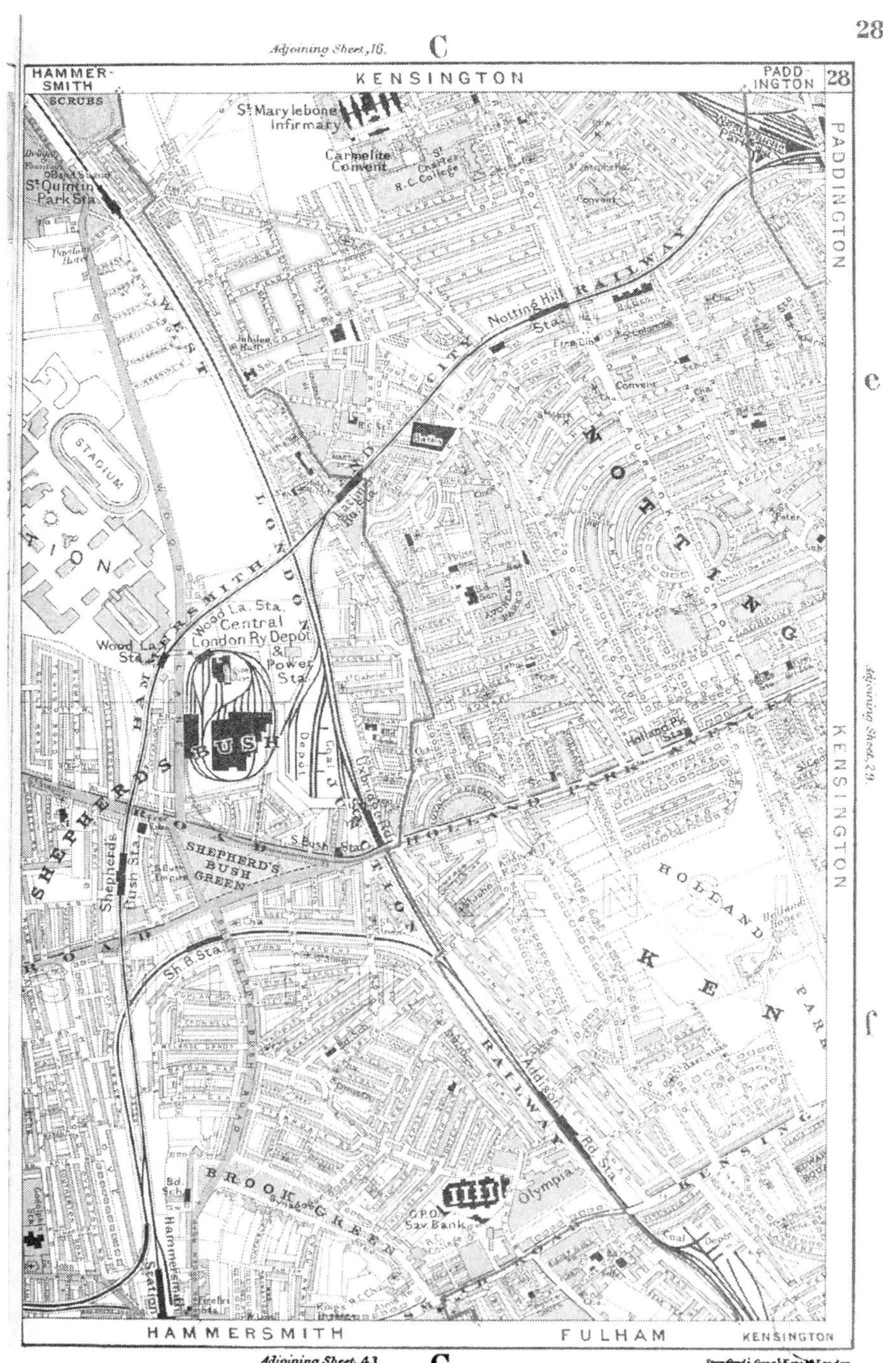
Adjoining Sheet, 16.
C
HAMMERSMITH
KENSINGTON
PADDINGTON
28
PADDINGTON
KENSINGTON
Adjoining Sheet, 29.
HAMMERSMITH
FULHAM
KENSINGTON
Adjoining Sheet, 43.
C
St Marylebone Infirmary
Carmelite Convent
St Quintin Park Sta.
STADIUM
Notting Hill Sta.
Wood La. Sta.
Central London Ry. Depot & Power Sta.
Wood La. Sta.
BUSH
SHEPHERDS
SHEPHERD'S BUSH GREEN
Shepherds Bush Sta.
Sh. B. Sta.
Holland Pk. Sta.
HOLLAND PARK
HOLLAND
Holland House
BROOK GREEN
Olympia
G.P.O. Sav. Bank
Addison Rd. Sta.
Hammersmith Station
Stanford's Geog! Estabt London

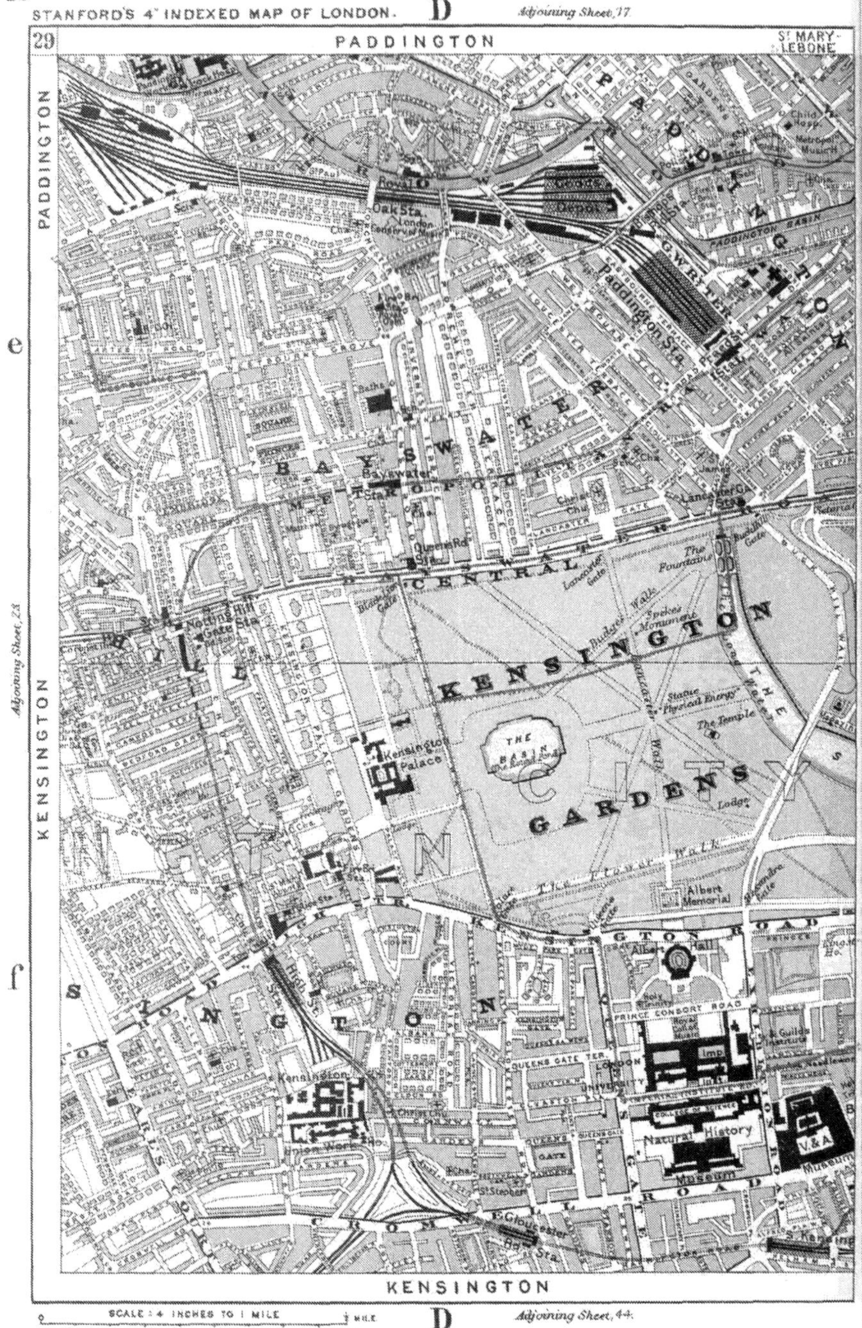
29
STANFORD'S 4" INDEXED MAP OF LONDON.
D
Adjoining Sheet, 17
PADDINGTON
St. MARY-LEBONE
PADDINGTON
KENSINGTON
e
f
Adjoining Sheet, 28
Royal Oak Sta.
Goods Depot
Paddington Sta.
PADDINGTON BASIN
BAYSWATER
Bayswater Sta.
Queens Rd. Sta.
Notting Hill Gate Sta.
Lancaster Ga. Sta.
CENTRAL
KENSINGTON
The Fountains
Speke's Monument
Statue "Physical Energy"
The Temple
THE BASIN
Kensington Palace
GARDENS
Albert Memorial
KENSINGTON ROAD
Albert Hall
PRINCE CONSORT ROAD
QUEENS GATE TER.
LONDON UNIVERSITY
Natural History Museum
V&A
Kensington
Gloucester Rd. Sta.
KENSINGTON
SCALE: 4 INCHES TO 1 MILE
D
Adjoining Sheet, 44.

Adjoining Sheet, 18. E

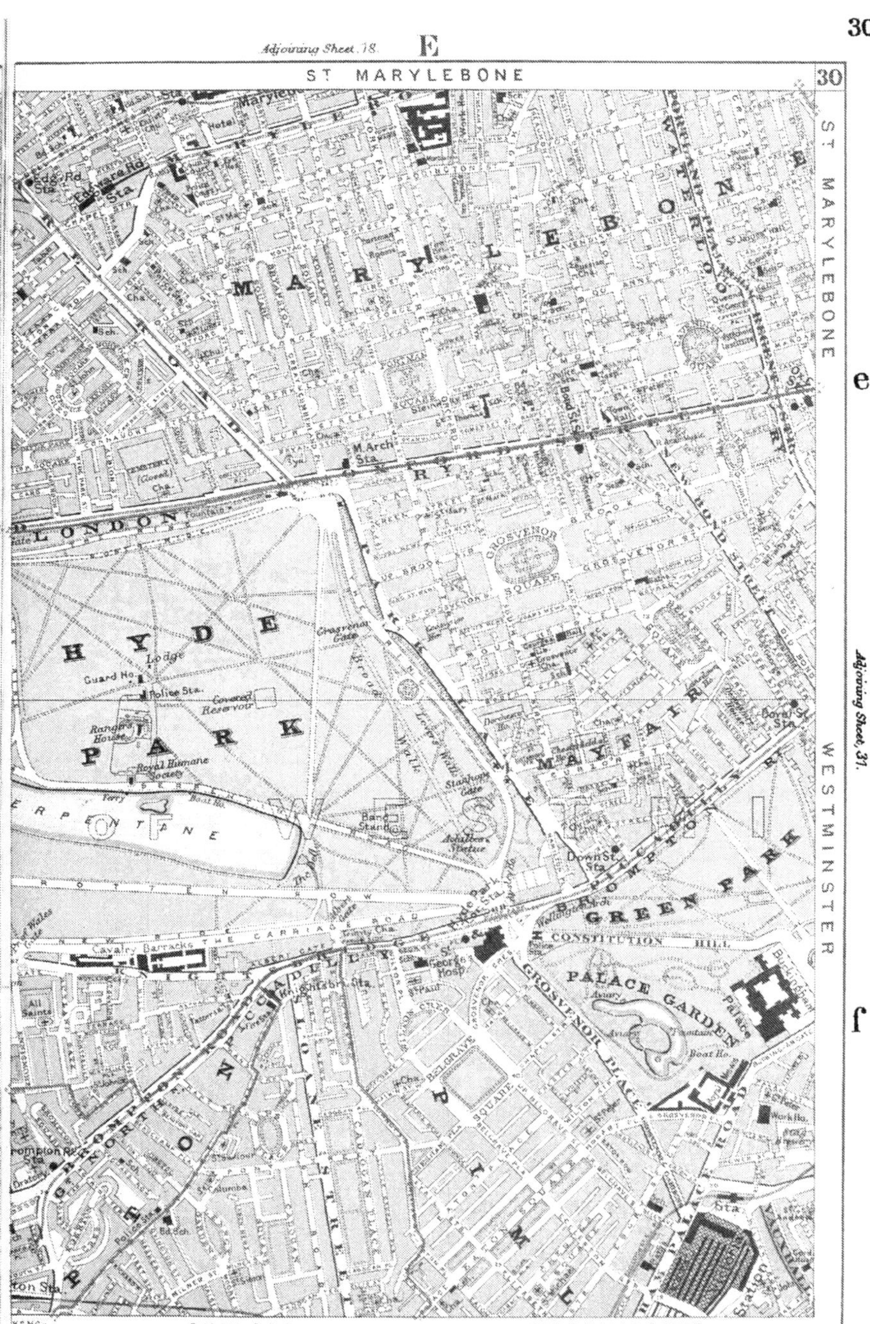

e

Adjoining Sheet, 31.

f

Adjoining Sheet, 45. E

Stanford's Geog.l Estab.t London.

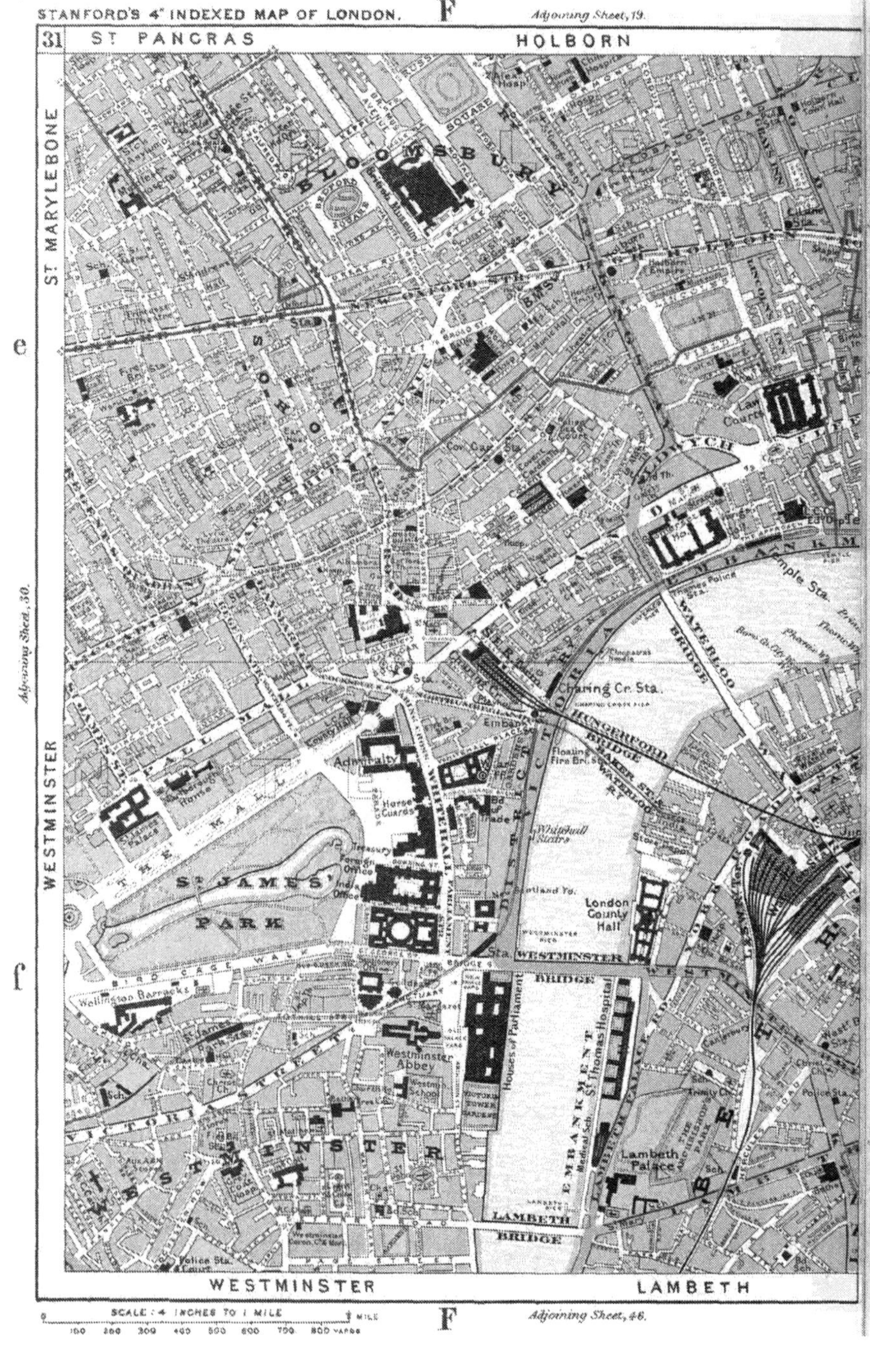
STANFORD'S 4" INDEXED MAP OF LONDON.
F
Adjoining Sheet, 19.
31
ST PANCRAS
HOLBORN
ST MARYLEBONE
e
f
Adjoining Sheet, 30.
WESTMINSTER
BLOOMSBURY
SOHO
Charing Cr. Sta.
HUNGERFORD BRIDGE
WATERLOO BRIDGE
Admiralty
Horse Guards
ST JAMES' PARK
THE MALL
Wellington Barracks
WESTMINSTER BRIDGE
London County Hall
Houses of Parliament
Westminster Abbey
St Thomas Hospital
EMBANKMENT
Lambeth Palace
LAMBETH BRIDGE
WESTMINSTER
LAMBETH
SCALE: 4 INCHES TO 1 MILE
F
Adjoining Sheet, 46.

Adjoining Sheet, 20. G

e

Adjoining Sheet, 31.

f

Adjoining Sheet, 47. G

Stanford's Geog. Estab. London

STANFORD'S 4" INDEXED MAP OF LONDON H *Adjoining Sheet 21.*

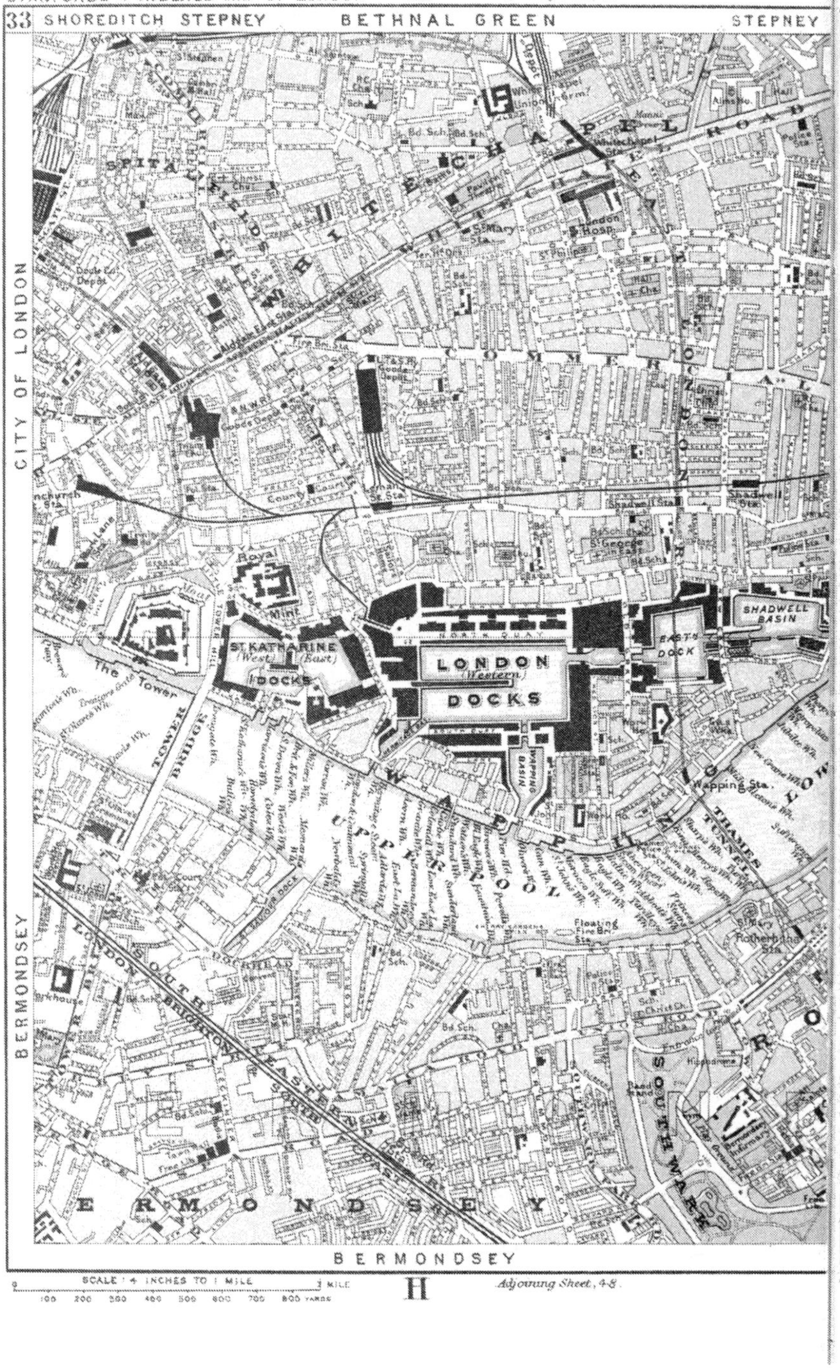

33 SHOREDITCH STEPNEY BETHNAL GREEN STEPNEY

CITY OF LONDON

e

Adjoining Sheet, 32.

f

BERMONDSEY

BERMONDSEY

SCALE 4 INCHES TO 1 MILE ½ MILE

0 100 200 300 400 500 600 700 800 YARDS

H *Adjoining Sheet, 48.*

Adjoining Sheet, 22. J

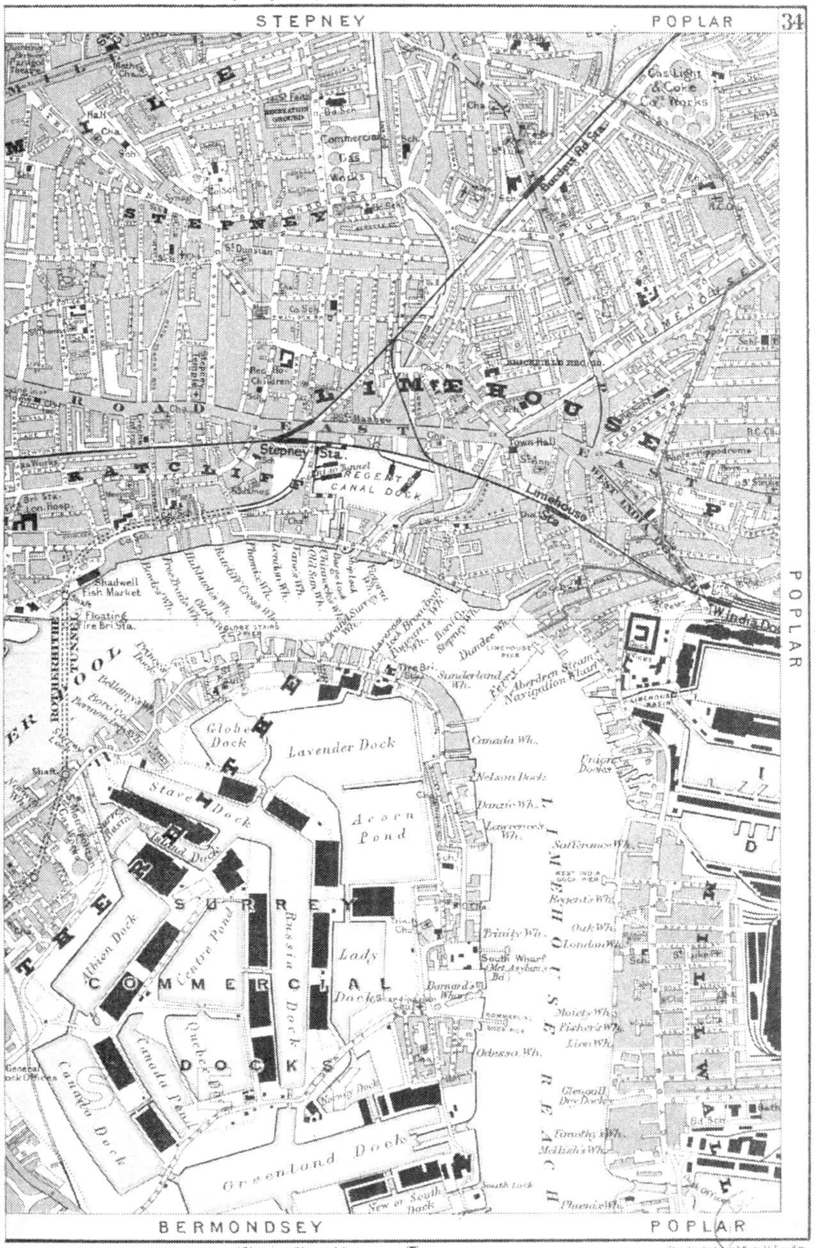

STEPNEY POPLAR 34

Gas Light & Coke Co.s Works

Burdett Rd. Sta.

Commercial Gas Works

St. Dunstan

STEPNEY

LIMEHOUSE

EAST

ROAD

RATCLIFF

Stepney Sta.

REGENT CANAL DOCK

Limehouse Sta.

Town Hall

Shadwell Fish Market

Floating Fire Bri. Sta.

W. India Dock

Dundee Wh.

Sunderland Wh.

Aberdeen Steam Navigation Wharf

Canada Wh.

Nelson Dock

Dantzic Wh.

Lawrence's Wh.

Globe Dock

Lavender Dock

Stave Dock

Acorn Pond

Island Dock

Albion Dock

Centre Pond

Russia Dock

Lady Dock

Quebec Pond

Canada Pond

Canada Dock

Greenland Dock

New or South Dock

South Lock

SURREY COMMERCIAL DOCKS

LIMEHOUSE REACH

Trinity Wh.

South Wharf

Barnard's Wharf

Odessa Wh.

Friars Docks

Sufferance Wh.

Regent's Wh.

Oak Wh.

London Wh.

Moiety Wh.

Fisher's Wh.

Lion Wh.

Glengall Dry Docks

Timothy Wh.

Mellish's Wh.

Phoenix Wh.

POPLAR

Adjoining Sheet, 35.

BERMONDSEY POPLAR

Adjoining Sheet, 49. J

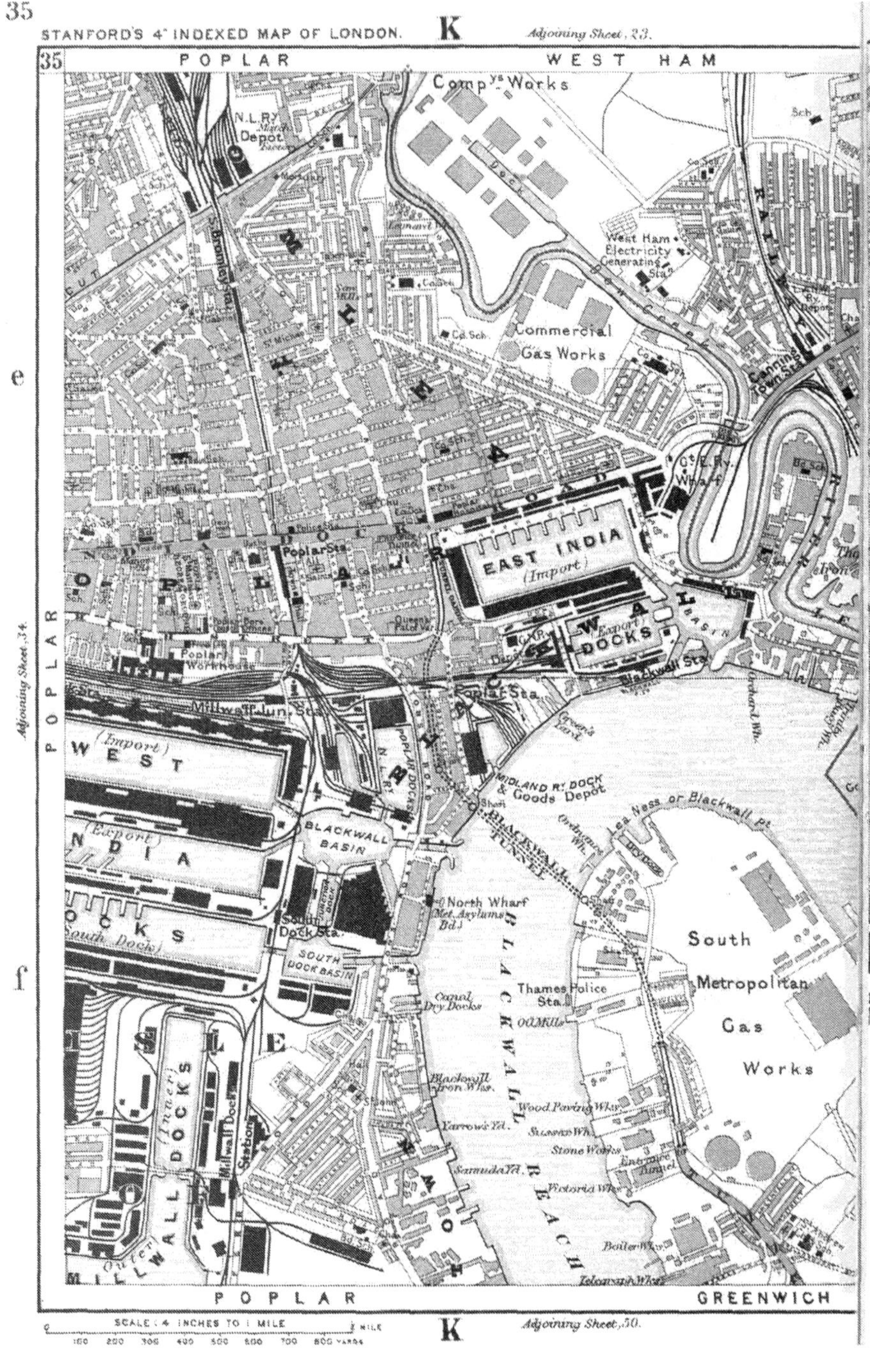
35
STANFORD'S 4" INDEXED MAP OF LONDON.
K
Adjoining Sheet, 23.
POPLAR
WEST HAM
Compys Works
N.L.RY Depot
West Ham Electricity Generating Sta.
Commercial Gas Works
Canning Town Sta.
Gt. E. Ry. Wharf
EAST INDIA (Import)
(Export) DOCKS
Blackwall Sta.
Poplar Sta.
Poplar Workhouse
Millwall Jun. Sta.
(Import) WEST INDIA (Export) DOCKS (South Dock)
POPLAR DOCK
MIDLAND Ry. DOCK & Goods Depot
BLACKWALL TUNNEL
Lea Ness or Blackwall Pt.
BLACKWALL BASIN
North Wharf
Met. Asylums Bd.
South Dock Sta.
SOUTH DOCK BASIN
BLACKWALL REACH
Canal Dry Docks
Thames Police Sta.
Oil Mills
South Metropolitan Gas Works
Blackwall Iron Whs.
Wood Paving Whf.
Yarrow's Yd.
Stone Works
Samuda Yd.
Victoria Whf.
Entrance Tunnel
Boiler Whf.
Telegraph Whf.
(Inner) DOCKS
(Outer) MILLWALL
Millwall Dock Station
Adjoining Sheet, 34.
POPLAR
e
f
POPLAR
GREENWICH
SCALE: 4 INCHES TO 1 MILE
K
Adjoining Sheet, 30.

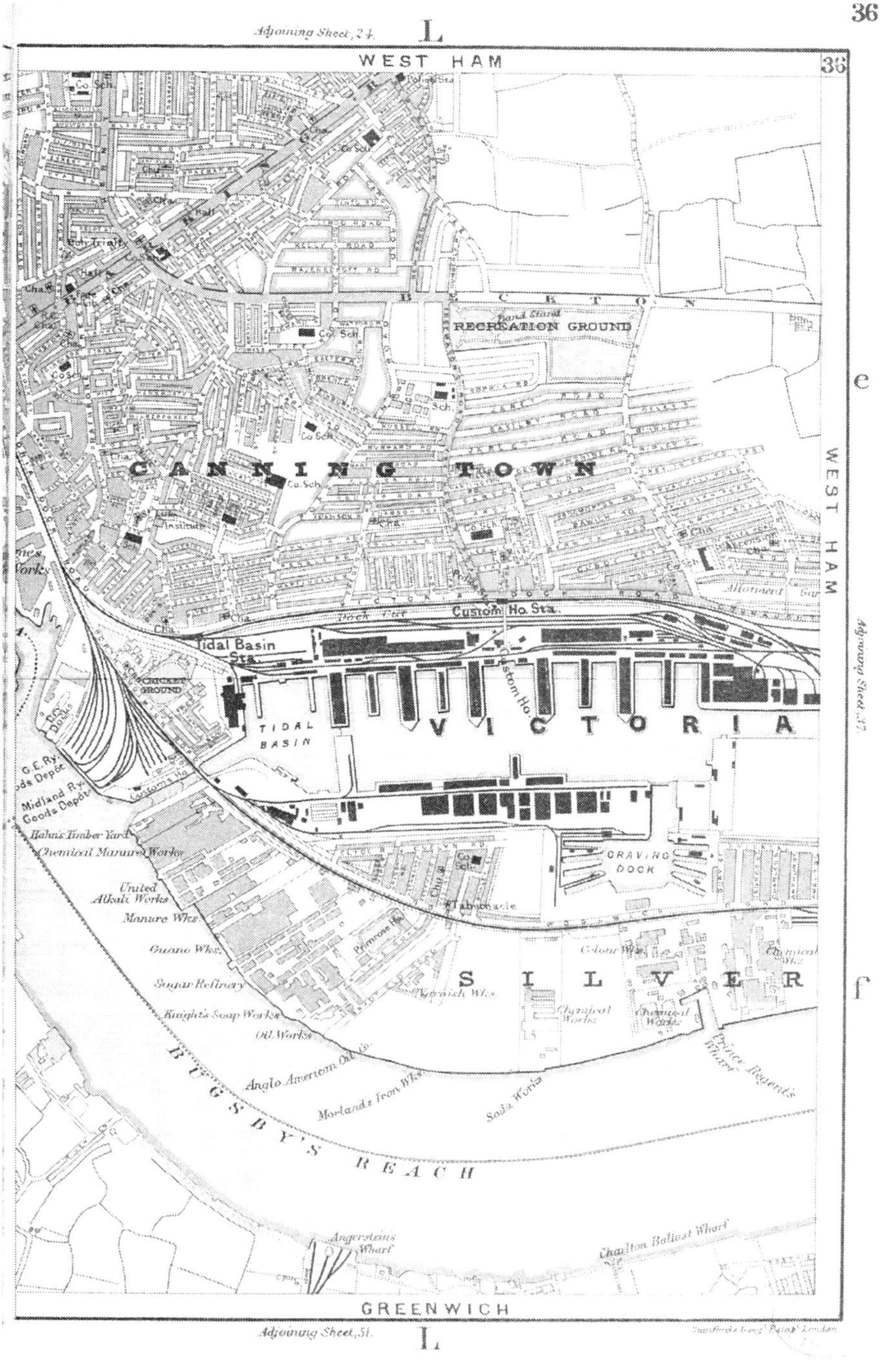

Adjoining Sheet, 24. L

WEST HAM 36

BECKTON

Band Stand

RECREATION GROUND

CANNING TOWN

Custom Ho. Sta.

Tidal Basin Sta.

Dock Cut

Custom Ho.

CRICKET GROUND

TIDAL BASIN

VICTORIA

G.E. Ry. Goods Depot

Midland Ry. Goods Depot

Hahn's Timber Yard

Chemical Manure Works

United Alkali Works

Manure Wks.

Guano Wks.

Sugar Refinery

Knight's Soap Works

Oil Works

Anglo American Oil Co.

Morland's Iron Wks.

Soda Works

Primrose Rd.

Tabernacle

CRAVING DOCK

Allotment

Colour Wks.

Chemical Works

Chemical Wks.

SILVER

Prince Regent's Wharf

BUGSBY'S REACH

Angersteins Wharf

Charlton Ballast Wharf

WEST HAM

Adjoining Sheet, 37.

e

f

GREENWICH

Adjoining Sheet, 31. L

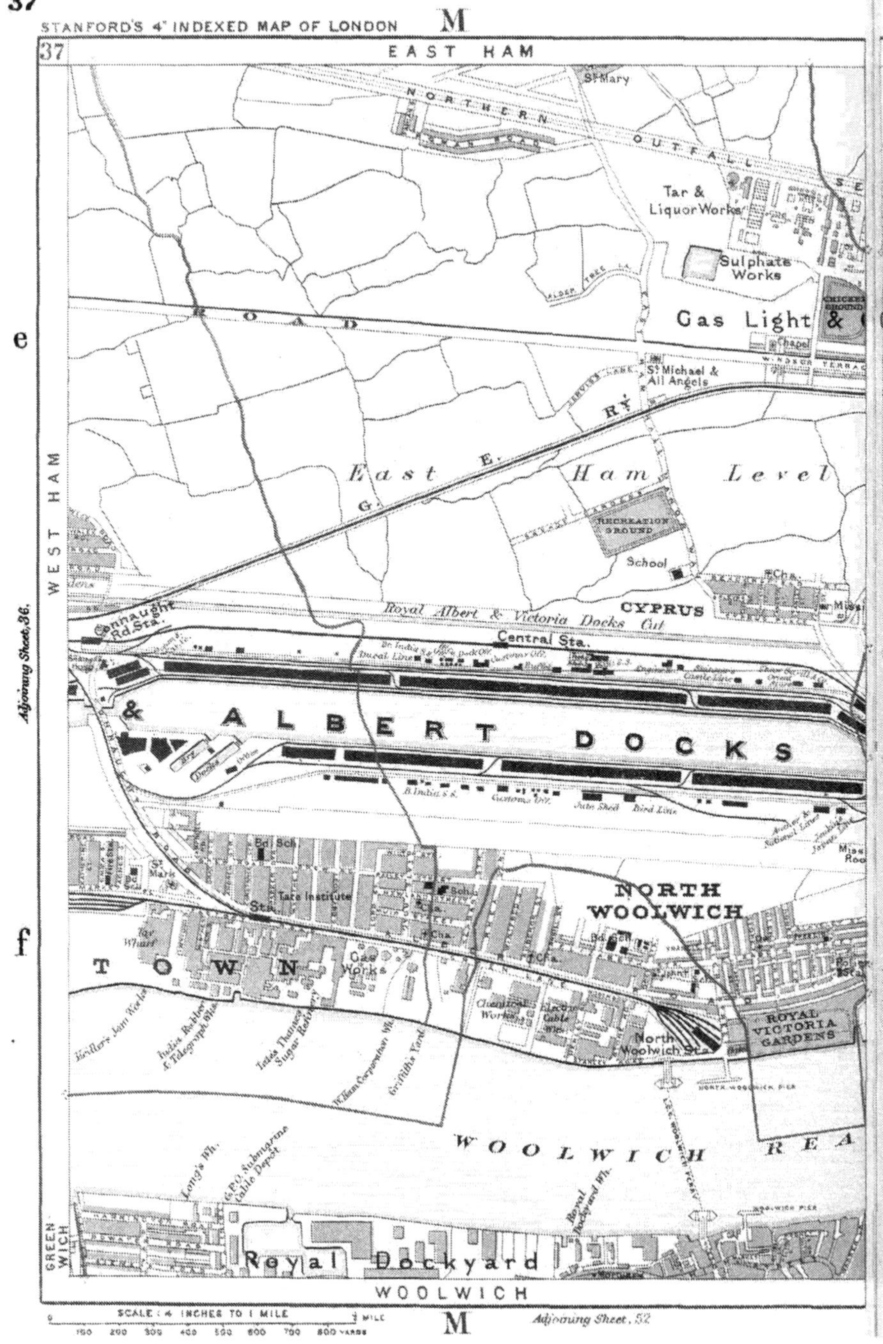
STANFORD'S 4" INDEXED MAP OF LONDON
M
37
EAST HAM
St Mary
NORTHERN OUTFALL SEWER
Tar & Liquor Works
Sulphate Works
ROAD
Gas Light & Co.
Chapel
St Michael & All Angels
RY
East Ham Level
G. E.
RECREATION GROUND
School
Cha.
CYPRUS
Royal Albert & Victoria Docks Cut
Connaught Rd. Sta.
Central Sta.
& ALBERT DOCKS
Bd. Sch.
St Mark
Tate Institute
Sta.
NORTH WOOLWICH
TOWN
Gas Works
Chemical Works
North Woolwich Sta.
ROYAL VICTORIA GARDENS
India Rubber & Telegraph Wks.
Tate's Thames Sugar Refinery
Griffith's Yard
WOOLWICH REACH
Long's Wh.
G.P.O. Submarine Cable Depot
Royal Dockyard Wh.
Royal Dockyard
WEST HAM
GREENWICH
WOOLWICH
e
f
Adjoining Sheet, 36.
Adjoining Sheet, 52
SCALE: 4 INCHES TO 1 MILE
M

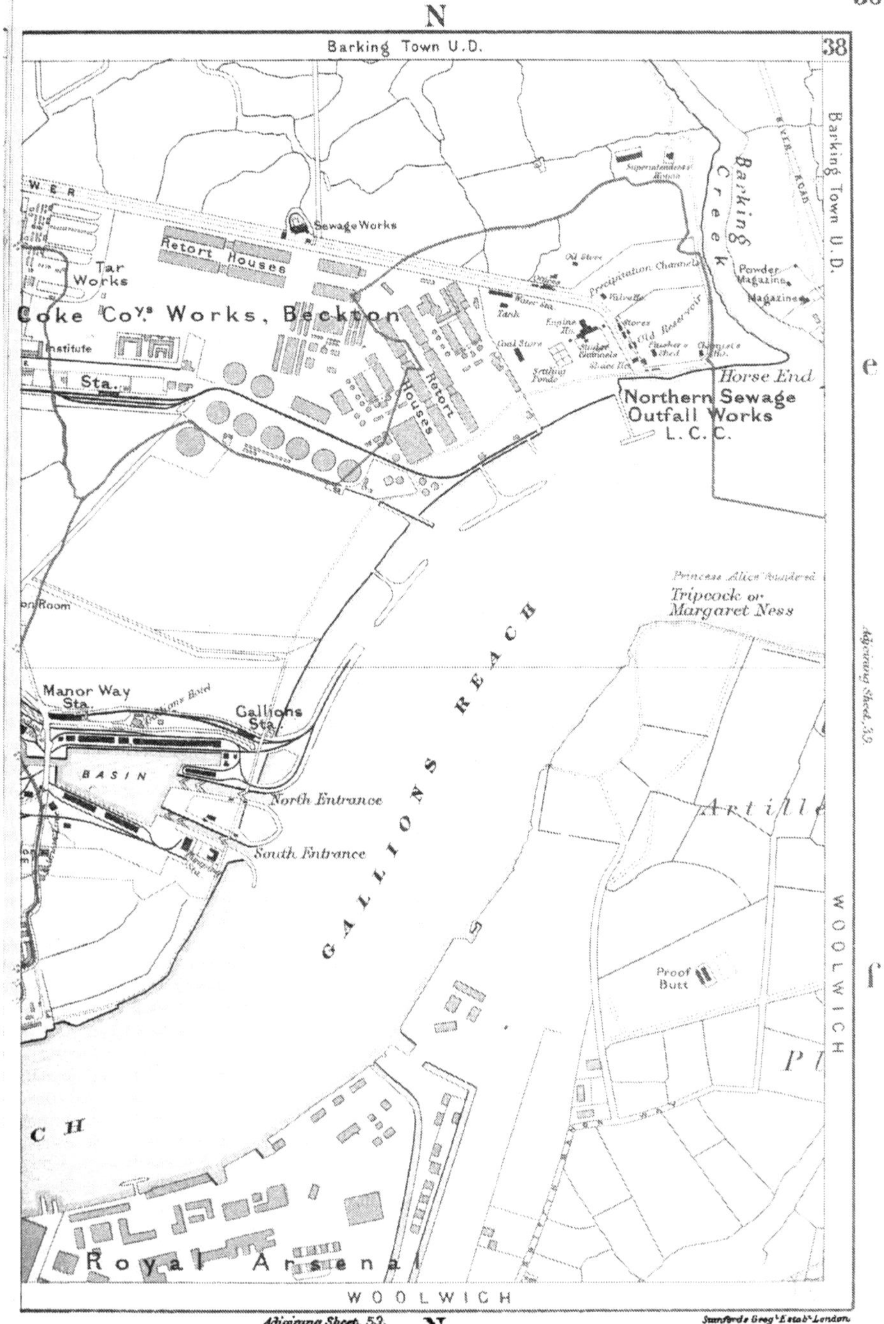
N
Barking Town U.D.
38
Barking Town U.D.
Barking Creek
Sewage Works
Retort Houses
Tar Works
Coke Co.ys Works, Beckton
Institute
Sta.
Powder Magazine
Magazine
Precipitation Channels
Old Reservoir
Horse End
Northern Sewage Outfall Works L. C. C.
Retort Houses
Princess Alice Foundered
Tripcock or Margaret Ness
GALLIONS REACH
Manor Way Sta.
Gallions Sta.
BASIN
North Entrance
South Entrance
Proof Butt
WOOLWICH
Royal Arsenal
WOOLWICH
Adjoining Sheet, 53.
Adjoining Sheet, 39.
N
Stanford's Geog.l Estab.t London.

e

f

39

Barking Town U.D.

Pond

CHOATS MANOR WAY

Barking Town U.D.

Sch.

CREEKMOUTH

Chemical Works

Guano Works

Ripple Marsh

Powder Magazine

RIVER THAMES

Riv. Roding

Barking or False Pt.

BARKING REACH

Cross Ness

Powder Magazine

Picquet Stables

ry Practice Ground

Rifle Ranges

Splinter Proof Shelter Shed

CROSS NESS MANOR WAY

umstead Marshes

Linton Shed

Road to Crossness Southern Outfall Sewage Works, and course of the Southern Outfall Sewer

WOOLWICH

WOOLWICH

Adjoining Sheet, 38.

e

f

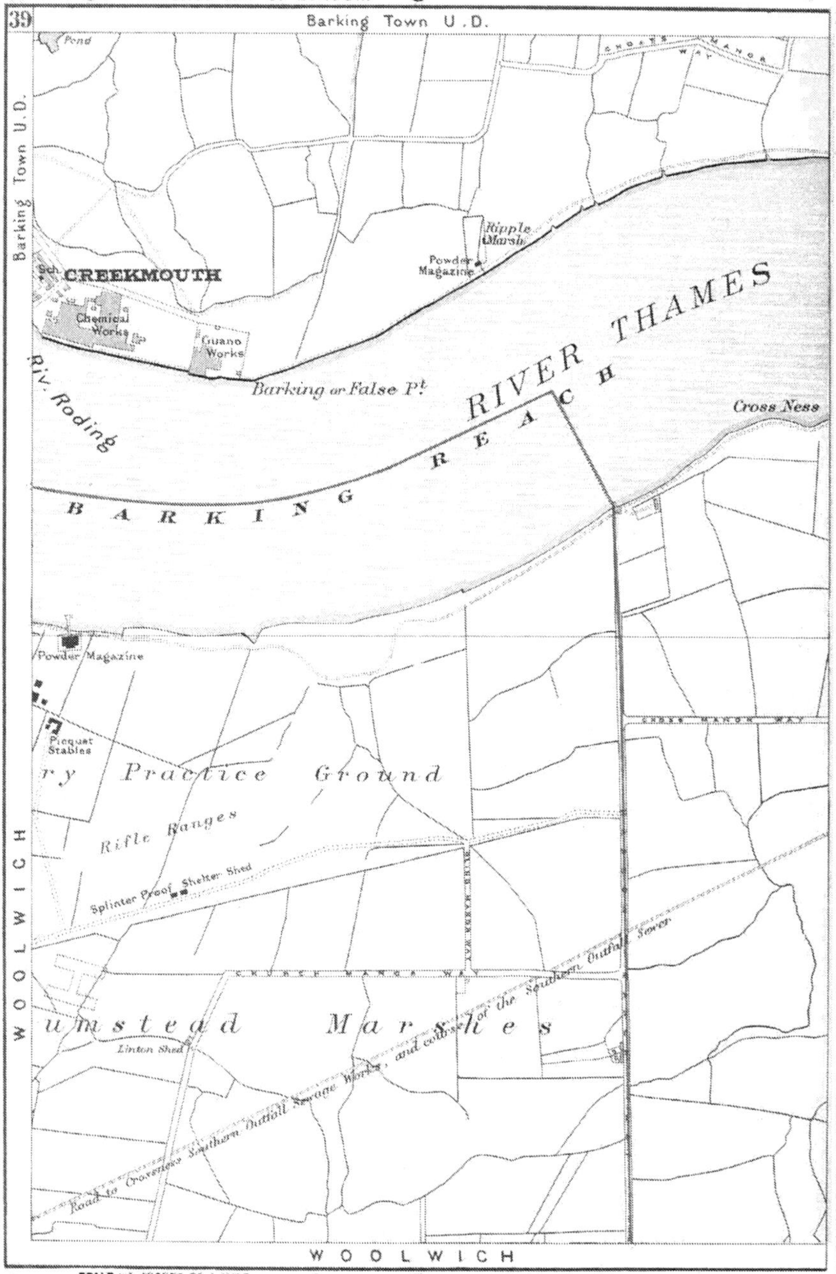

SCALE: 4 INCHES TO 1 MILE

0 100 200 300 400 500 600 700 800 YARDS ¼ MILE

Adjoining Sheet, 54.

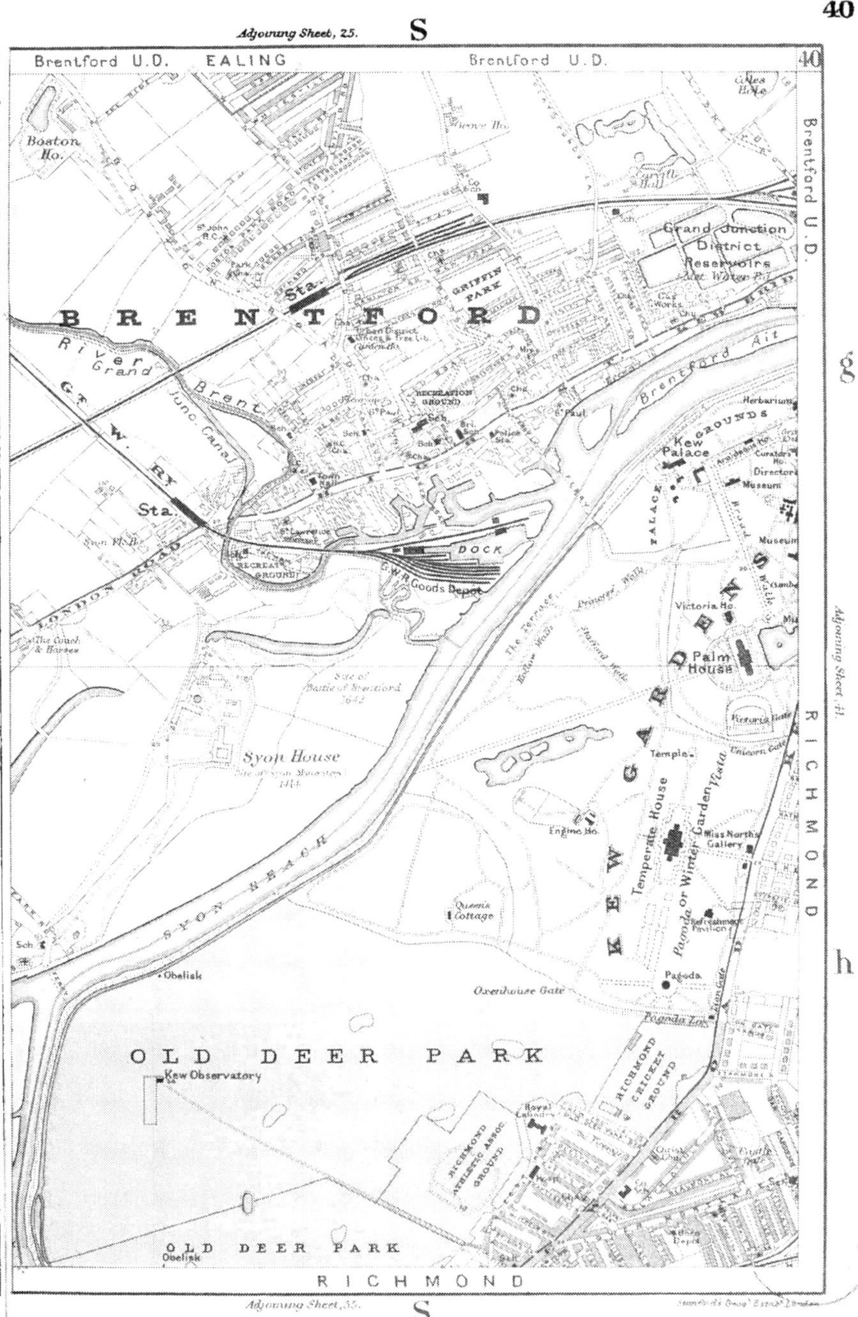
Adjoining Sheet, 25.
S
Brentford U.D.
EALING
Brentford U.D.
40
Boston Ho.
BRENTFORD
River Brent
Grand Junc. Canal
GT. W. RY.
Sta.
GRIFFIN PARK
RECREATION GROUND
Grand Junction District Reservoirs
Brentford Ait
Kew Palace
Herbarium
Museum
Victoria Ho.
Palm House
KEW GARDENS
Temple
Temperate House
Pagoda or Winter Garden Vista
Miss North's Gallery
Engine Ho.
Queen's Cottage
Refreshment Pavilion
Pagoda
Oxenhouse Gate
Pagoda Lo.
DOCK
G.W.R. Goods Depot
LONDON ROAD
Syon House
SYON REACH
Obelisk
OLD DEER PARK
Kew Observatory
RICHMOND CRICKET GROUND
RICHMOND ATHLETIC ASSOC. GROUND
OLD DEER PARK
Obelisk
RICHMOND
Brentford U.D.
g
h
Adjoining Sheet, 41.
RICHMOND
Adjoining Sheet, 55.
S

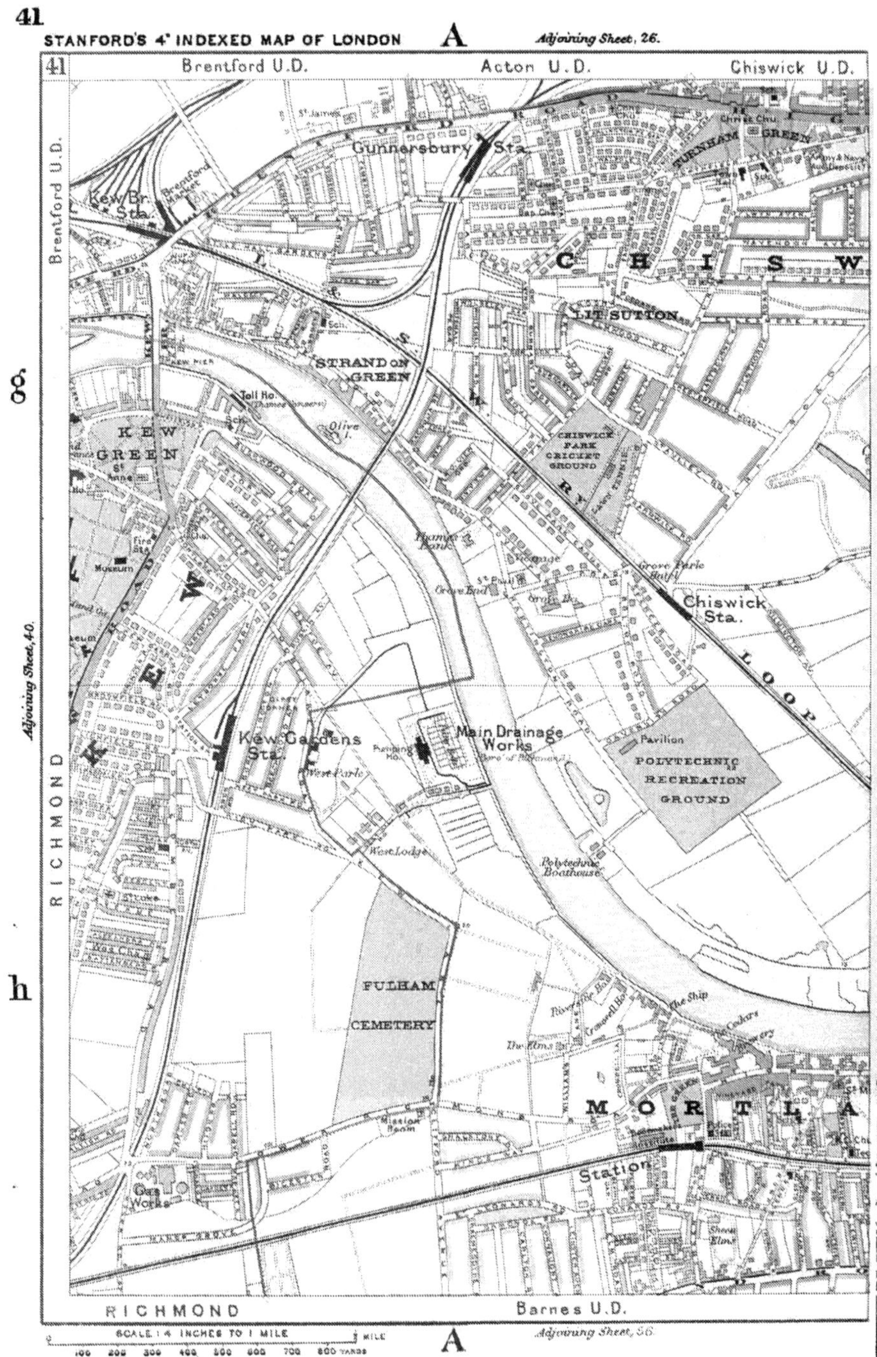
STANFORD'S 4" INDEXED MAP OF LONDON
A
Adjoining Sheet, 26.
41
Brentford U.D.
Acton U.D.
Chiswick U.D.
Gunnersbury Sta.
Kew Br. Sta.
TURNHAM GREEN
C H I S W
LIT SUTTON
STRAND ON GREEN
Toll Ho.
Olive I.
KEW GREEN
Museum
CHISWICK PARK CRICKET GROUND
Grove Park
Chiswick Sta.
LOOP
Grove End
Kew Gardens Sta.
Main Drainage Works
Pavilion
POLYTECHNIC RECREATION GROUND
West Park
West Lodge
Polytechnic Boathouse
FULHAM CEMETERY
The Ship
The Elms
Mission Room
M O R T L A
Station
Gas Works
Sheen Elms
RICHMOND
Barnes U.D.
Brentford U.D.
Adjoining Sheet, 40.
g
h
SCALE: 4 INCHES TO 1 MILE
100 200 300 400 500 600 700 800 YARDS
A
Adjoining Sheet, 56.

Adjoining Sheet, 27. B

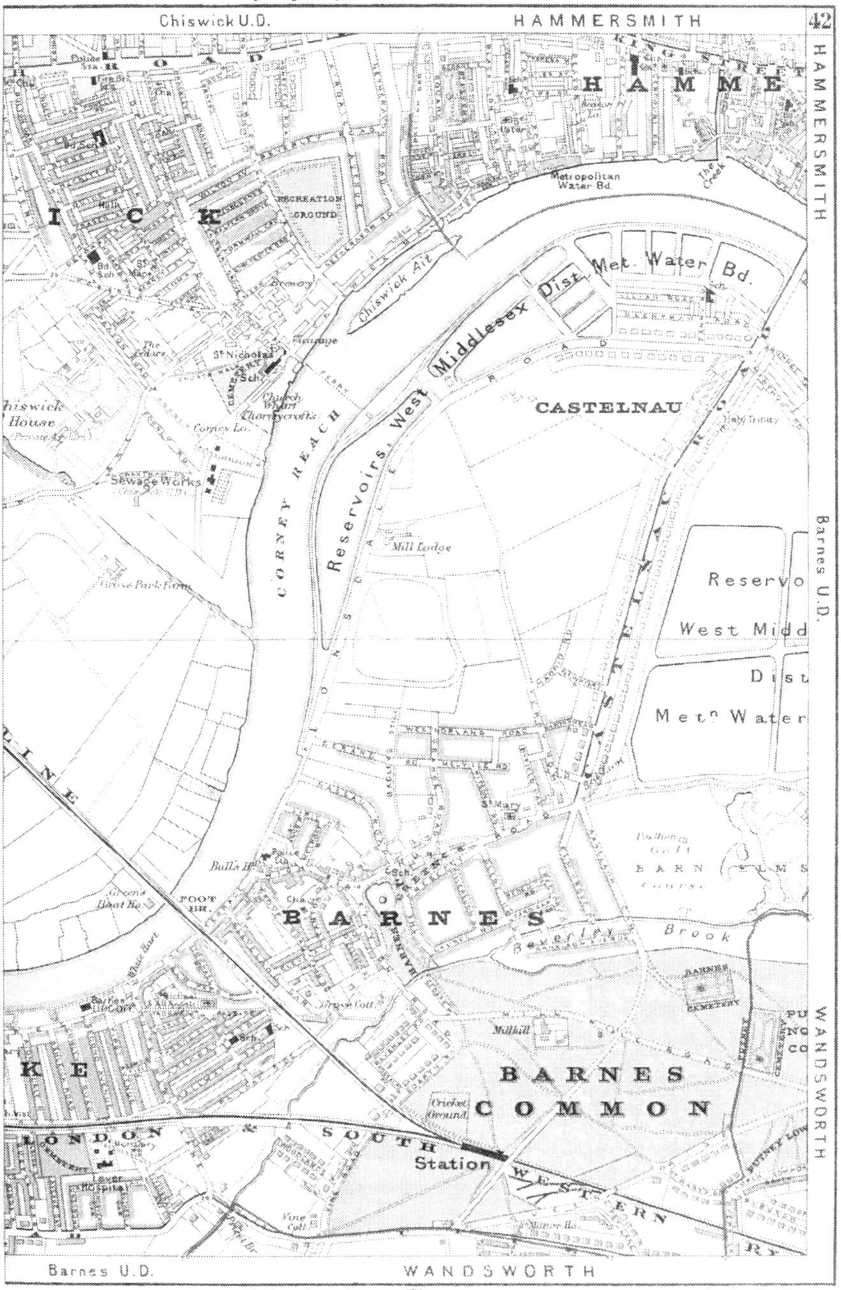

Chiswick U.D. HAMMERSMITH 42

HAMMERSMITH

CHISWICK
HAMMERSMITH
Metropolitan Water Bd.
The Creek
Recreation Ground
Chiswick Ait
Met. Water Bd.
West Middlesex Dist.
Reservoirs
St Nicholas
Church Wharf
Thornycrofts
Chiswick House
Corney Lo.
Sewage Works
Corney Reach
Castelnau
Mill Lodge
Grove Park Farm
Reservoirs West Middlesex Dist. Metn Water
Barnes U.D.
Adjoining Sheet, 43
St Mary
Bull's Hd
Green's Boat Ho.
Foot Br.
Barnes
Barn Elms
Beverley Brook
Barnes Cemetery
Millhill
Barnes Common
Cricket Ground
London & South Western Ry
Station
Vine Cott.
Manor Ho.
Putney Lower Common
WANDSWORTH

g

h

Barnes U.D. WANDSWORTH

Adjoining Sheet, 57. B

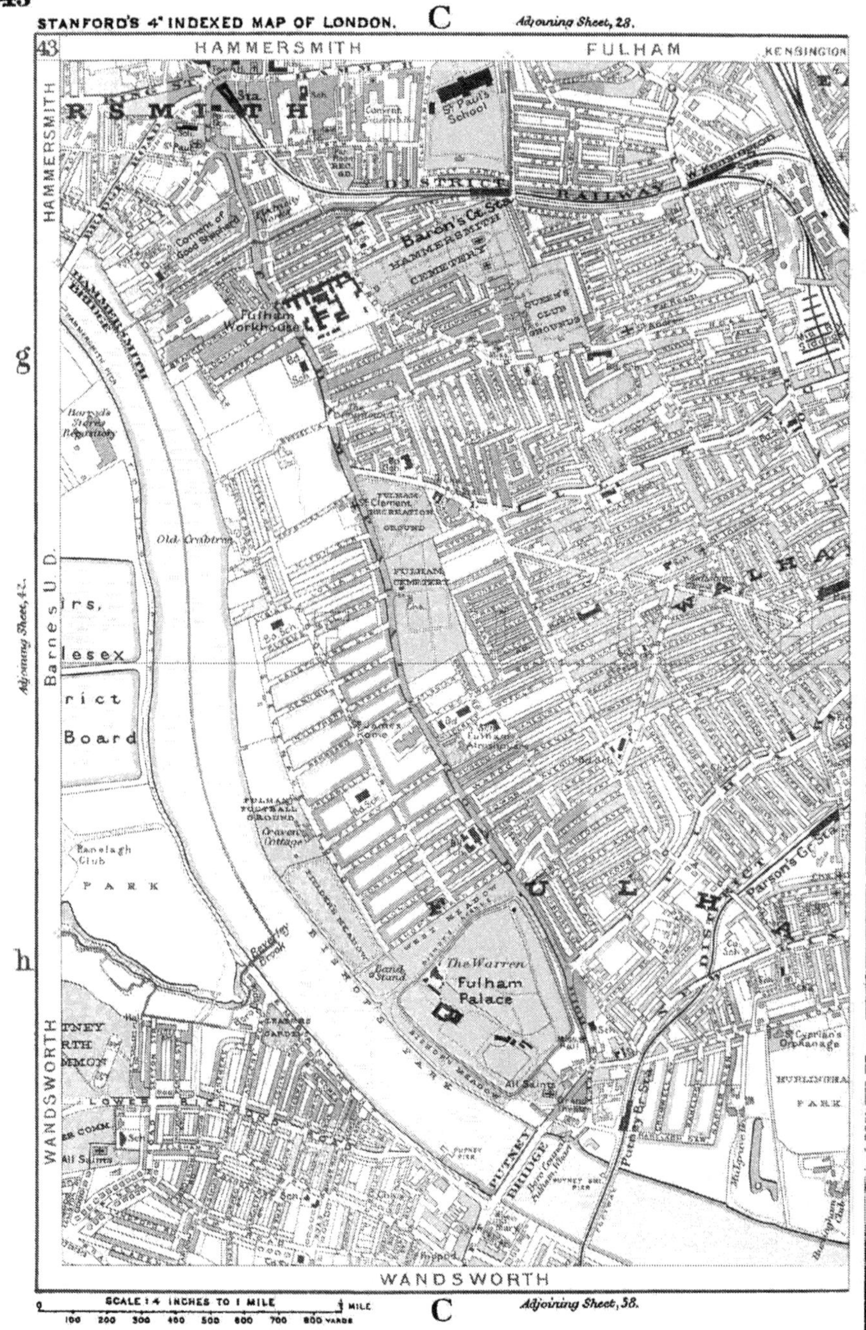
43
STANFORD'S 4" INDEXED MAP OF LONDON.
C
Adjoining Sheet, 28.
HAMMERSMITH
FULHAM
KENSINGTON
HAMMERSMITH
Barnes U.D.
WANDSWORTH
Adjoining Sheet, 42.
g
h
St Paul's School
DISTRICT RAILWAY
Convent of Good Shepherd
HAMMERSMITH BRIDGE
Baron's Ct. Sta.
HAMMERSMITH CEMETERY
QUEEN'S CLUB GROUNDS
Fulham Workhouse
Old Crabtree
FULHAM CEMETERY
Ranelagh Club
PARK
FULHAM FOOTBALL GROUND
Craven Cottage
The Warren
Fulham Palace
BISHOP'S PARK
All Saints
PUTNEY BRIDGE
Putney Br. Sta.
Parson's Gr. Sta.
St Cyprian's Orphanage
HURLINGHAM PARK
WANDSWORTH
SCALE: 4 INCHES TO 1 MILE
0 100 200 300 400 500 600 700 800 YARDS
¼ MILE
C
Adjoining Sheet, 38.

Adjoining Sheet, 29. D

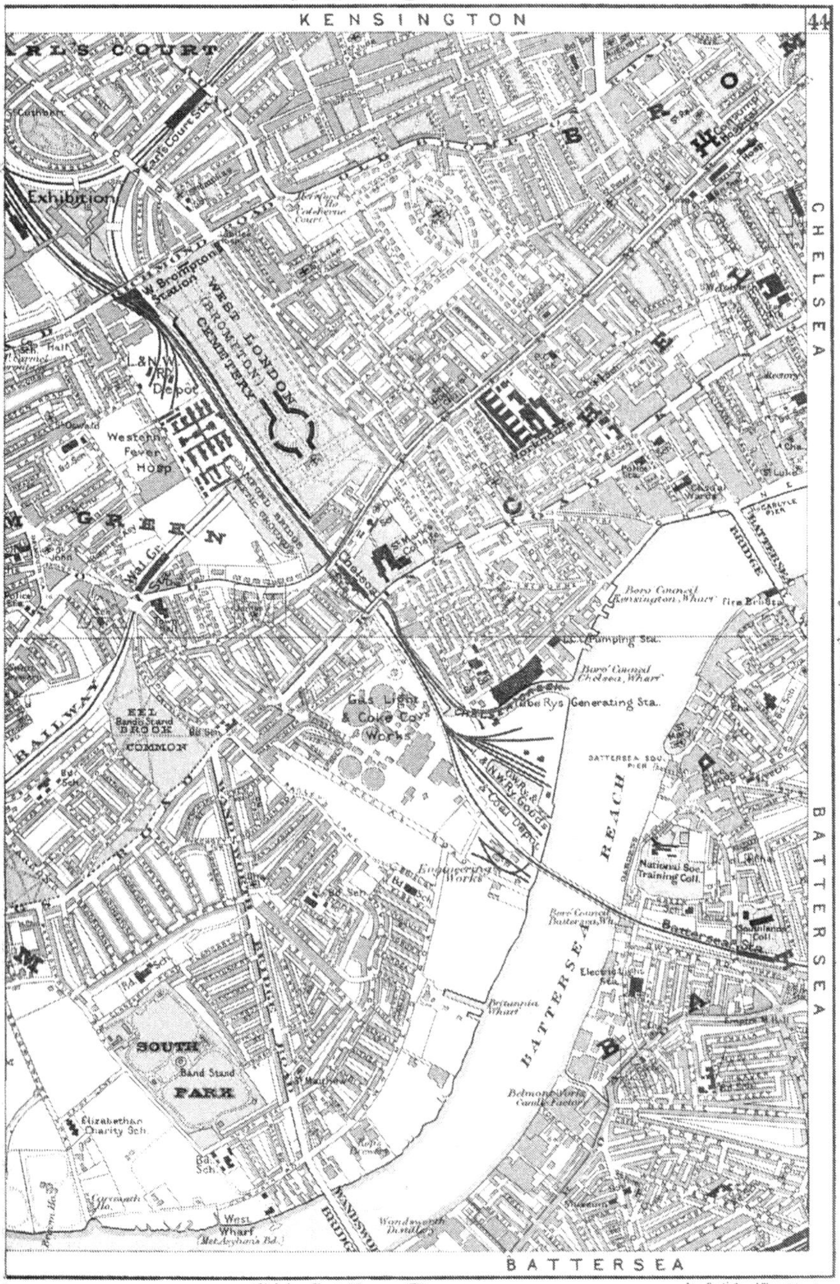

KENSINGTON

44

EARLS COURT

Exhibition

W. Brompton Station

WEST LONDON (BROMPTON) CEMETERY

L&NW Ry Depot

Western Fever Hosp

GREEN

Workhouse

CHELSEA

St Marks Coll.

Boro' Council Kensington Wharf

L.C.C. Pumping Sta.

Boro' Council Chelsea Wharf

Tube Rys Generating Sta.

Gas Light & Coke Co. Works

EEL BROOK COMMON

Band Stand

RAILWAY

L&NW Ry & GWR Goods & Coal Depot

Engineering Works

BATTERSEA REACH

National Soc. Training Coll.

Boro' Council Battersea Wh.

Electric Light Sta.

Britannia Wharf

BATTERSEA

Belmont Works Candle Factory

SOUTH PARK

Band Stand

Elizabethan Charity Sch.

West Wharf (Met. Asylums Bd.)

Wandsworth Distillery

WANDSWORTH BRIDGE

BATTERSEA BRIDGE

Adjoining Sheet, 45.

g

h

BATTERSEA

Adjoining Sheet, 59. D

Stanford's Geog^l Estab^t London

STANFORD'S 4" INDEXED MAP OF LONDON

E

Adjoining Sheet 30.

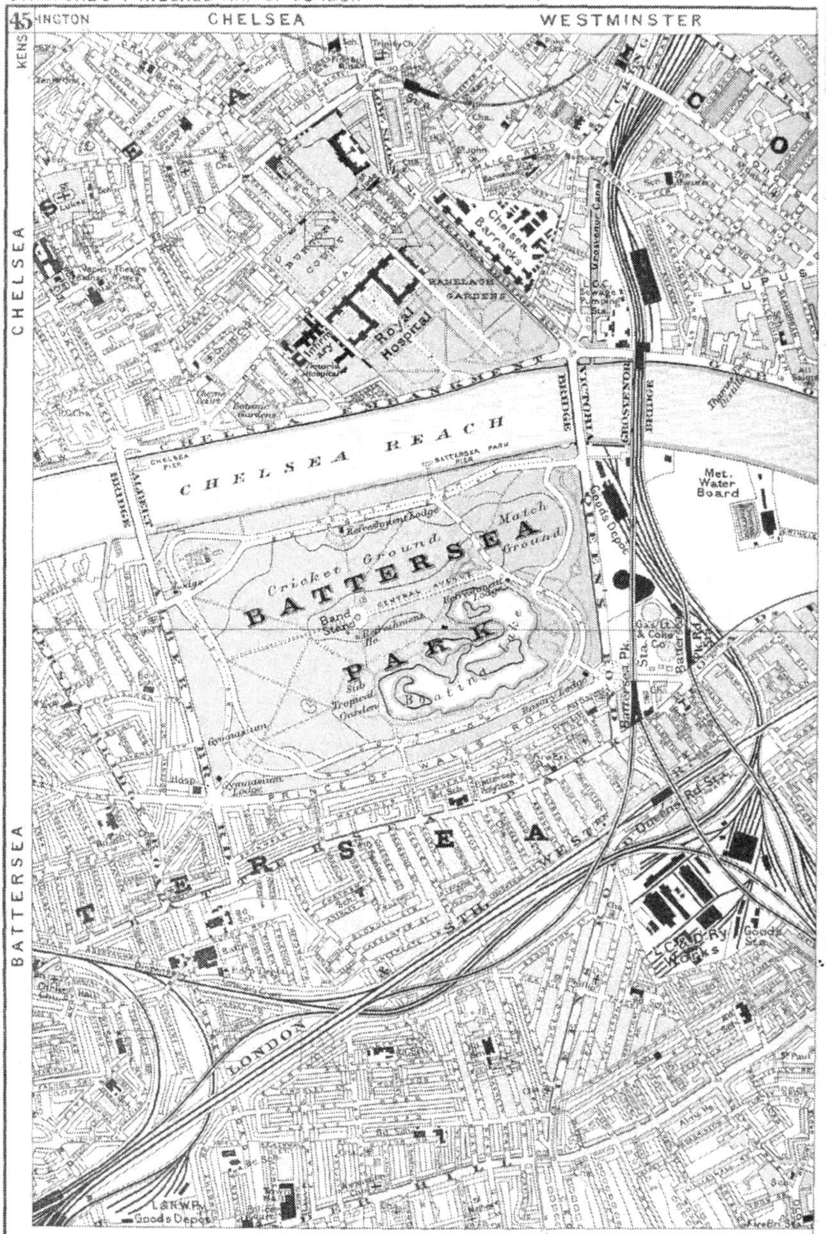

45 KENSINGTON

CHELSEA

WESTMINSTER

CHELSEA

g.

Adjoining Sheet 44.

BATTERSEA

h

BATTERSEA

WANDSWORTH

SCALE: 4 INCHES TO 1 MILE

0 100 200 300 400 500 600 700 800 YARDS ¼ MILE

E

Adjoining Sheet 60.

Adjoining Sheet, 31. F

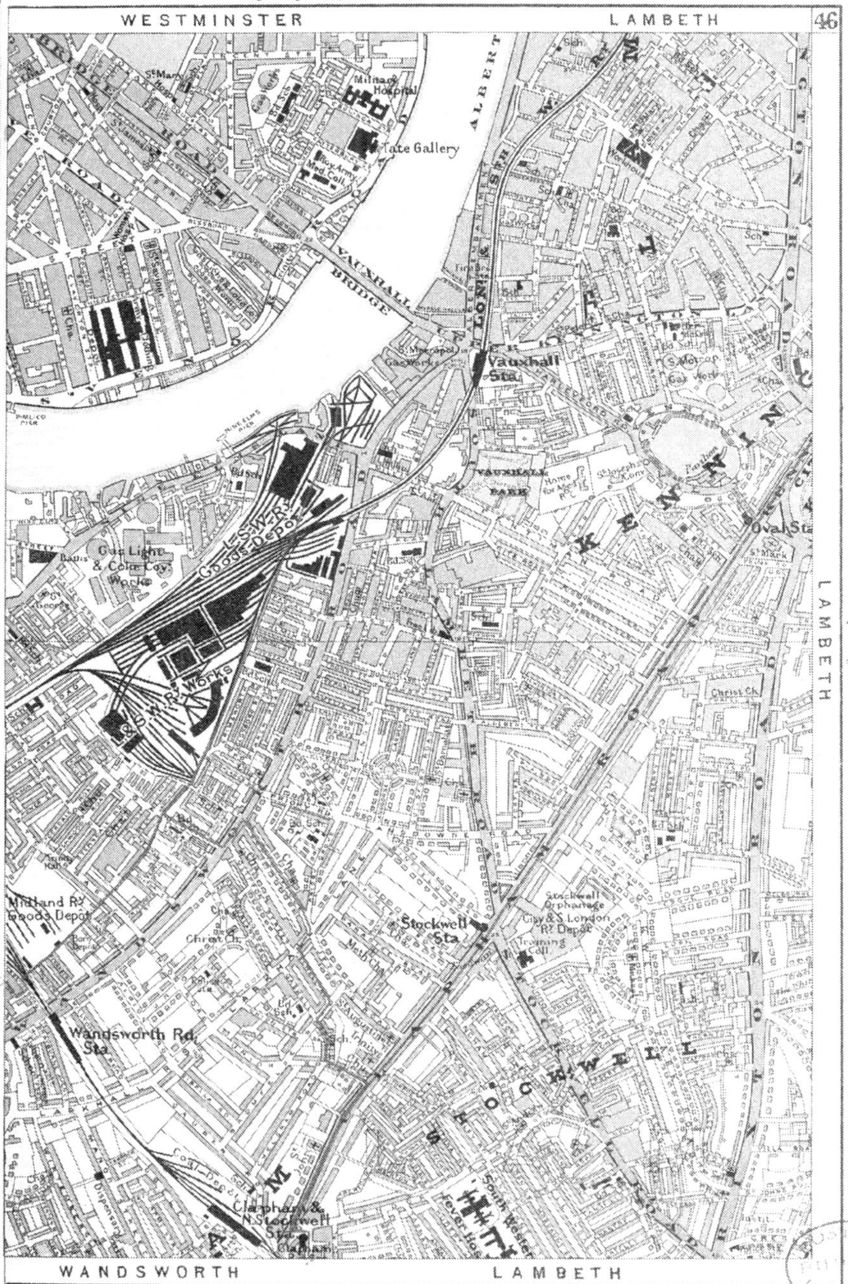

Adjoining Sheet, 61. F

Stanford's Geog.^l Estab.^t London

G

h

STANFORD'S 4" INDEXED MAP OF LONDON. G *Adjoining Sheet, 32.*

47 LAMBETH SOUTHWARK BERMONDSEY

g

Adjoining Sheet, 46. LAMBETH

h

LAMBETH CAMBERWELL

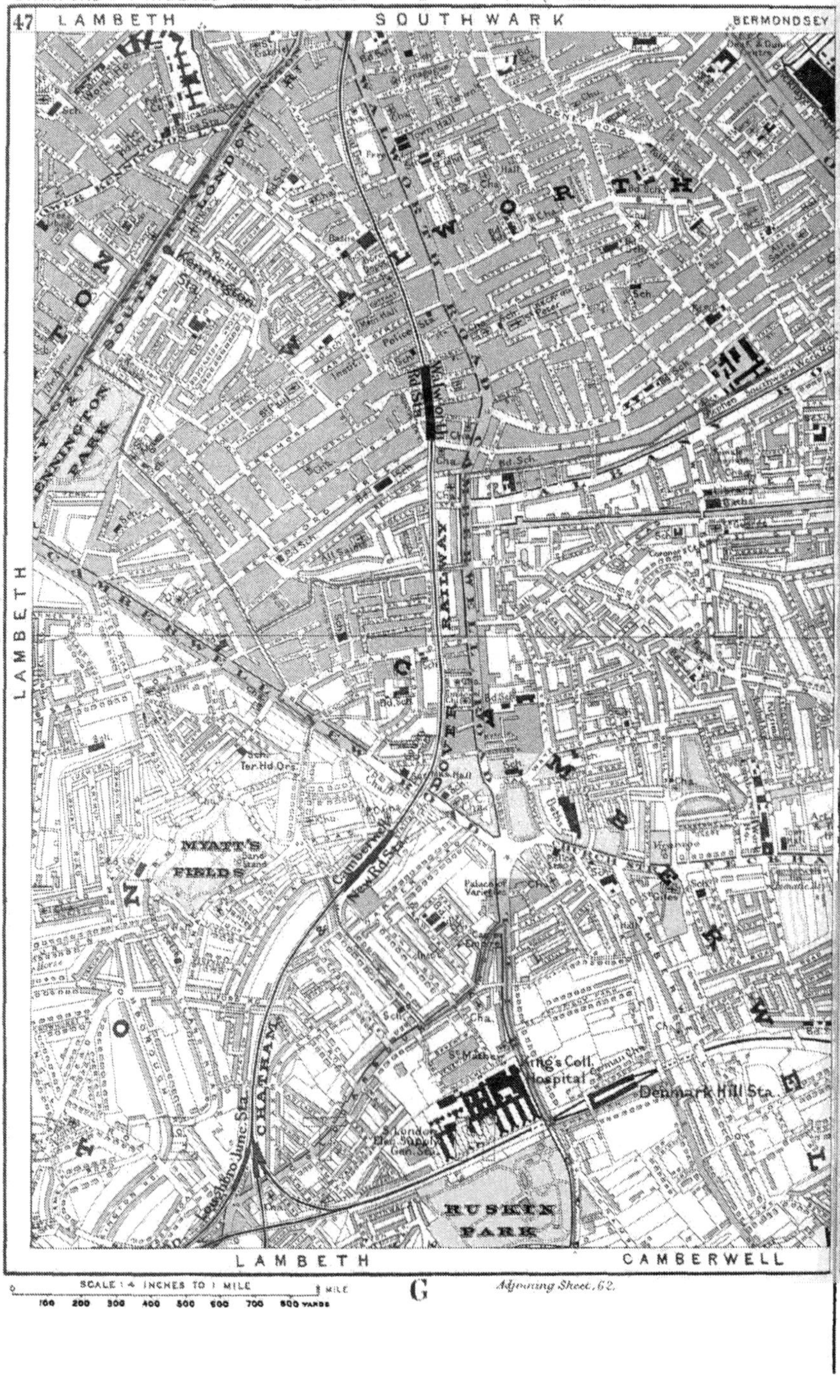

SCALE: 4 INCHES TO 1 MILE ½ MILE
0 100 200 300 400 500 600 700 800 YARDS

G *Adjoining Sheet, 62.*

Adjoining Sheet 33

H

BERMONDSEY

48

PARK

Lady Gomm's Hospital

Southwark Park Sta.

South Bermondsey Sta.

SOUTHWARK PARK ROAD

St. Augustine

S. E. Ry.

GRAND SURREY CANAL

South Metropolitan Gas Works

Cricket Ground

Licensed Victuallers Asylum

St. Gasts. Fever Hosp.

Old Kent Rd. Sta.

Queens Rd. Sta.

Police Sta.

St. Giles Camb.

Peckham Rye Sta.

Nunhead Sta.

NUNHEAD

DEPTFORD

Adjoining Sheet 49

h

CAMBERWELL

CAMBERWELL

Adjoining Sheet 63

H

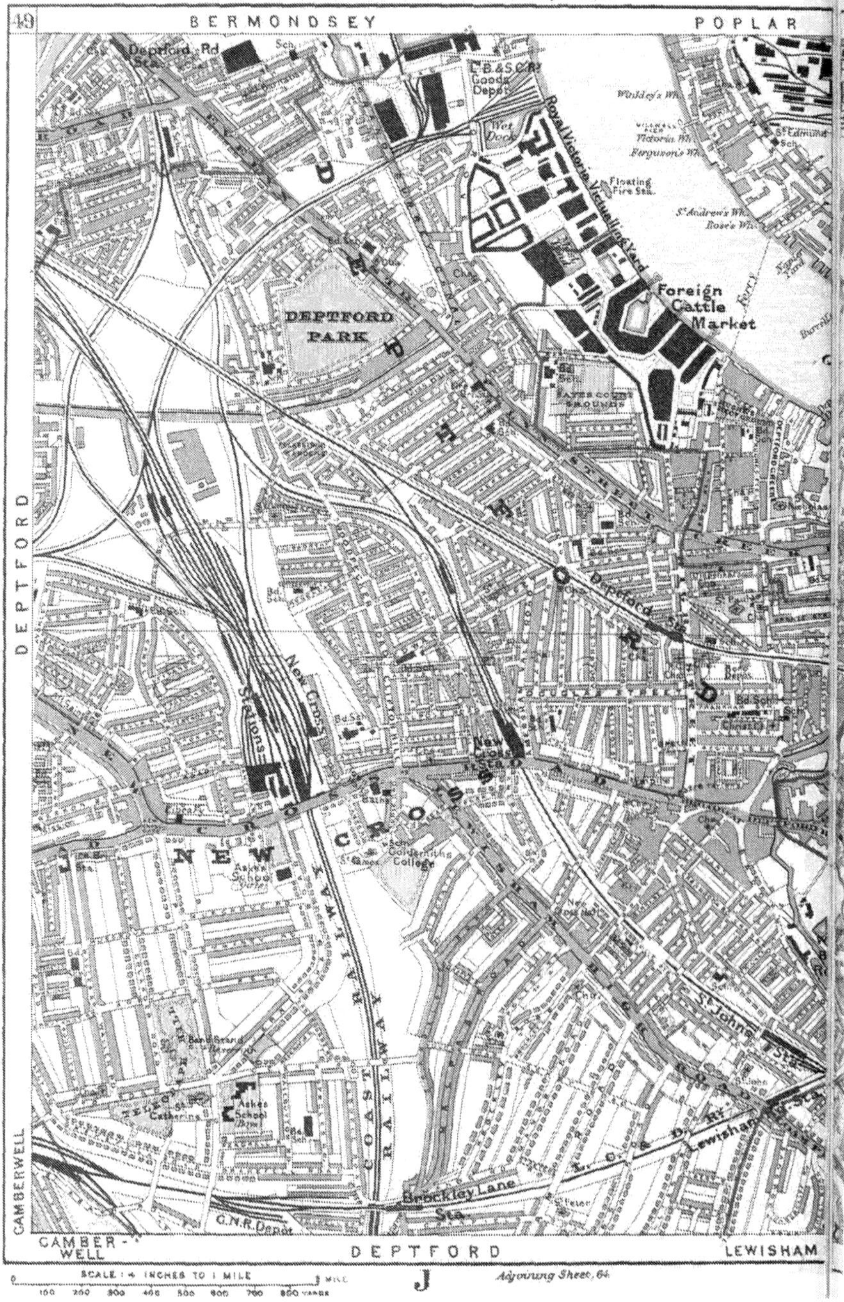

49 BERMONDSEY POPLAR

DEPTFORD PARK

Foreign Cattle Market

Royal Victoria Victualling Yard

New Cross Stations

New Cross Sta.

Deptford Sta.

Goldsmiths College

St. Johns Sta.

Lewisham Jn. Sta.

Brockley Lane Sta.

G.N.R. Depot

NEW CROSS

DEPTFORD

CAMBERWELL

CAMBER-WELL DEPTFORD LEWISHAM

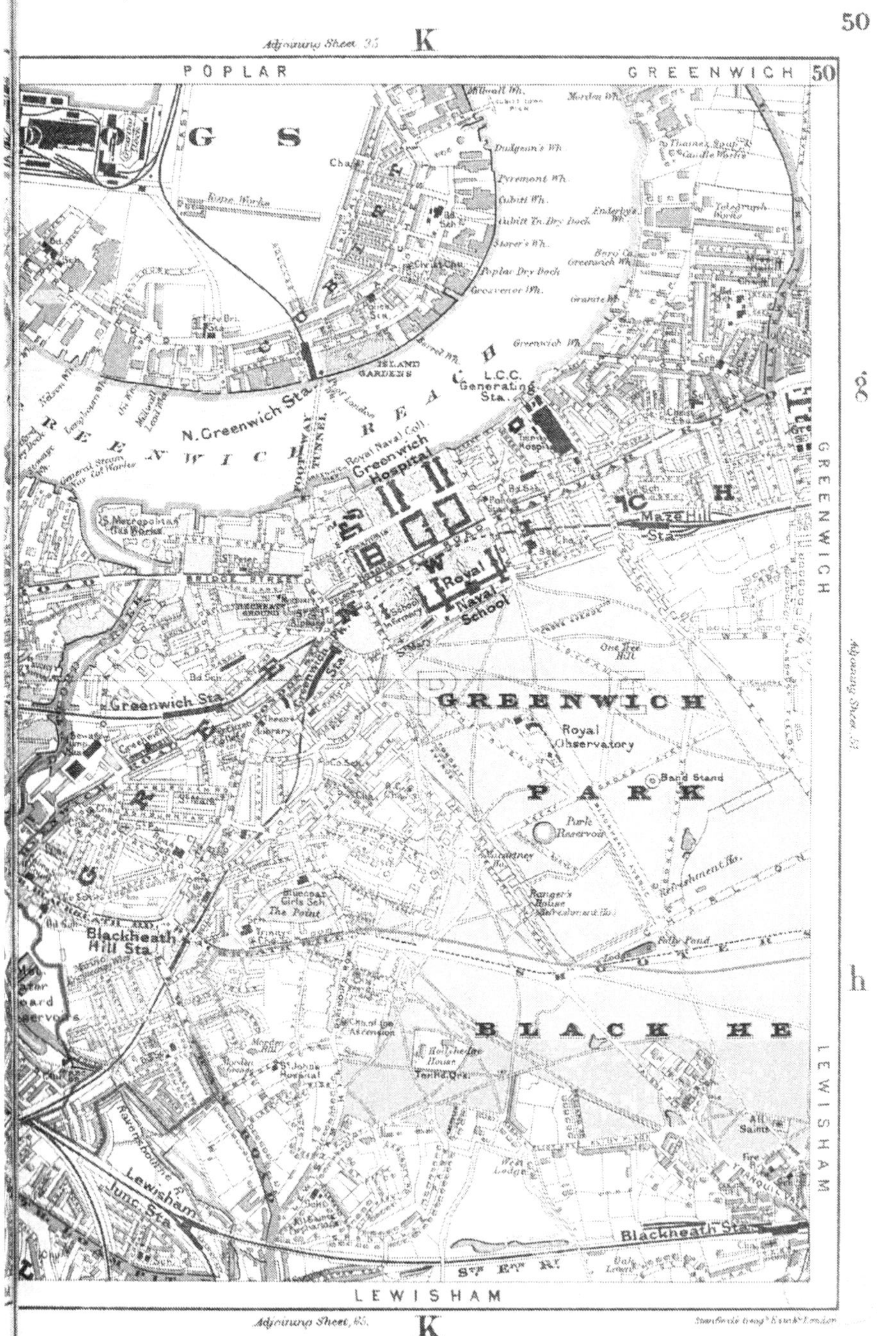
Adjoining Sheet 35
K
POPLAR
GREENWICH
50
Millwall Wh.
Morden Wh.
Dudgeon's Wh.
Pyrmont Wh.
Cubitt Wh.
Cubitt Tn. Dry Dock
Storer's Wh.
Poplar Dry Dock
Grosvenor Wh.
Enderby's Wh.
Granite Wh.
Greenwich Wh.
Rope Works
ISLAND GARDENS
L.C.C. Generating Sta.
N. Greenwich Sta.
GREENWICH REACH
Greenwich Hospital
Royal Naval School
Maze Hill Sta.
BRIDGE STREET
Greenwich Sta.
GREENWICH PARK
Royal Observatory
Band Stand
Park Reservoir
One Tree Hill
Ranger's House
Refreshment Ho.
Blackheath Hill Sta.
The Point
BLACK HE
Ch. of the Ascension
St. John's Hospital
West Lodge
All Saints
Lewisham Junc. Sta.
Blackheath Sta.
GREENWICH
LEWISHAM
Adjoining Sheet 51
h
LEWISHAM
Adjoining Sheet 65
K

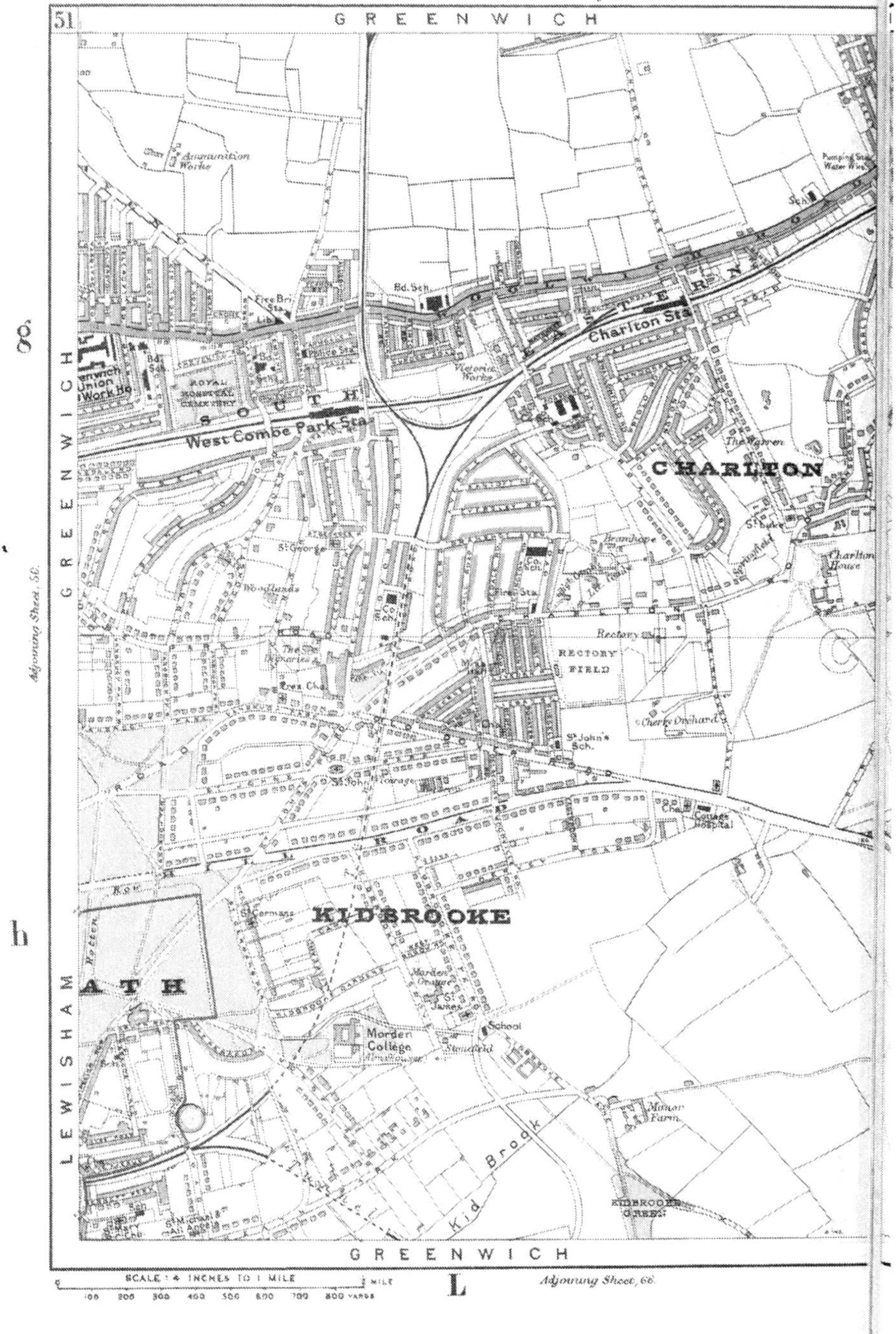

51 GREENWICH

g

h

Adjoining Sheet 50.

GREENWICH

LEWISHAM

Ammunition Works

Pumping Sta. Water Wks.

Sch.

Bd. Sch.

Fire Bri. Sta.

Lib.

Police Sta.

Bd. Sch.

Greenwich Union Work Ho.

ROYAL HOSPITAL CEMETERY

SOUTH EASTERN

Charlton Sta.

Victoria Works

West Combe Park Sta.

The Green

CHARLTON

St. Luke

Charlton House

St. George

Woodlands

Co. Sch.

Fire Sta.

Rectory

RECTORY FIELD

Cherry Orchard

St. John's Sch.

St. John

Pres. Cha.

Cottage Hospital

ROAD

KIDBROOKE

St. Germans

ATH

Morden Grange

St. James

School

Morden College

Manor Farm

Kid Brook

KIDBROOKE GREEN

St. Michael & All Angels

GREENWICH

SCALE: 4 INCHES TO 1 MILE

0 100 200 300 400 500 600 700 800 YARDS ½ MILE

L *Adjoining Sheet 66.*

Adjoining Sheet, 37. M

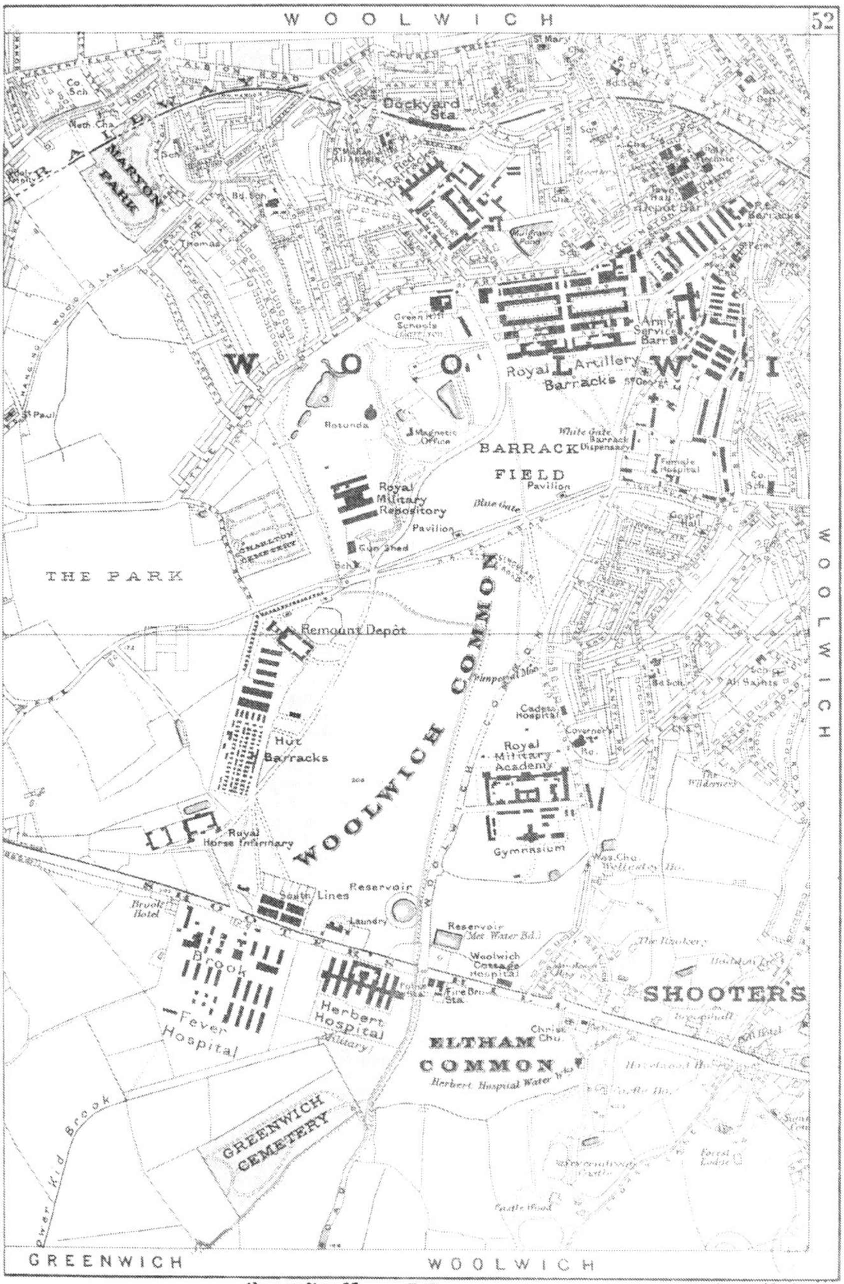

WOOLWICH 52

Dockyard Sta.

Red Barracks

MARYON PARK

Co. Sch.

St Thomas

Depot Bar

Barracks

Green Hill Schools

Army Service Barr

W O O L W I

Royal Artillery Barracks

Rotunda

Magnetic Office

BARRACK FIELD

Pavilion

White Gate Barrack Dispensary

Female Hospital

Royal Military Repository

Pavilion

Blue Gate

Gun Shed

Gospel Hall

THE PARK

CHARLTON CEMETERY

Remount Depot

WOOLWICH COMMON

Cadets Hospital

All Saints

Hut Barracks

Royal Military Academy

The Wilderness

Royal Horse Infirmary

Gymnasium

Wes. Cha.

Wellesley Ho.

South Lines

Reservoir

Brook Hotel

Laundry

Reservoir

Met. Water Bd.

The Rookery

Brook

Woolwich Cottage Hospital

Fire Bri. Sta.

SHOOTERS

Fever Hospital

Herbert Hospital

Military

Christ Chu.

ELTHAM COMMON

Herbert Hospital Water Wks.

GREENWICH CEMETERY

Kid Brook

Forest Lodge

Castle Wood

g

h

WOOLWICH

Adjoining Sheet, 53.

GREENWICH WOOLWICH

Adjoining Sheet, 67. M *Stanford's Geog.l Estab.t London.*

STANFORD'S 4" INDEXED MAP OF LONDON. N *Adjoining Sheet, 38*

53 WOOLWICH

g

h

Adjoining Sheet, 52

WOOLWICH

Plumstead Sta.

PLUMSTEAD

PLUMSTEAD COMMON

Infirmary

Bathing Lake

Clothworkers Wood

Shrewsbury Ho.

HILL

SHOULDER OF MUTTON GREEN

WOOLWICH Bexley U.D.

SCALE : 4 INCHES TO 1 MILE

N *Adjoining Sheet, 68.*

Adjoining Sheet 55

O

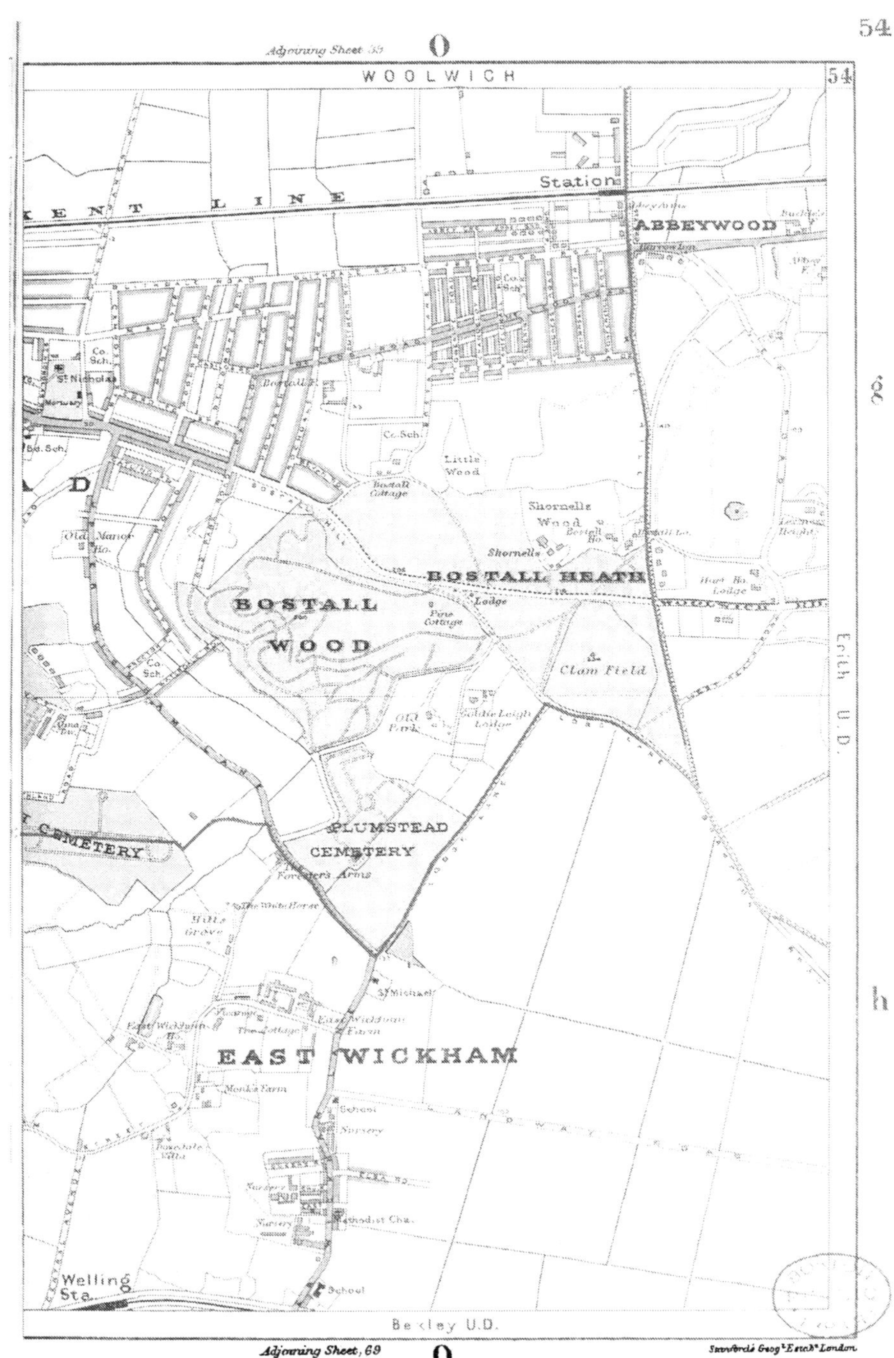

Adjoining Sheet, 69

O

Stanford's Geog. Estab. London

g

h

STANFORD'S 4" INDEXED MAP OF LONDON. S *Adjoining Sheet, 40*

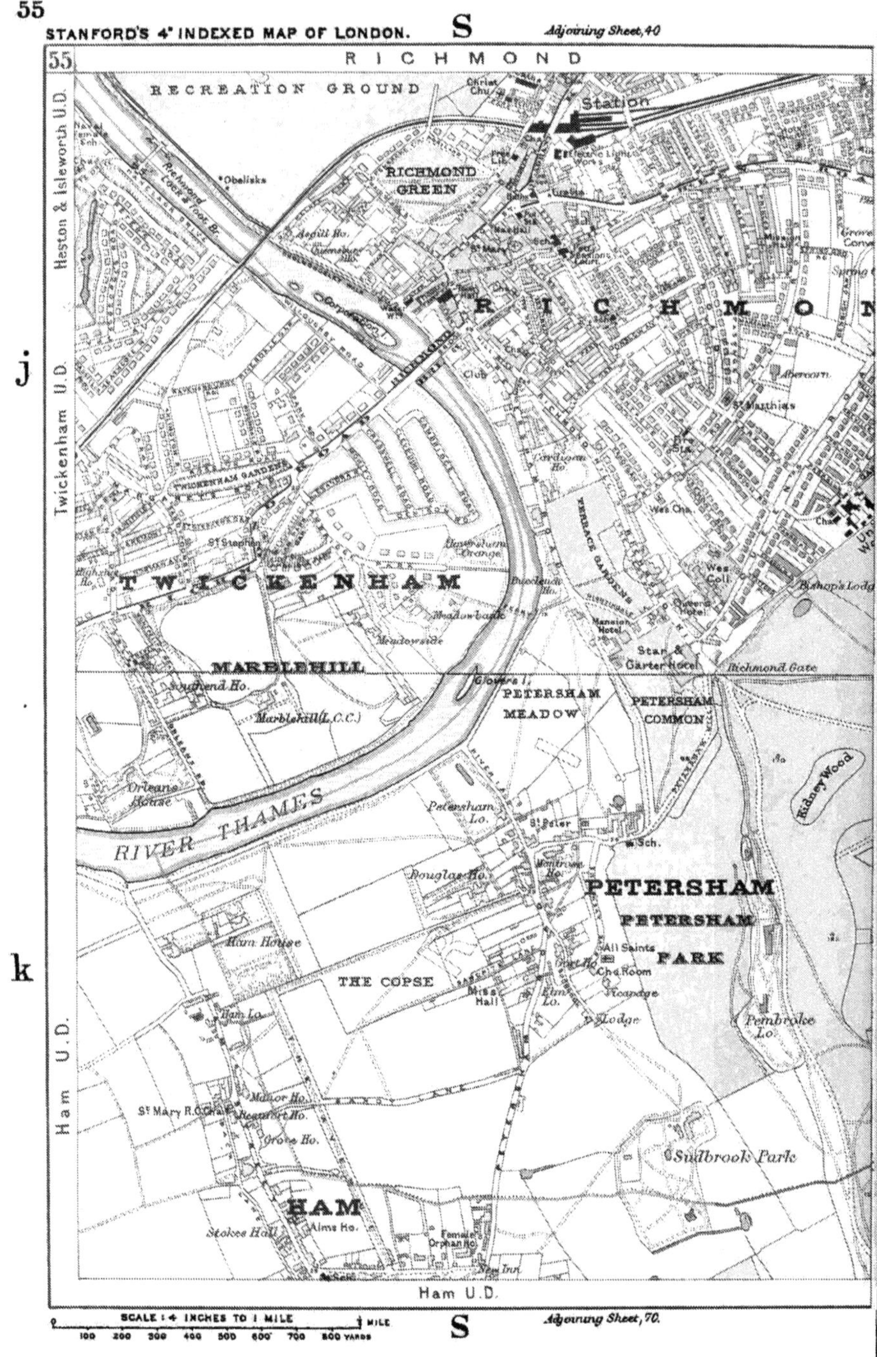

j

k

SCALE: 4 INCHES TO 1 MILE
0 100 200 300 400 500 600 700 800 YARDS ½ MILE

S *Adjoining Sheet, 70.*

Adjoining Sheet, 41

A

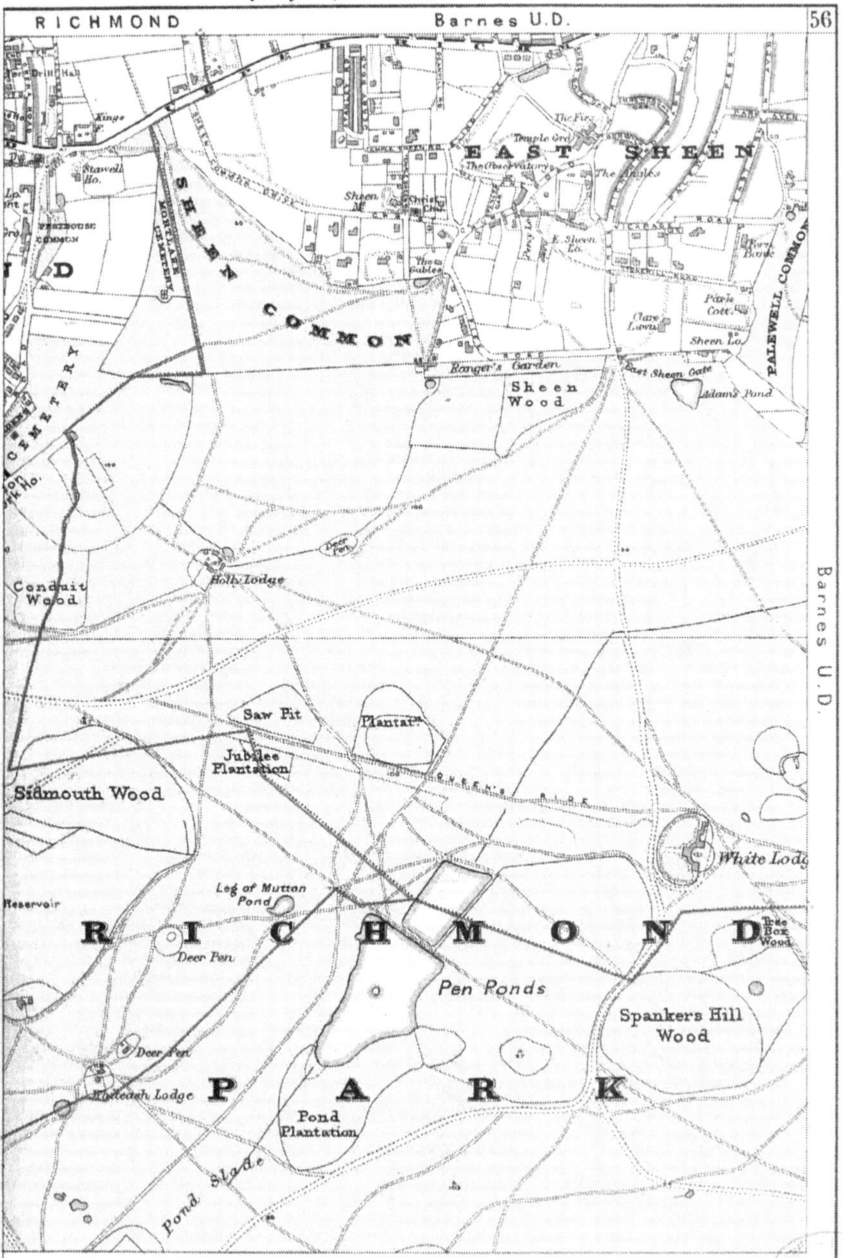

RICHMOND

Barnes U.D.

56

EAST SHEEN

SHEEN COMMON

MORTLAKE CEMETERY

CEMETERY

PESTHOUSE COMMON

PALEWELL COMMON

Drill Hall

Kings

Stawell Ho.

The Firs

Temple Grd.

The Observatory

The Ashes

Sheen M.

Christ Ch.

The Gables

E. Sheen Lo.

Fern Bank

Park Cott.

Clare Lawn

Sheen Lo.

Ranger's Garden

East Sheen Gate

Adams Pond

Sheen Wood

j

Deer Pen

Holly Lodge

Conduit Wood

Barnes U.D.

Adjoining Sheet, 57.

Saw Pit

Plantatn.

Jubilee Plantation

Sidmouth Wood

QUEEN'S RIDE

White Lodge

Leg of Mutton Pond

Reservoir

RICHMOND

Deer Pen

Tree Box Wood

k

Pen Ponds

Spankers Hill Wood

Deer Pen

Whiteash Lodge

PARK

Pond Plantation

Pond Slade

Ham U.D.

Adjoining Sheet, 71.

Stanford's Geog.l Estab.t London.

A

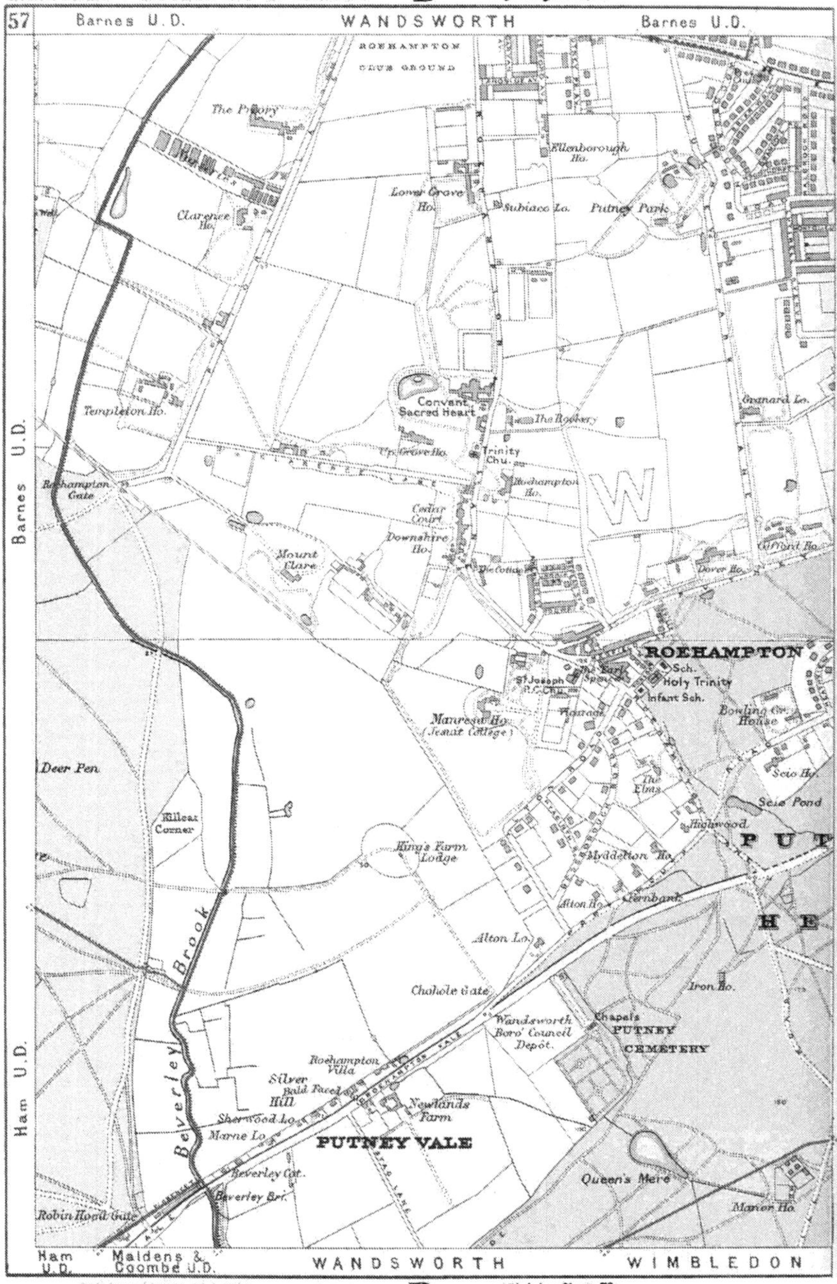

57

Barnes U.D. WANDSWORTH Barnes U.D.

ROEHAMPTON CLUB GROUND

The Priory

Clarence Ho.

Lower Grove Ho.

Subiaco Lo.

Ellenborough Ho.

Putney Park

Templeton Ho.

Convent Sacred Heart

The Rookery

Granard Lo.

Up. Grove Ho.

Trinity Chu.

Roehampton Ho.

W

Roehampton Gate

Cedar Court

Downshire Ho.

Mount Clare

Clifford Ho.

Dover Ho.

ROEHAMPTON

Sch.

Holy Trinity Infant Sch.

St Joseph R.C. Chu.

Manresa Ho. (Jesuit College)

Bowling Gr. House

Scio Ho.

Scio Pond

Deer Pen

Killcat Corner

The Elms

Highwood

King's Farm Lodge

Myddelton Ho.

P U T

Alton Ho.

Fernbank

H E

Alton Lo.

Beverley Brook

Cholole Gate

Iron Ho.

Wandsworth Boro' Council Depôt.

Chapels PUTNEY CEMETERY

Roehampton Villa

Silver Hill

Bald Faced Stag

Newlands Farm

Sherwood Lo.

Marne Lo.

PUTNEY VALE

Beverley Cot.

Beverley Bri.

Queen's Mere

Robin Hood Gate

Manor Ho.

Barnes U.D.

Ham U.D.

Adjoining Sheet, 56.

j

k

Ham U.D. Maldens & Coombe U.D. WANDSWORTH WIMBLEDON

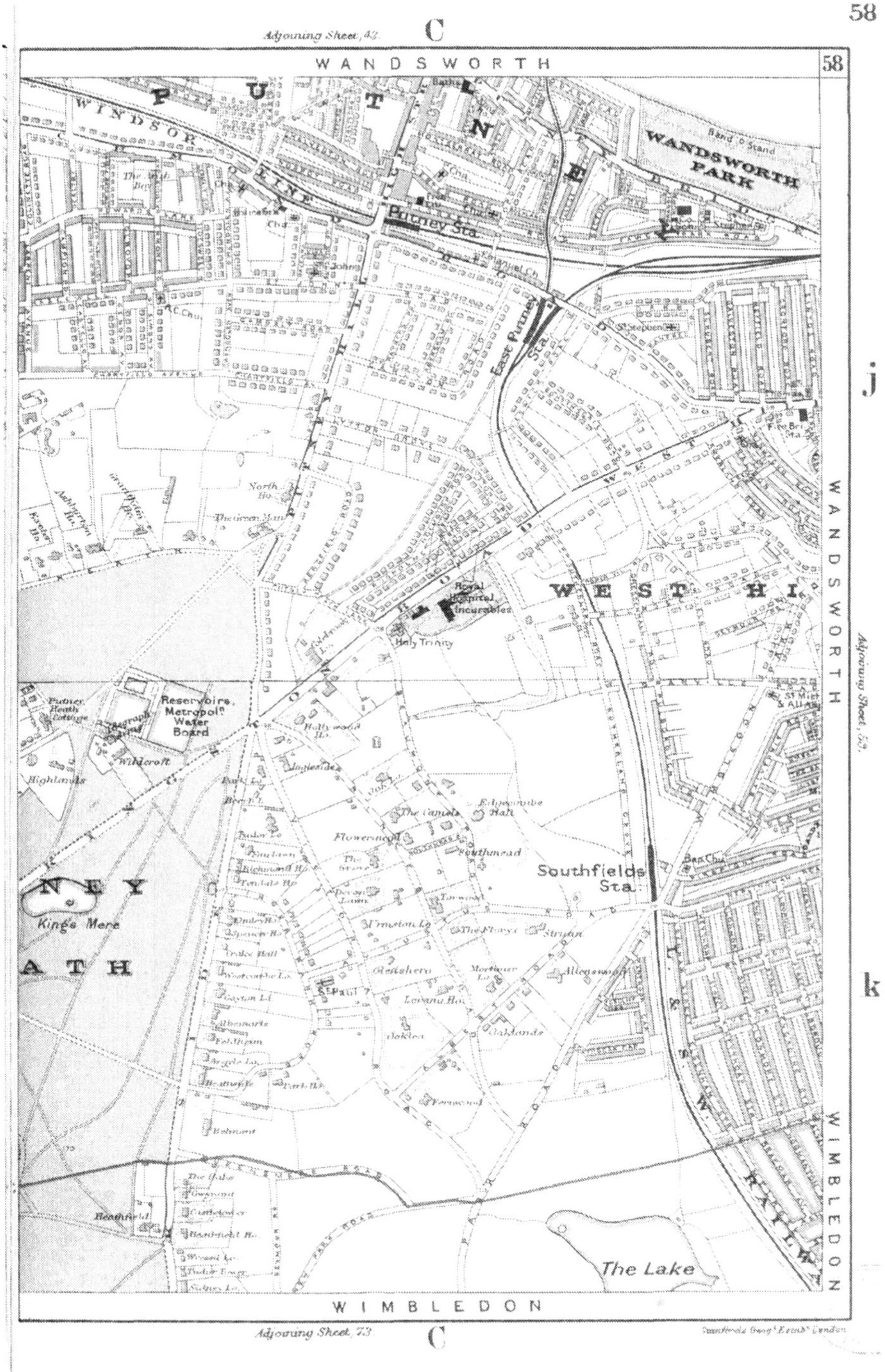

Adjoining Sheet, 43.

C

WANDSWORTH 58

PUTNEY

WINDSOR

WANDSWORTH PARK

Band Stand

Putney Sta.

East Putney Sta.

St. Stephen

Fire Bri. Sta.

WEST HILL

Royal Hospital, Incurables

Holy Trinity

Reservoirs, Metropol[n] Water Board

Putney Heath Cottage

Wildcroft

Highlands

Hollywood Ho.

Ingleside

Edgecombe Hall

The Canals

Flowermead

Southmead

Southfields Sta.

Bap. Chu.

St. Mich. & All Angels

PUTNEY HEATH

King's Mere

Allenswood

Oaklands

Fernwood

Belmont

Heathfield

The Lake

j

k

WANDSWORTH

Adjoining Sheet, 59.

WIMBLEDON

WIMBLEDON

Adjoining Sheet, 73

C

Stanford's Geog.l Estab.t London

STANFORD'S 4" INDEXED MAP OF LONDON. D *Adjoining Sheet, 44.*

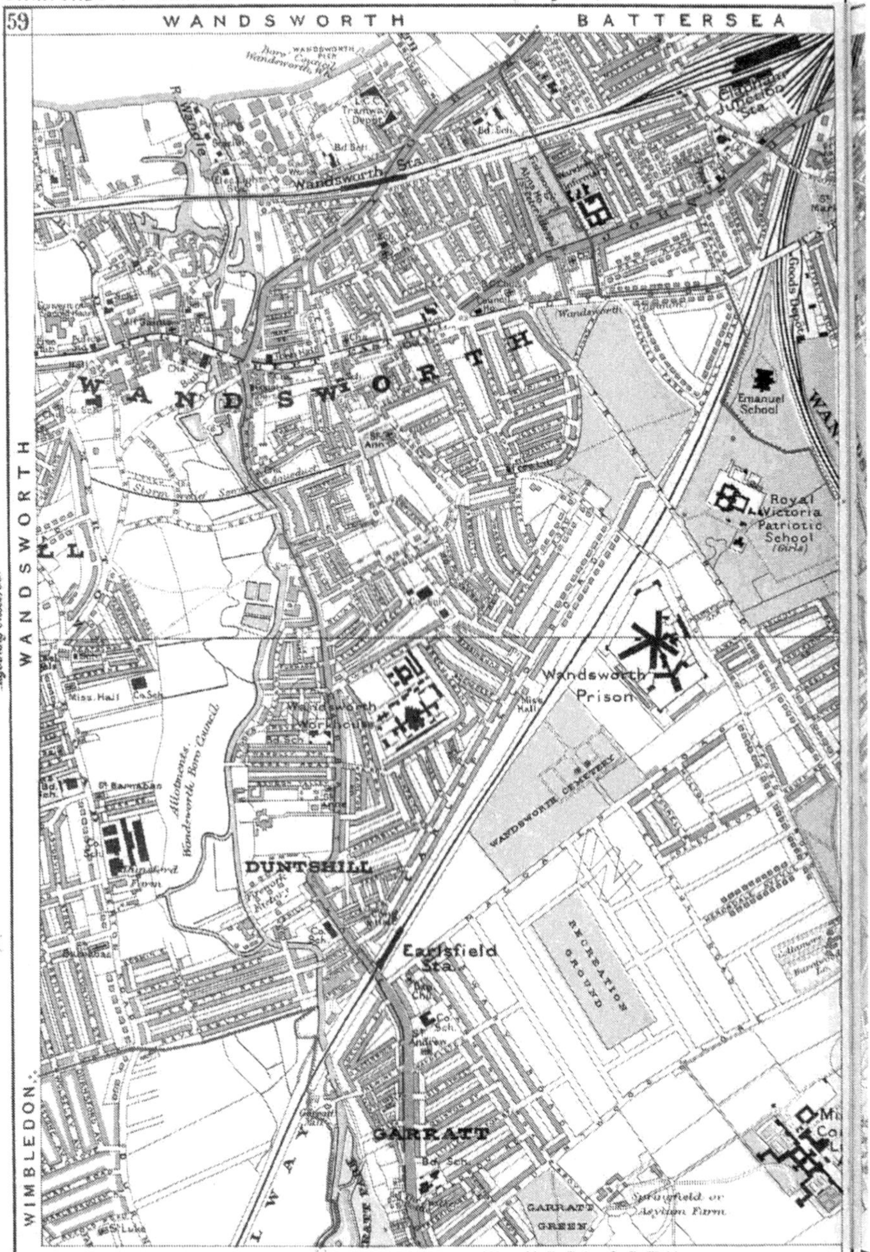

59 WANDSWORTH BATTERSEA

j

k

Adjoining Sheet, 58. WANDSWORTH

WIMBLEDON

WIMBLEDON WANDSWORTH

SCALE: 4 INCHES TO 1 MILE

0 100 200 300 400 500 600 700 800 YARDS ¼ MILE

D *Adjoining Sheet, 74*

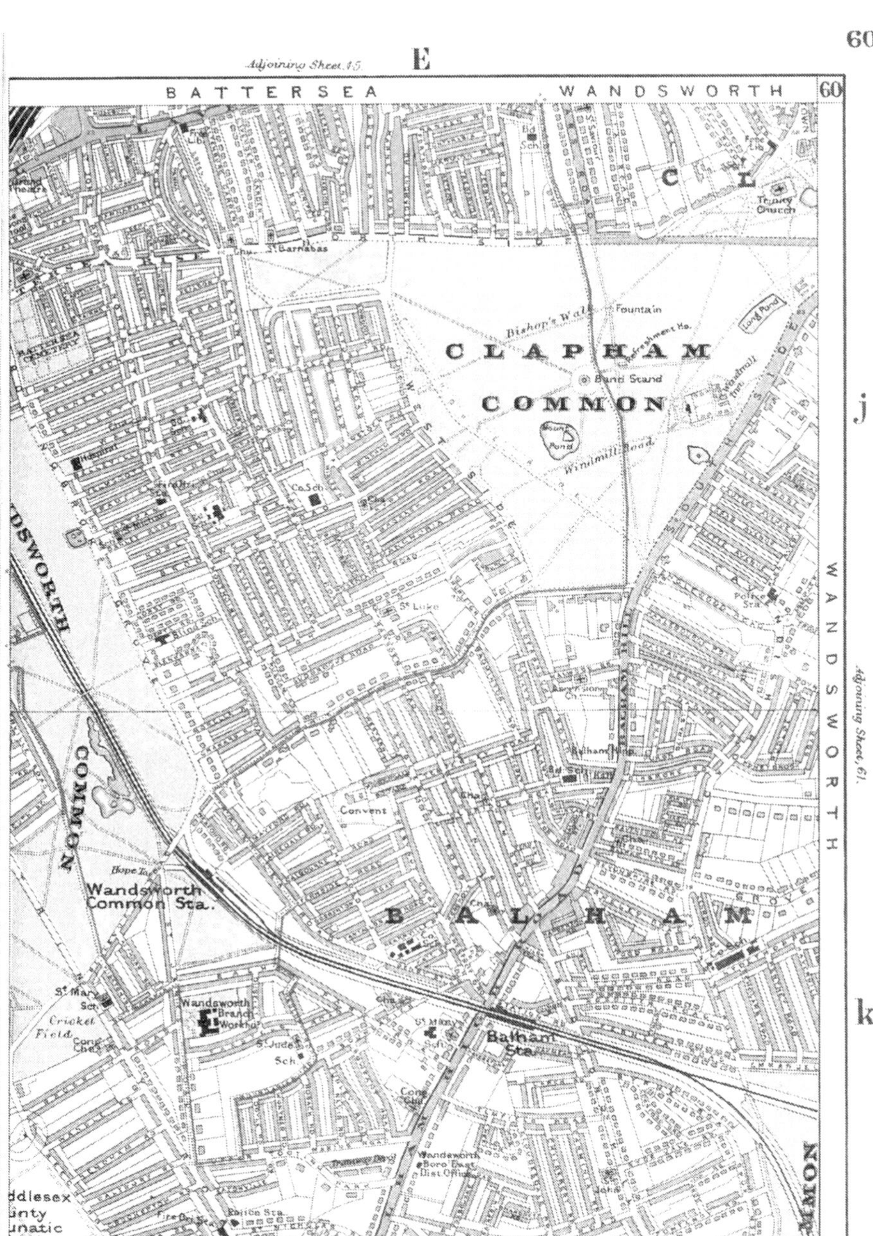
Adjoining Sheet 45.
E
BATTERSEA
WANDSWORTH
60
CLAPHAM
COMMON
Bishop's Walk
Fountain
Band Stand
Windmill Road
Long Pond
Trinity Church
St. Barnabas
Co. Sch.
St. Luke
Convent
Ascension Ch.
Balham Hill
Police Sta.
WANDSWORTH
Adjoining Sheet 61.
j
k
COMMON
Hope Ter.
Wandsworth Common Sta.
BALHAM
Balham Sta.
St. Mary
St. Jude Sch.
Wandsworth Branch Workho.
Cricket Field
Cong. Chap.
Wandsworth Boro East Dist. Office
St. John
Police Sta.
Fire Bri. Sta.
Holy Trinity Ch.
Middlesex County Lunatic Asylum
UPPER TOOTING
COMMON
WANDSWORTH
Adjoining Sheet 75.
E
Stanford's Geog. Estab. London.

STANFORD'S 4" INDEXED MAP OF LONDON. F *Adjoining Sheet, 46*

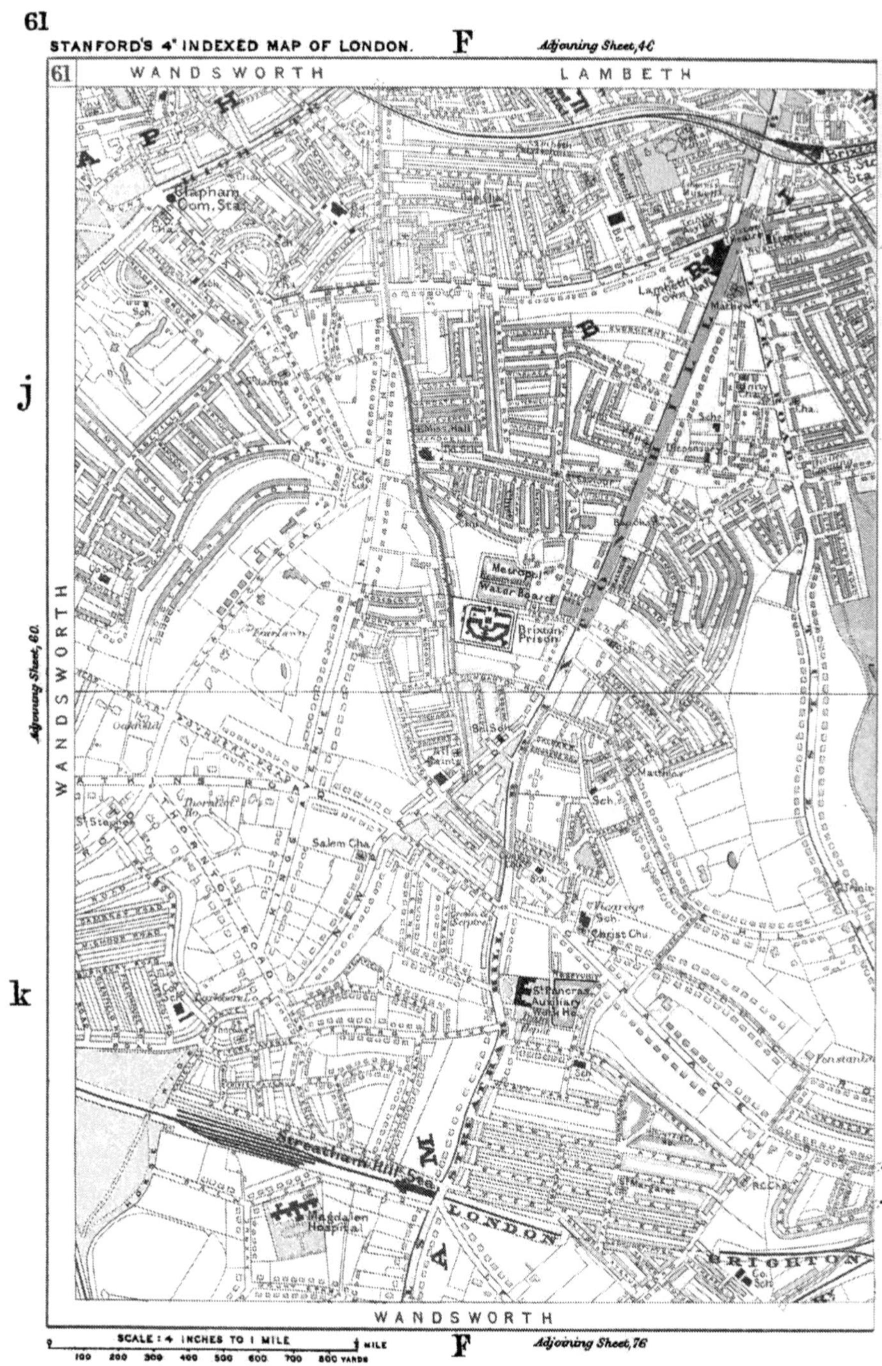

SCALE: 4 INCHES TO 1 MILE

F *Adjoining Sheet, 76*

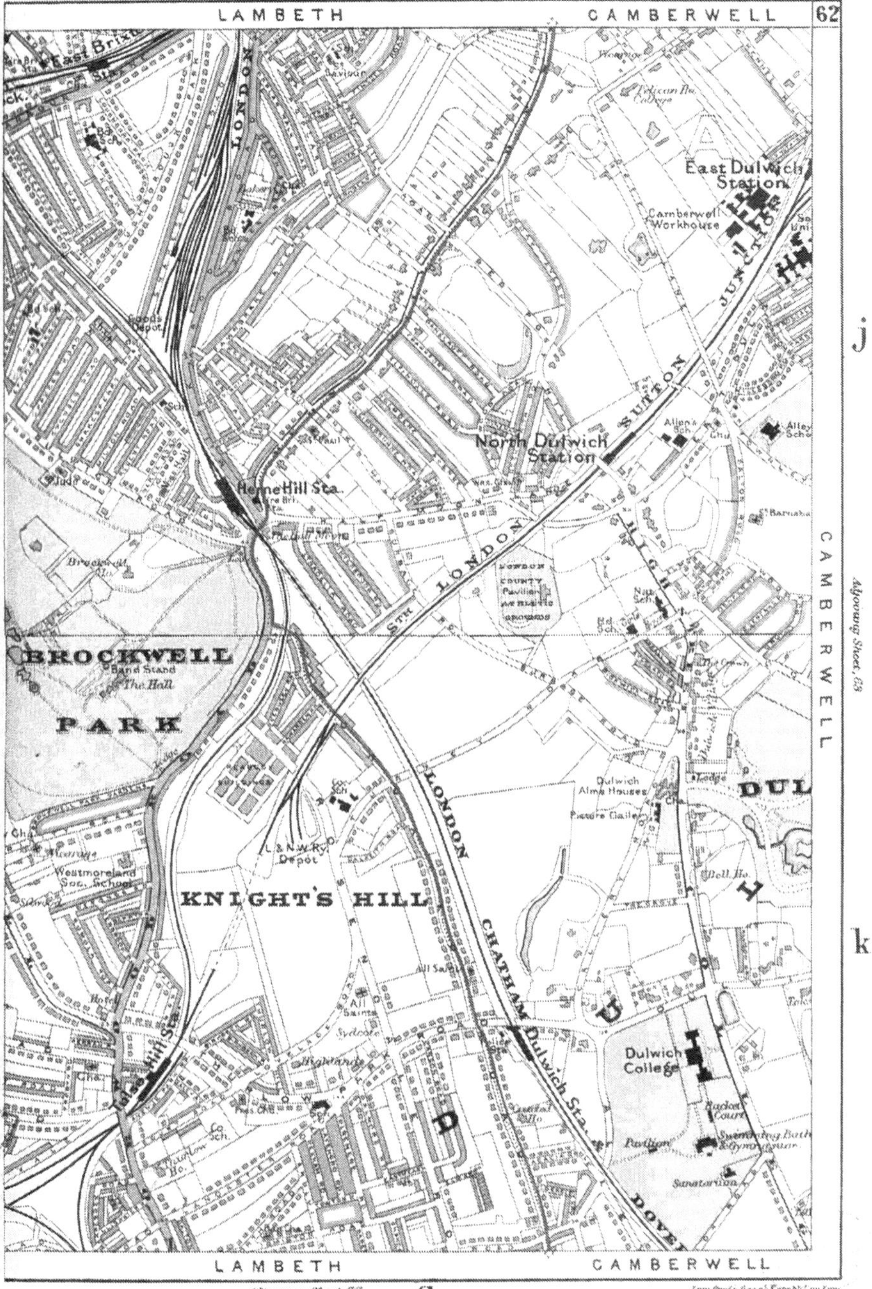
Adjoining Sheet, 47.
G
LAMBETH
CAMBERWELL
62
East Dulwich Station
Camberwell Workhouse
North Dulwich Station
Herne Hill Sta.
BROCKWELL
PARK
KNIGHT'S HILL
Westmoreland Soc. School
L & N.W. Ry. Depot
Dulwich Alms Houses
Dulwich College
CAMBERWELL
Adjoining Sheet, 63
LAMBETH
CAMBERWELL
Adjoining Sheet, 77
G
j
k

STANFORD'S 4" INDEXED MAP OF LONDON. **H** *Adjoining Sheet, 48.*

j

k

Adjoining Sheet, 62.

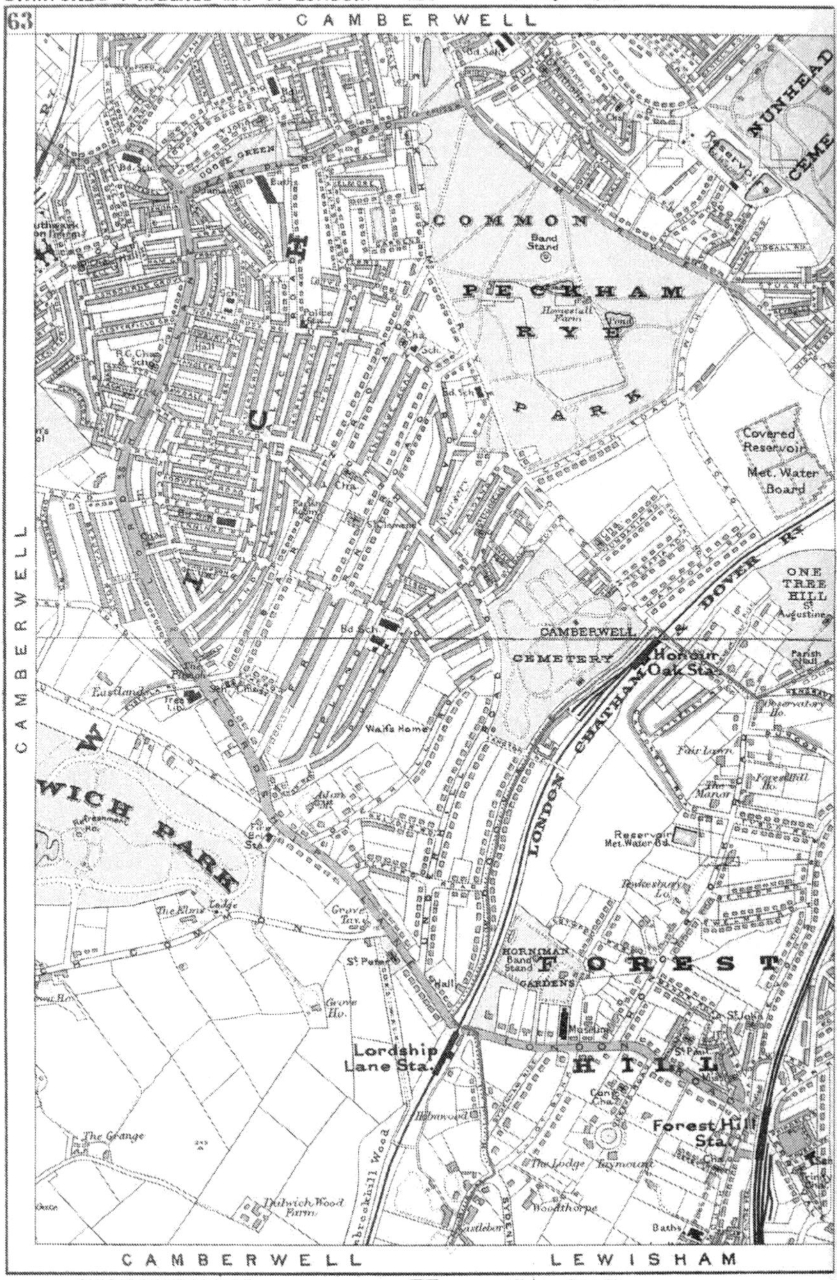
63
CAMBERWELL
CAMBERWELL
CAMBERWELL
LEWISHAM
NUNHEAD CEME
COMMON
PECKHAM RYE PARK
Band Stand
Homestall Farm
Pond
Covered Reservoir
Met. Water Board
DOVER RD.
ONE TREE HILL
St. Augustine
CAMBERWELL CEMETERY
Honour Oak Sta.
LONDON CHATHAM
Parish Hall
Observatory Ho.
Fair Lawn
The Manor
Forest Hill Ho.
Reservoir Met. Water Bd.
Tewkesbury Lo.
FOREST HILL
HORNIMAN GARDENS
Band Stand
Museum
St. Paul
St. John
Forest Hill Sta.
Baths
Woodthorpe
The Lodge
Lordship Lane Sta.
Hall
St. Peter
Grove Tav.
Grove Ho.
Hillswood
Castlebar
Dulwich Wood Farm
The Grange
The Elms
Lodge
WICH PARK
Refreshment Ho.
Eastlands
Free Library
The Plough
Fire Brig. Sta.
Adam
Waifs Home
Bd. Sch.
St. Clement
GOOSE GREEN
Police Sta.
Sydenham Hill Wood

SCALE : 4 INCHES TO 1 MILE

0 100 200 300 400 500 600 700 800 YARDS 1 MILE

H *Adjoining Sheet, 78.*

Adjoining Sheet, 48. J

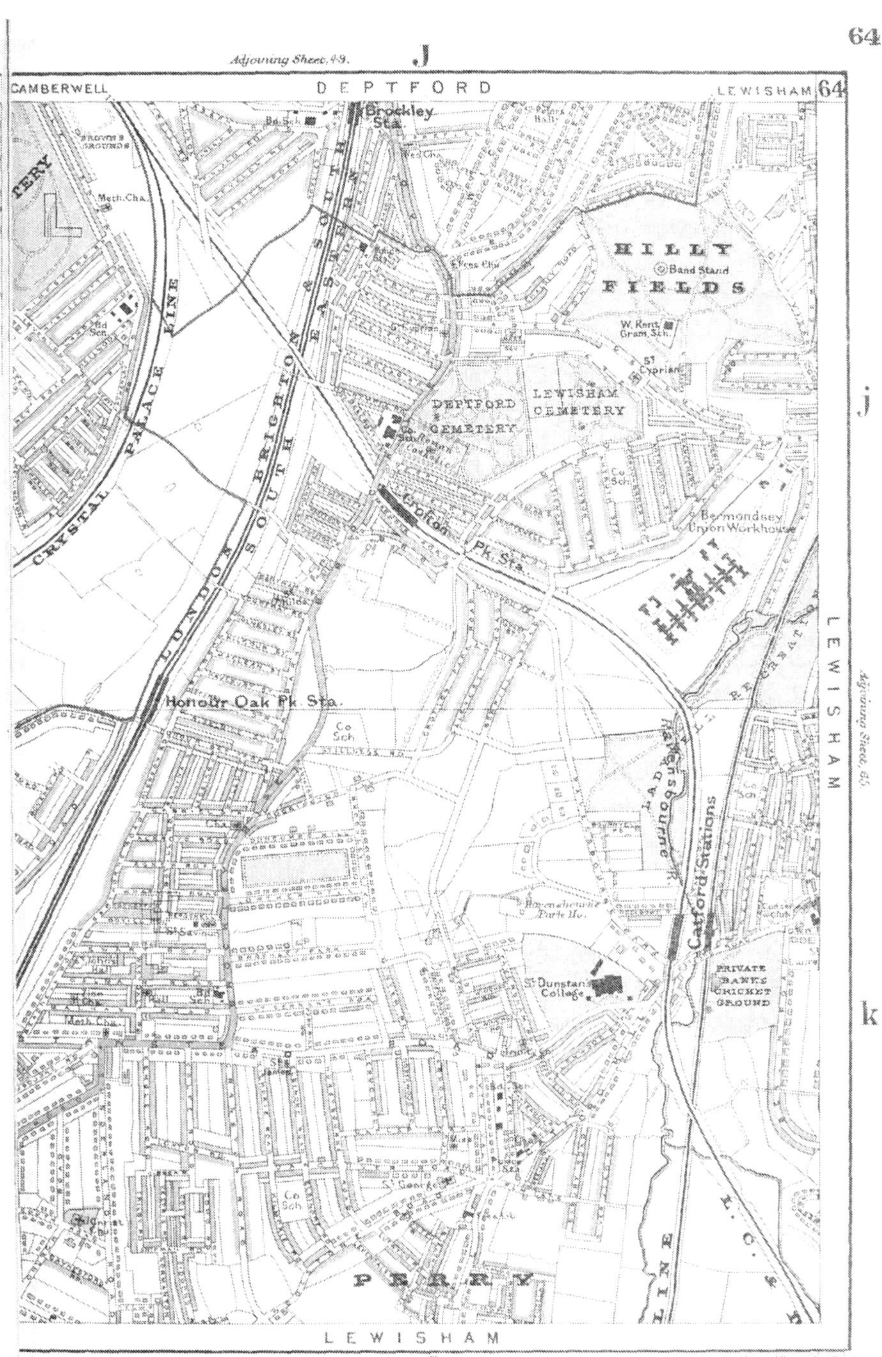

j

Adjoining Sheet, 65.

k

Adjoining Sheet, 79 J

STANFORD'S 4" INDEXED MAP OF LONDON. K *Adjoining Sheet, 51.*

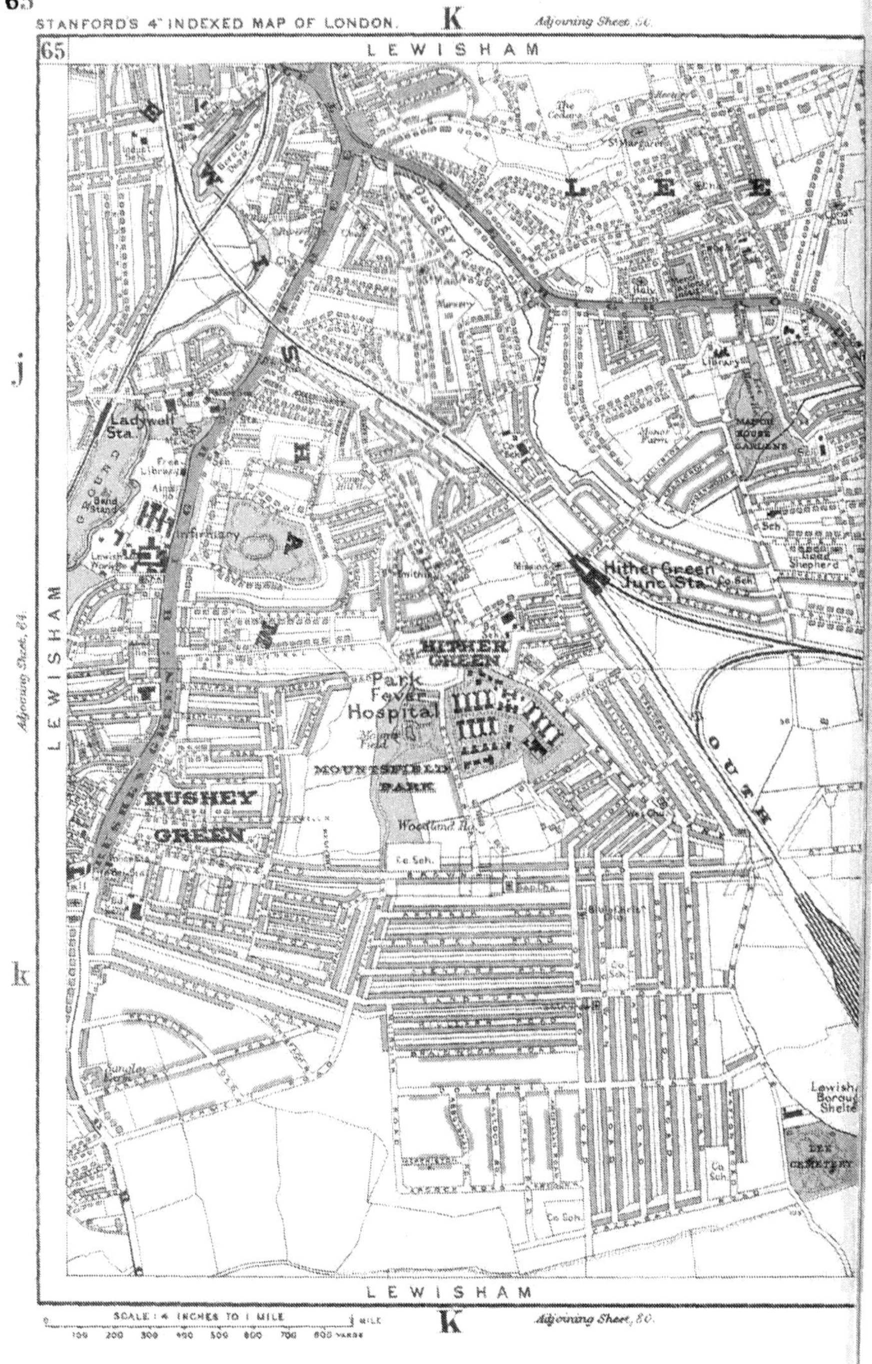

K *Adjoining Sheet, 80.*

Adjoining Sheet, 51. L

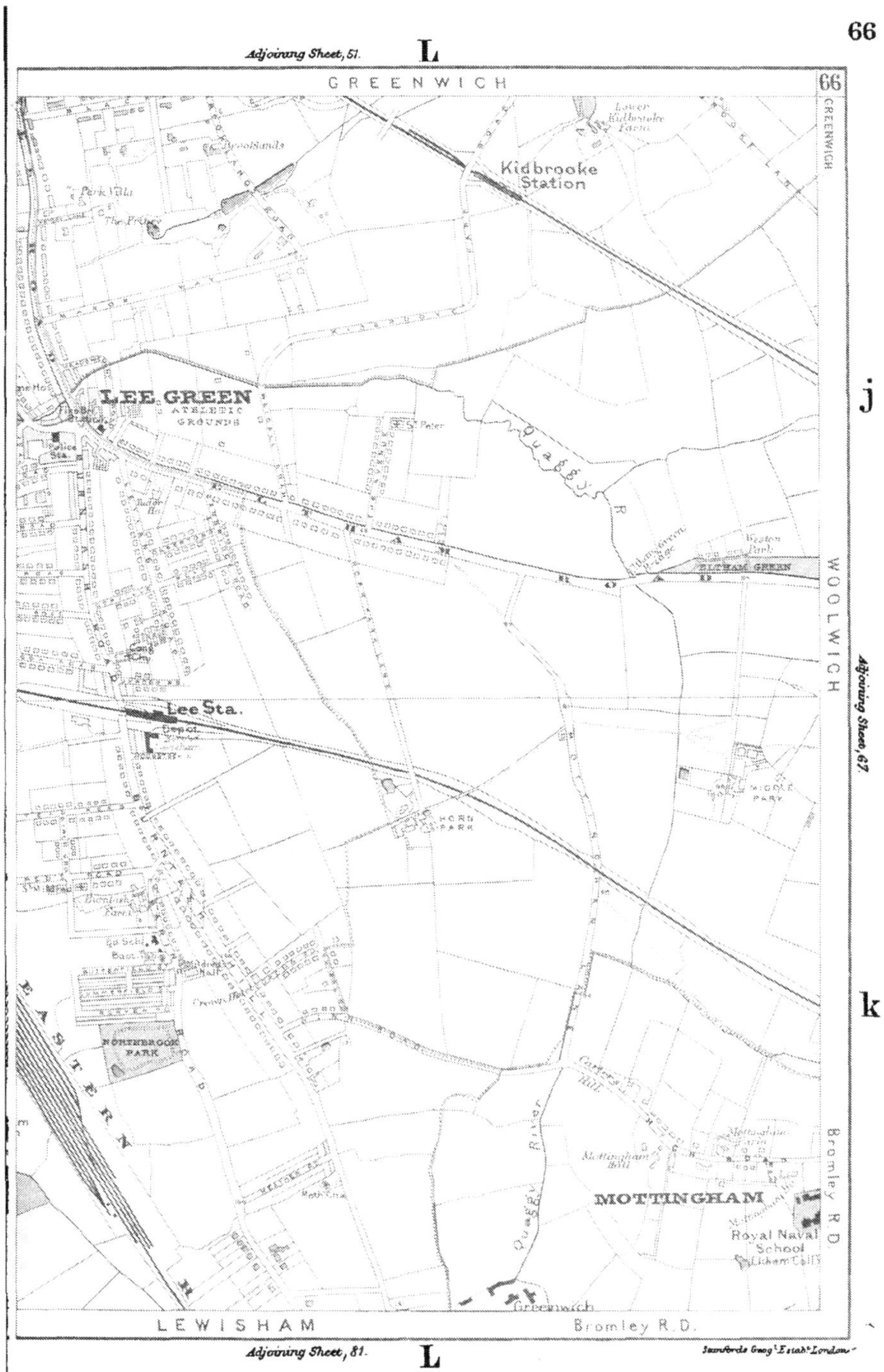

j

Adjoining Sheet, 67.

k

Adjoining Sheet, 81. L

Stanford's Geog. Estab. London.

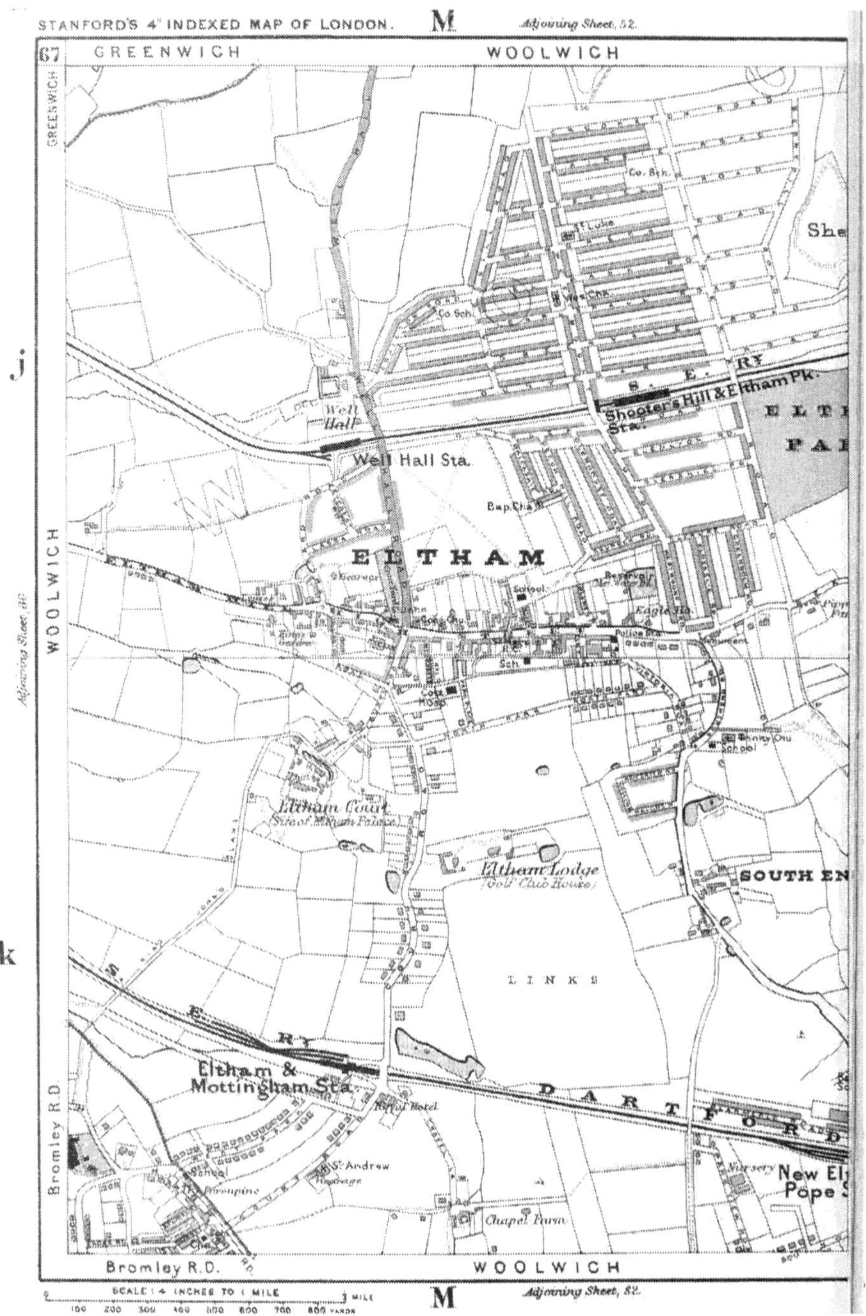
STANFORD'S 4" INDEXED MAP OF LONDON.
M
Adjoining Sheet, 52.
67
GREENWICH
WOOLWICH
WOOLWICH
Bromley R.D.
j
k
Shooter's Hill & Eltham Pk.
Well Hall Sta.
ELTHAM
Eltham Court
(Site of Eltham Palace)
Eltham Lodge
(Golf Club House)
SOUTH EN
LINKS
Eltham & Mottingham Sta.
DARTFORD
Chapel Farm
Nursery
New El
Pope
Bromley R.D.
WOOLWICH
SCALE: 4 INCHES TO 1 MILE
M
Adjoining Sheet, 82.

Adjoining Sheet, 53.

N

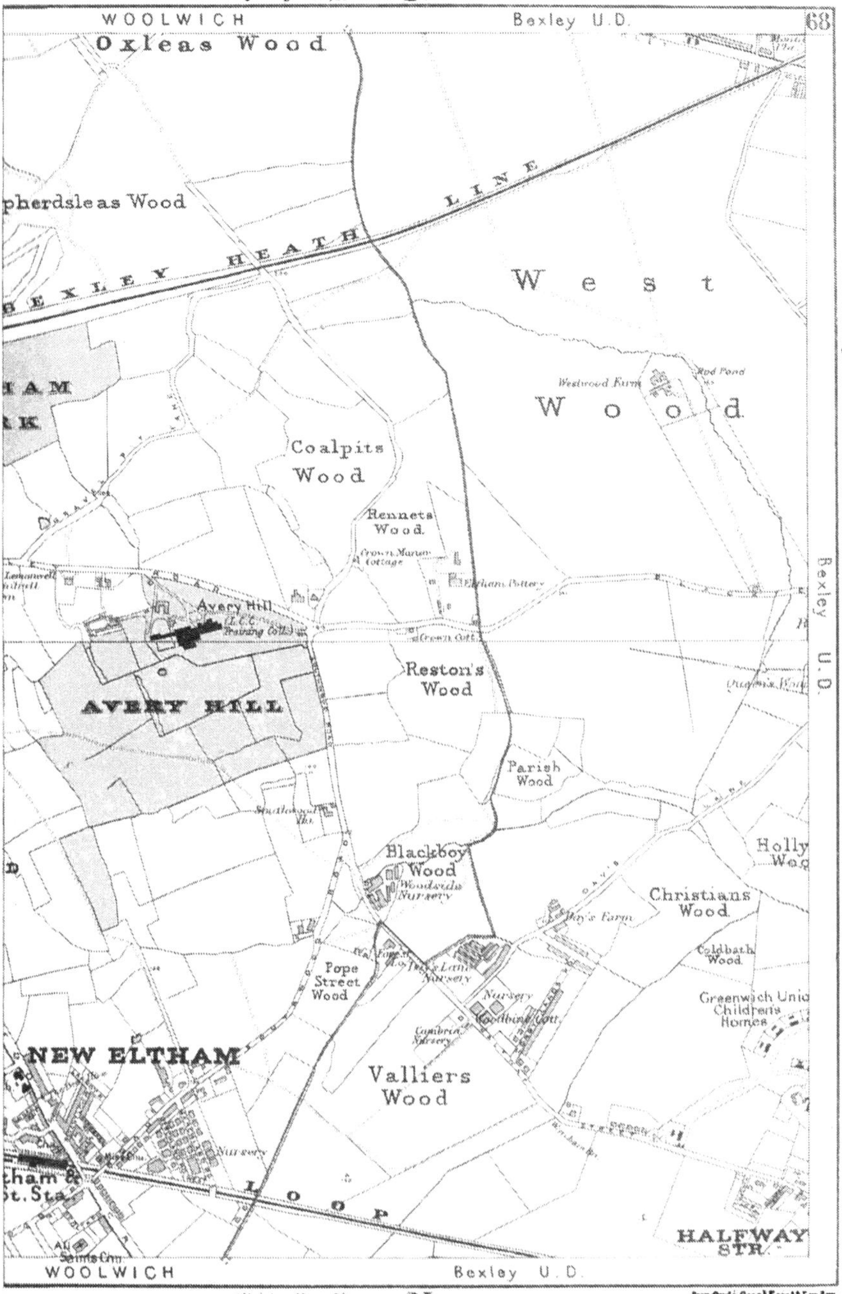

WOOLWICH

Bexley U.D.

68

Oxleas Wood

pherdsleas Wood

BEXLEY HEATH LINE

West Wood

Westwood Farm

Rod Pond

HAM
RK

Coalpits Wood

Rennets Wood

Crown Manor Cottage

Eltham Pottery

Avery Hill (L.C.C. Training Coll.)

Green Cott.

AVERY HILL

Reston's Wood

Parish Wood

Blackboy Wood

Woodside Nursery

Hay's Farm

Christians Wood

Holly Woo

Coldbath Wood

Pope Street Wood

Nursery

Woodbine Cott.

Cambria Nursery

Greenwich Unio Children's Homes

NEW ELTHAM

Valliers Wood

Nursery

LOOP

ltham & St. Sta.

All Saints Ch.

HALFWAY STR.

j

k

Bexley U.D.

Adjoining Sheet, 69.

WOOLWICH

Bexley U.D.

Adjoining Sheet 83.

N

Stanford's Geog^l Estab^t London

STANFORD'S 4" INDEXED MAP OF LONDON. O *Adjoining Sheet, 54.*

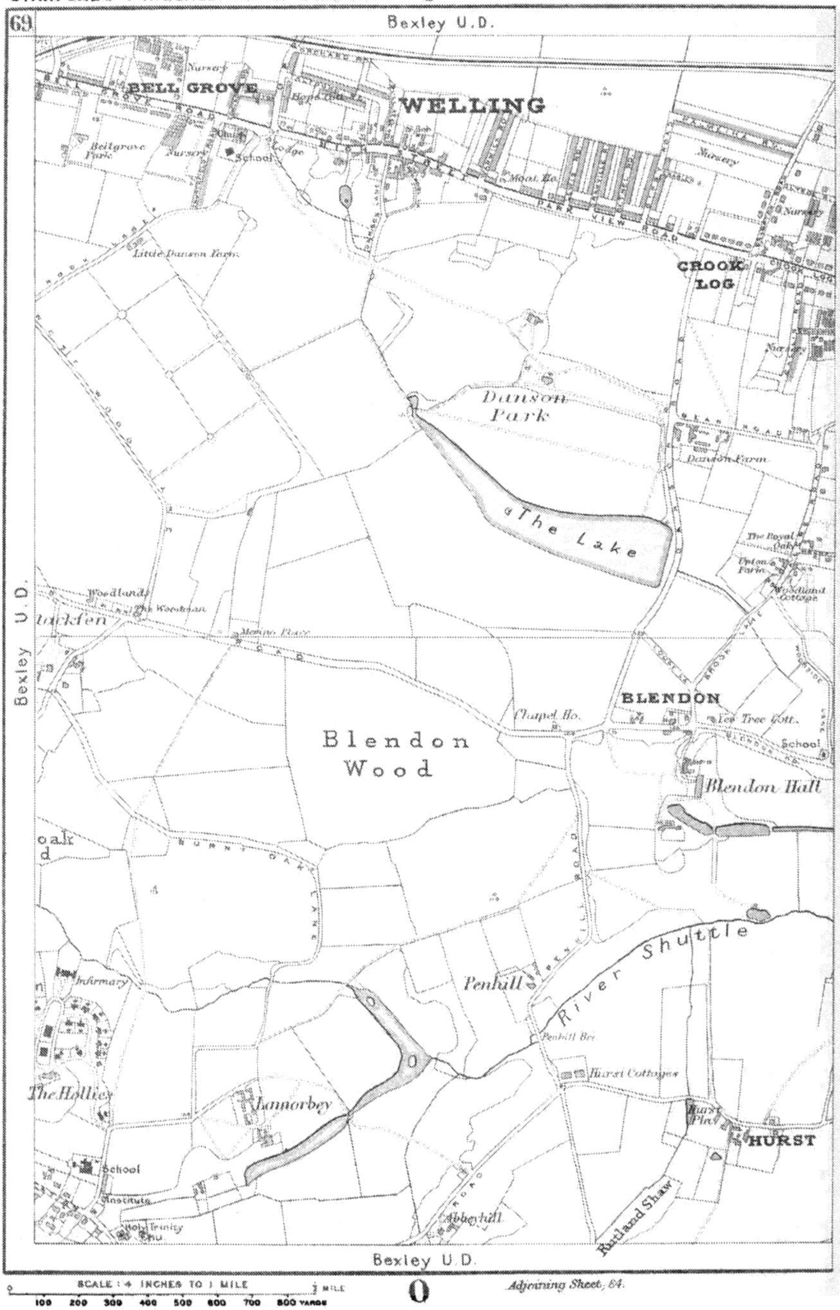

SCALE: 4 INCHES TO 1 MILE ½ MILE
0 100 200 300 400 500 600 700 800 YARDS

O *Adjoining Sheet, 84.*

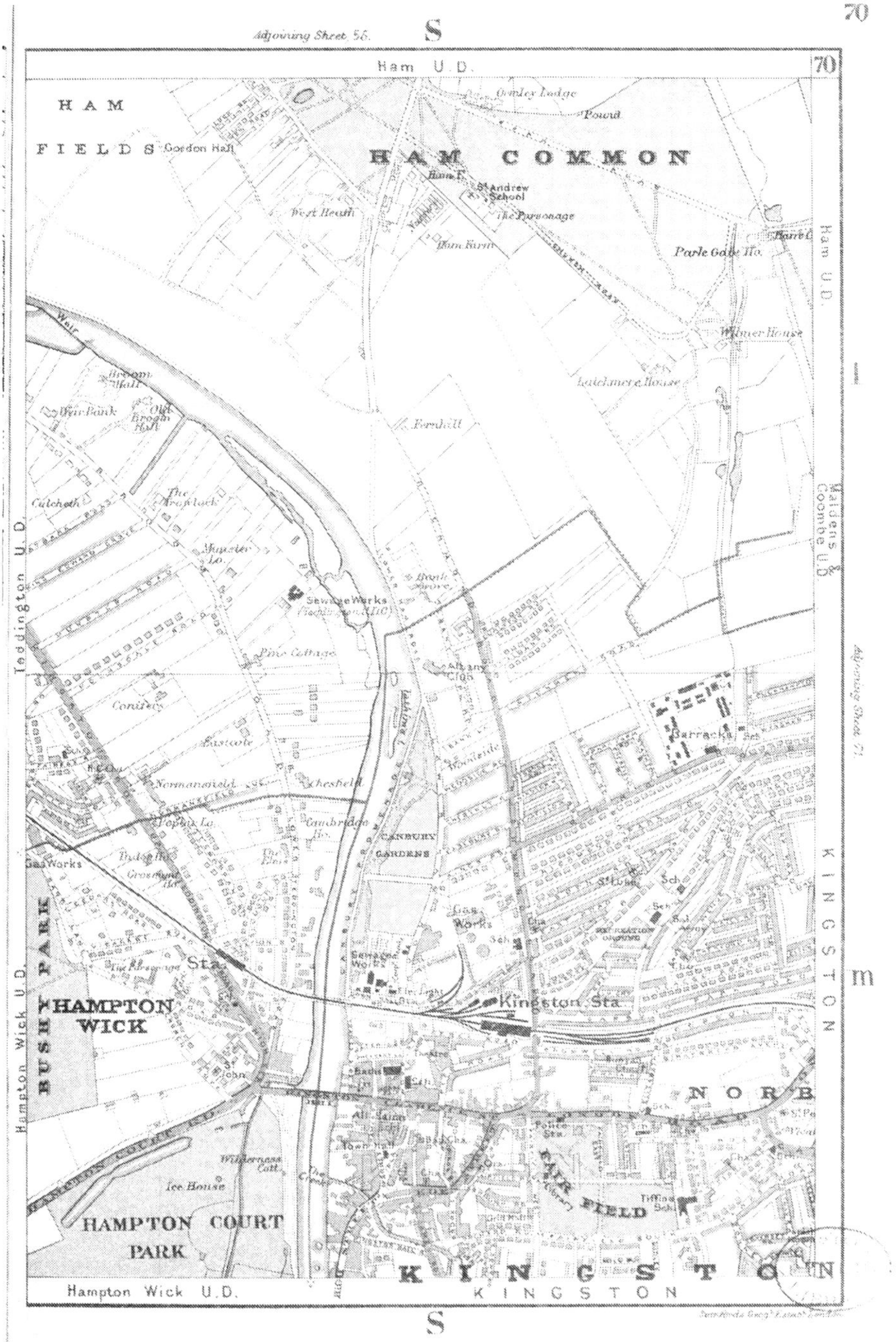

Adjoining Sheet 56.

S

Ham U.D.

70

HAM FIELDS

Gordon Hall

HAM COMMON

Ormeley Lodge

Pound

St. Andrew School

The Parsonage

West Heath

Ham Farm

Park Gate Ho.

Wilmer House

Latchmere House

Ferndale

Weir

Broom Hall

Old Broom Hall

Catchett

The Avonlock

Munster Lo.

Sewage Works (Teddington U.D.C.)

Pine Cottage

Conifers

Eastcote

Normansfield

Albany Club

Barracks

Woodside

Canbury Gardens

Gas Works

St. Luke

Recreation Ground

Sewage Works

Kingston Sta.

Sta.

HAMPTON WICK

BUSHY PARK

St. John

All Saints

Town Hall

Police Sta.

FAIR FIELD

Tiffins Sch.

Wilderness Cott.

Ice House

HAMPTON COURT PARK

NORBITON

KINGSTON

Teddington U.D.

Hampton Wick U.D.

Ham U.D.

Maldens & Coombe U.D.

Adjoining Sheet 71.

Hampton Wick U.D.

KINGSTON

S

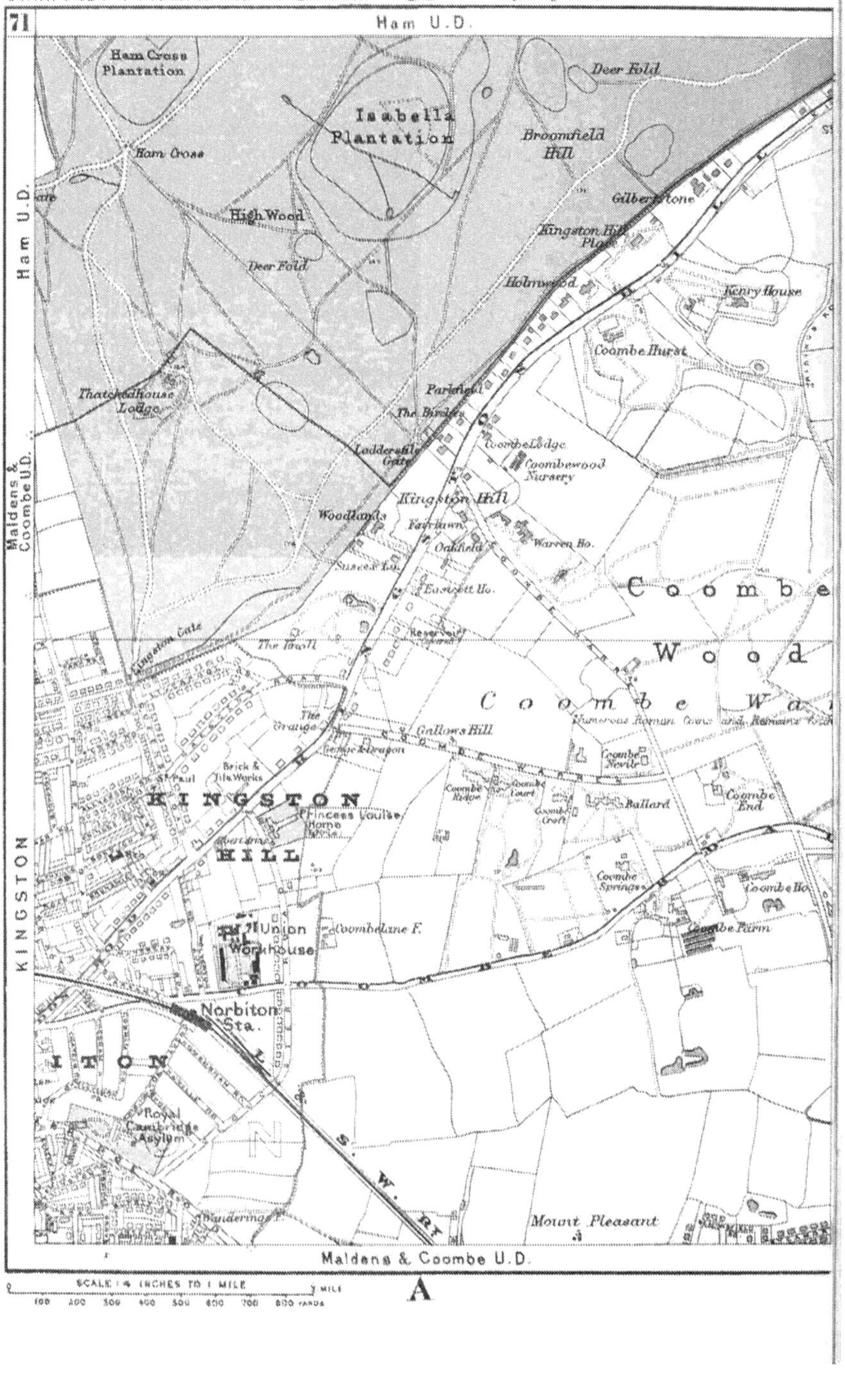

71 Ham U.D.

Ham Cross Plantation
Isabella Plantation
Deer Fold
Broomfield Hill
Ham Cross
High Wood
Deer Fold
Gilbert Stone
Kingston Hill Place
Holmwood
Kenry House
Coombe Hurst
Thatchedhouse Lodge
Parkfield
The Birches
Coombe Lodge
Ladderstile Gate
Coombewood Nursery
Kingston Hill
Woodlands
Fairlawn
Oakfield
Warren Ho.
Sussex Lo.
Eastcott Ho.
Coombe Wood
Reservoir
Kingston Gate
The Knoll
Coombe Warren
The Grange
Gallows Hill
Numerous Roman Coins and Remains found
George & Dragon
Coombe Nevile
Brick & Tile Works
St. Paul
Coombe Ridge
Coombe Court
Coombe Croft
Ballard
Coombe End
KINGSTON HILL
Princess Louise Home
Coombe Springs
Coombe Ho.
Union Workhouse
Coombelane F.
Coombe Farm
Norbiton Sta.
ITON
L. & S. W. Ry.
Royal Cambridge Asylum
Wandering F.
Mount Pleasant

Ham U.D.

Maldens & Coombe U.D.

KINGSTON

Adjoining Sheet, 70.

l

m

Maldens & Coombe U.D.

A

SCALE: 4 INCHES TO 1 MILE

100 200 300 400 500 600 700 800 YARDS ½ MILE

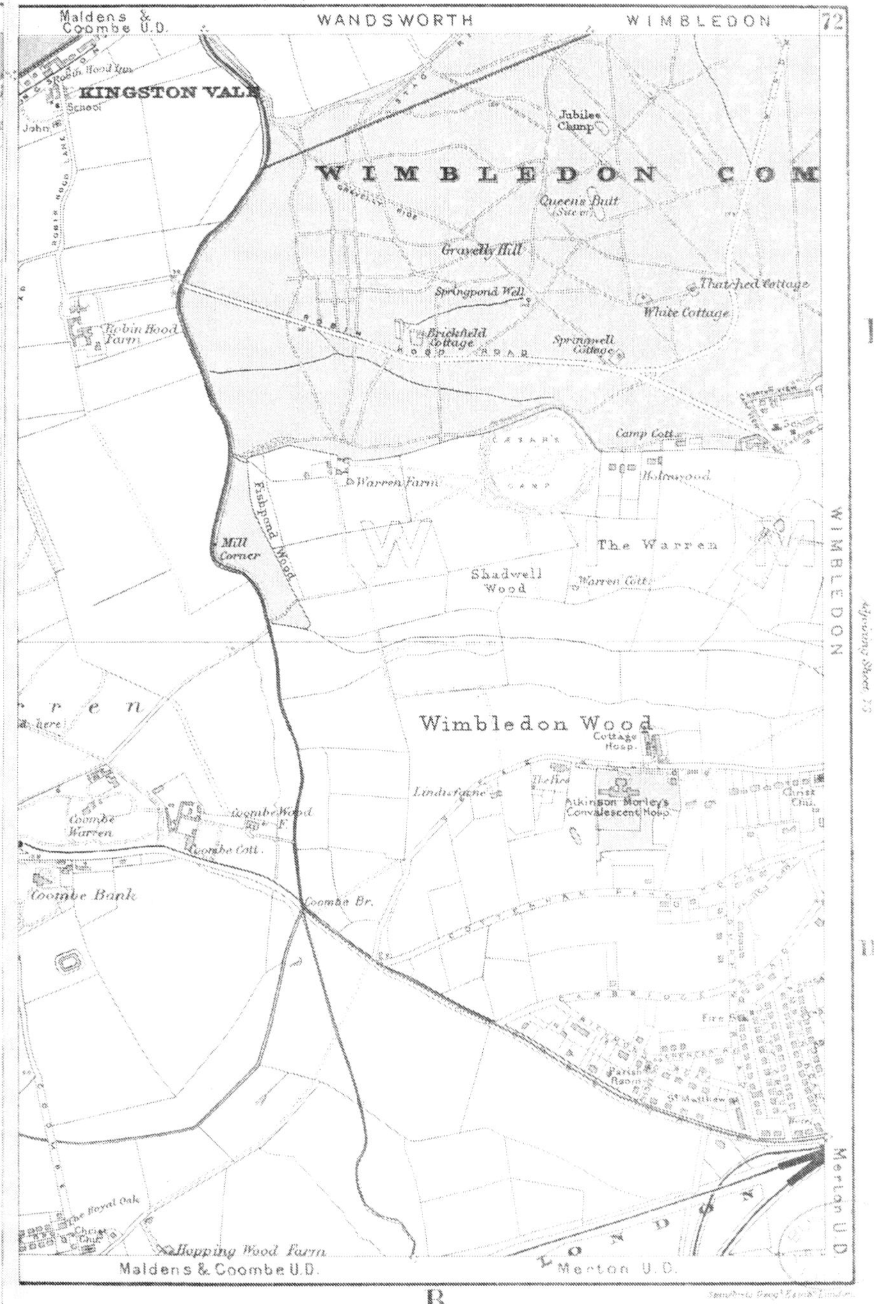
Adjoining Sheet, 57.
B
Maldens & Coombe U.D.
WANDSWORTH
WIMBLEDON
72
KINGSTON VALE
Robin Hood Inn
School
John
ROBIN HOOD LANE
Jubilee Clump
WIMBLEDON COM
Queens Butt (Site of)
Gravelly Hill
Springpond Well
Thatched Cottage
White Cottage
Robin Hood Farm
Brickfield Cottage
Springwell Cottage
Camp Cott.
CÆSAR'S CAMP
Warren Farm
Holmwood
Fishpond Wood
Mill Corner
The Warren
Shadwell Wood
Warren Cott.
WIMBLEDON
Adjoining Sheet, 73
Wimbledon Wood
Cottage Hosp.
Lindisfarne
Atkinson Morley's Convalescent Hosp.
Christ Ch.
Coombe Warren
Coombe Wood F.
Coombe Cott.
Coombe Bank
Coombe Br.
Fire Sta.
Parish Room
St. Matthew
The Royal Oak
Christ Ch.
Hopping Wood Farm
LONDON
Merton U.D.
Maldens & Coombe U.D.
Merton U.D.
B

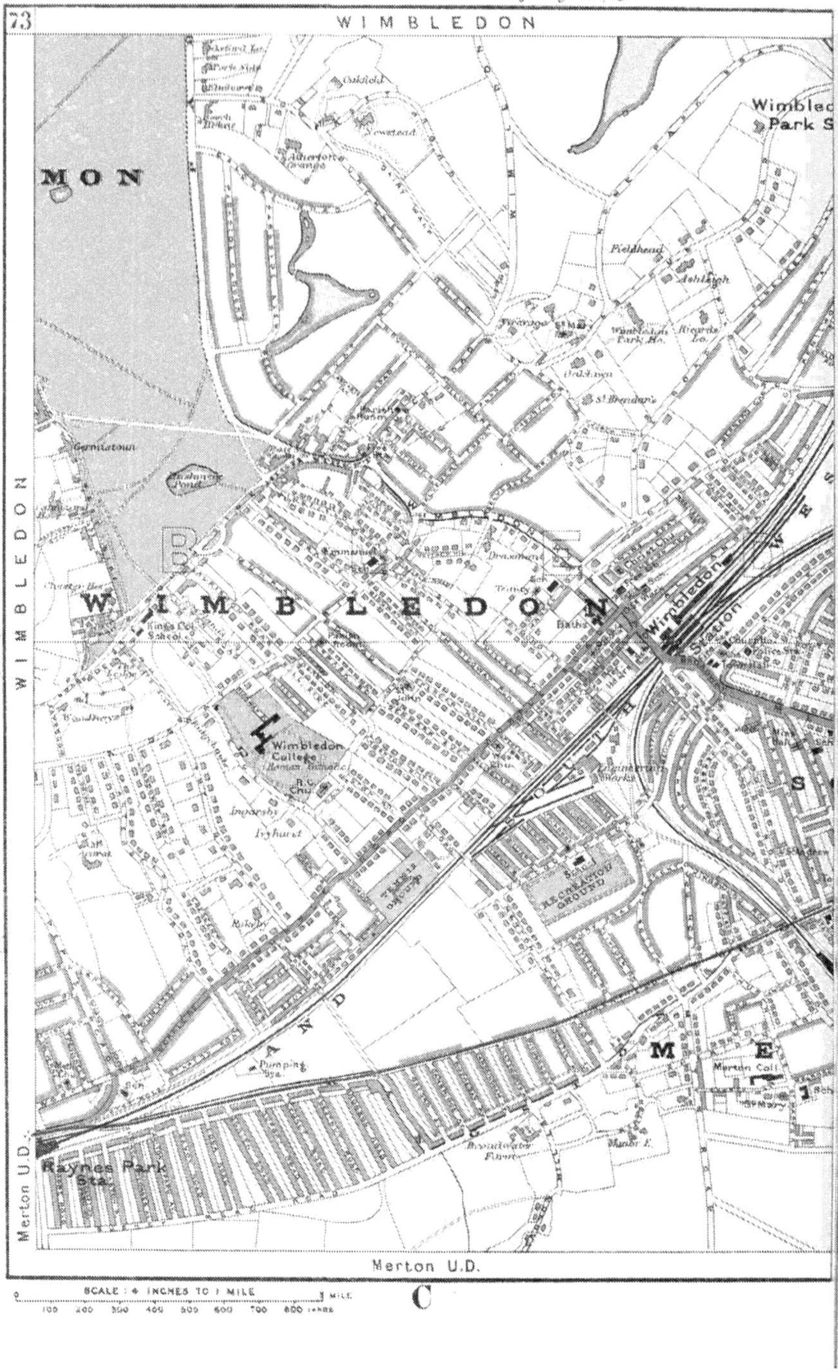
73
WIMBLEDON
MON
Wimbledon Park Sta.
Fieldhead
Ashleigh
Vicarage
Wimbledon Park Ho.
Oaklawn
St Brendan's
Adjoining Sheet, 72.
WIMBLEDON
W I M B L E D O N
Kings Coll. School
Wimbledon College
Roman Catholic
R.C. Chu.
Ivyhurst
Baths
Wimbledon Station
RECREATION GROUND
M E
Merton Coll.
Pumping Sta.
Raynes Park Sta.
Merton U.D.
Broadwater Farm
Manor F.
l
m

Merton U.D.

SCALE: 4 INCHES TO 1 MILE
0 100 200 300 400 500 600 700 800 YARDS 1 MILE

C

Adjoining Sheet, 59.

D

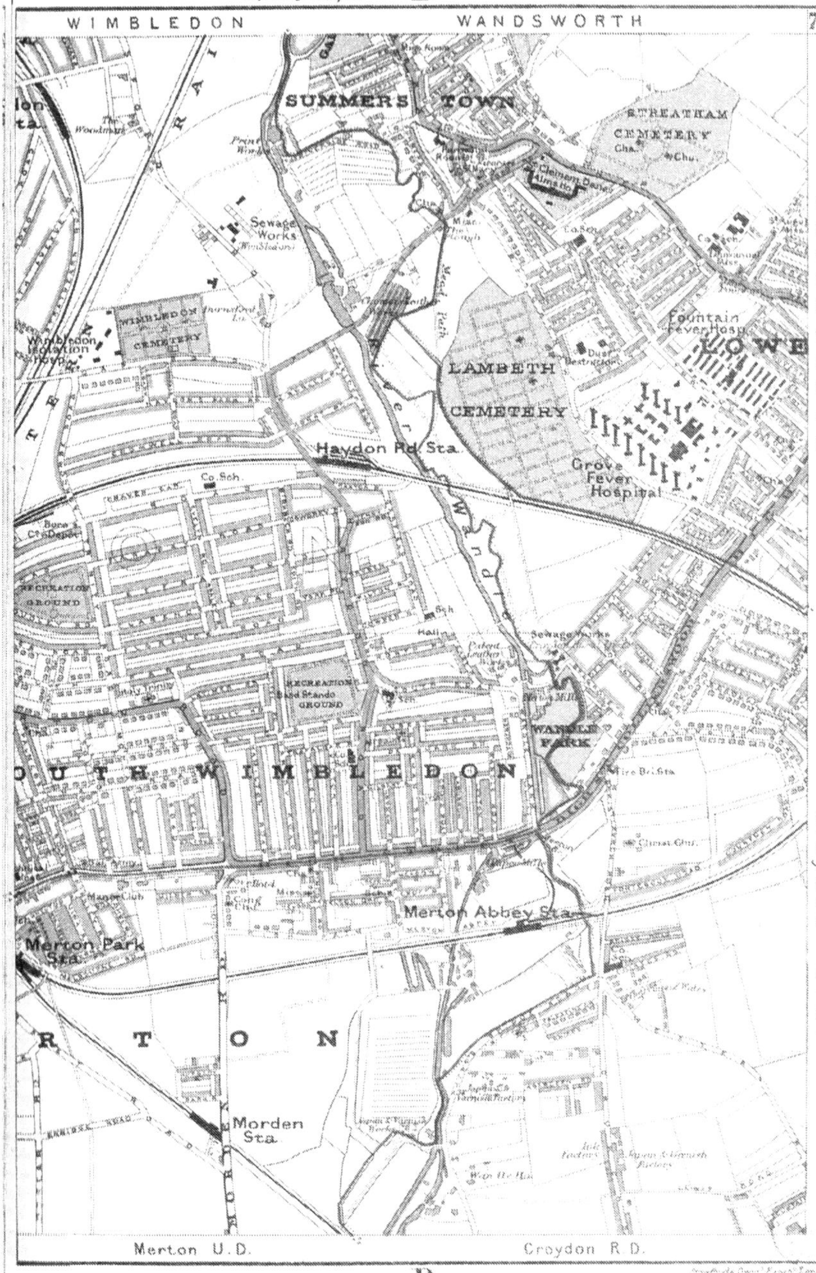
WIMBLEDON
WANDSWORTH
74
SUMMERS TOWN
STREATHAM CEMETERY
Cha.
Chu.
Sewage Works
WIMBLEDON CEMETERY
Wimbledon Isolation Hosp.
LAMBETH CEMETERY
Fountain Fever Hosp.
LOWE
Grove Fever Hospital
Haydon Rd. Sta.
Co. Sch.
RECREATION GROUND
Sewage Works
WANDLE PARK
SOUTH WIMBLEDON
Christ Chu.
Merton Abbey Sta.
Merton Park Sta.
RTON
Morden Sta.
WANDSWORTH
Croydon R.D.
Merton U.D.
Croydon R.D.

Adjoining Sheet 75.

l

m

D

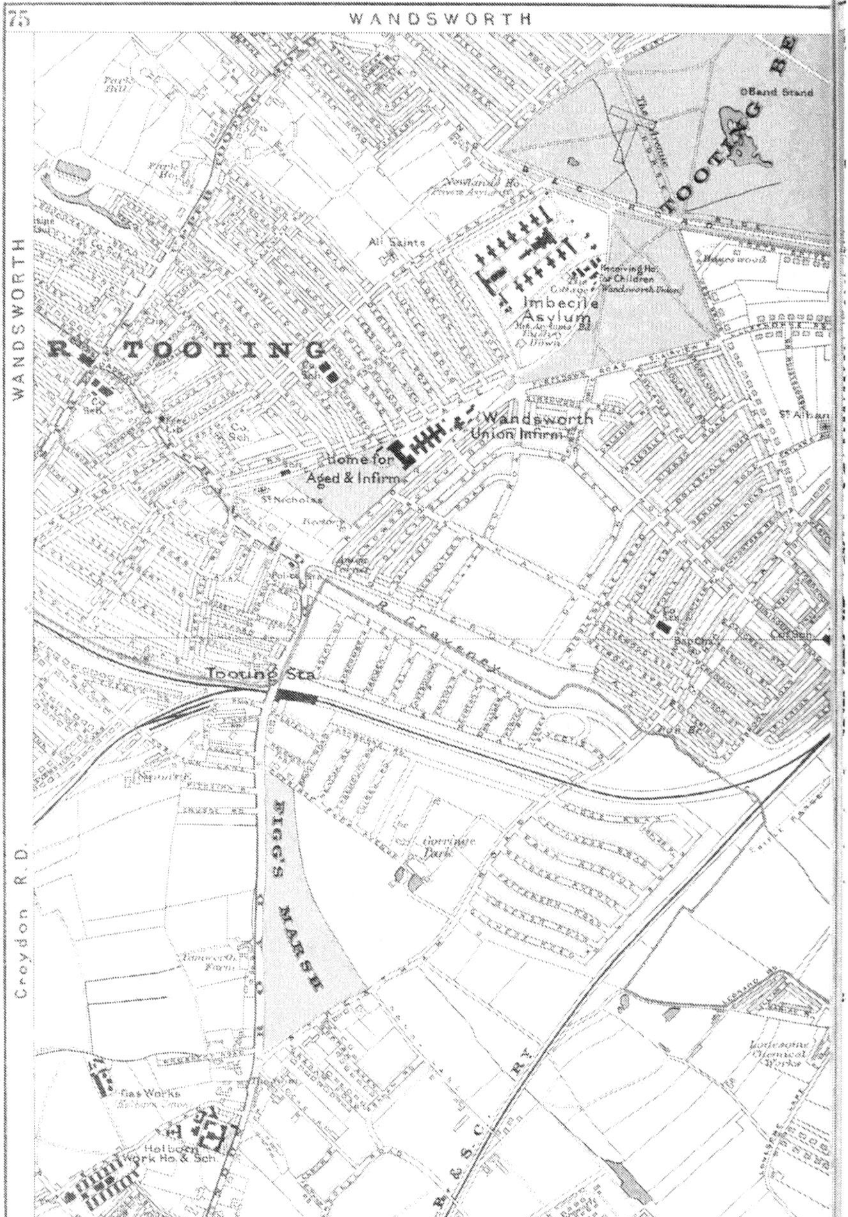
75
WANDSWORTH
WANDSWORTH
l
m
Adjoining Sheet, 74.
TOOTING
TOOTING BE
Band Stand
The Avenue
Newlands Ho.
All Saints
Receiving Ho. for Children Wandsworth Union
Imbecile Asylum
Wandsworth Union Infirm
Home for Aged & Infirms
St Nicholas
Rectory
Tooting Sta
Hayeswood
St Alban
Co. Sch.
Bap Ch.
PIGG'S MARSH
Gorringe Park
Tamworth Farm
Gas Works
Holborn Work Ho. & Sch.
Co. Sch.
L. B. & S. C. RY
Lonesome Chemical Works
Croydon R.D.
Croydon R.D.

SCALE : 4 INCHES TO 1 MILE
0 100 200 300 400 500 600 700 800 YARDS ½ MILE

E

Adjoining Sheet, 61.

F

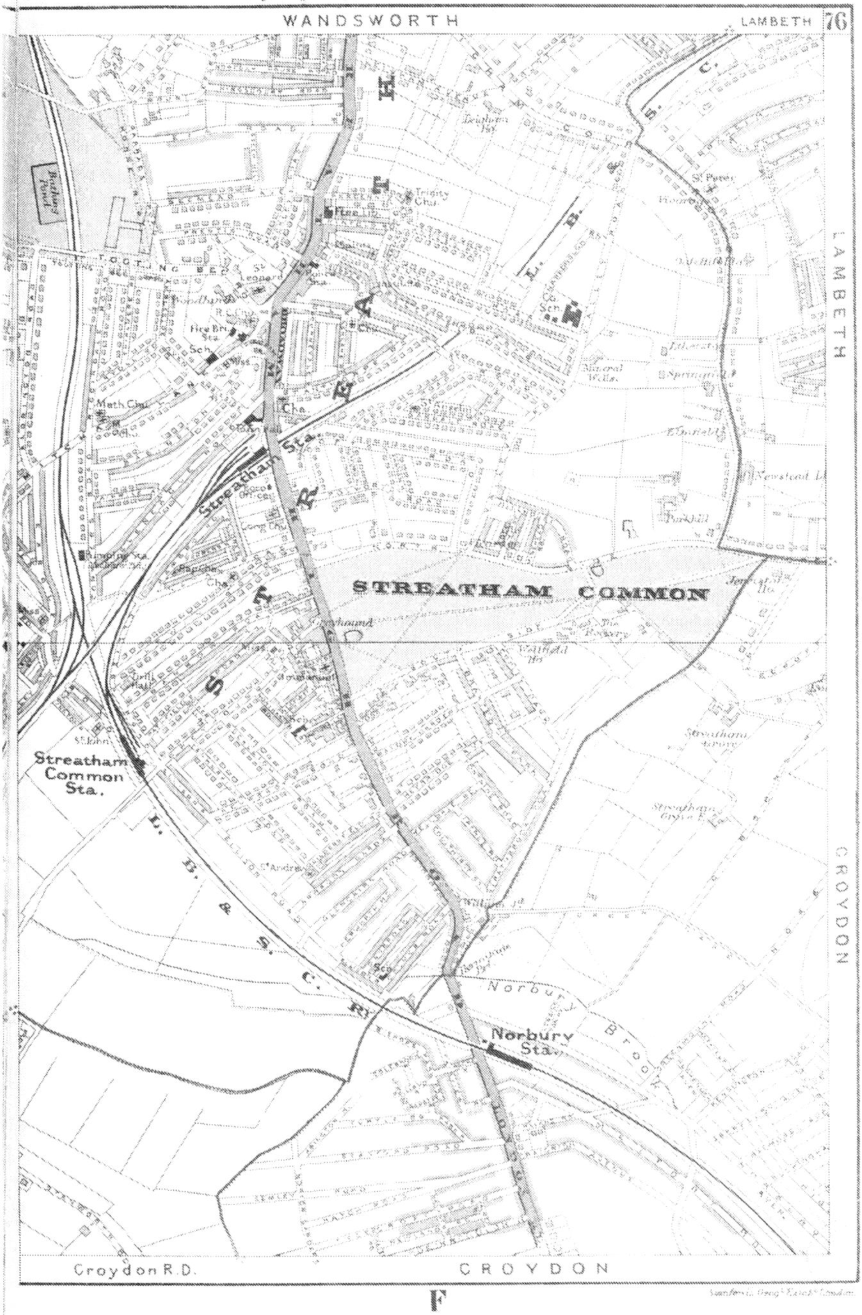

WANDSWORTH

LAMBETH 76

STREATHAM

STREATHAM COMMON

Streatham Sta.

Streatham Common Sta.

Norbury Sta.

Trinity Chu.

Free Lib.

TOOTING BEC

Mineral Wells

Streatham Grove

Norbury Brook

St Andrew

L. B. & S. C. Ry

LAMBETH

l

Adjoining Sheet, 77.

CROYDON

m

Croydon R.D.

CROYDON

F

Stanford's Geog.l Estab.t London.

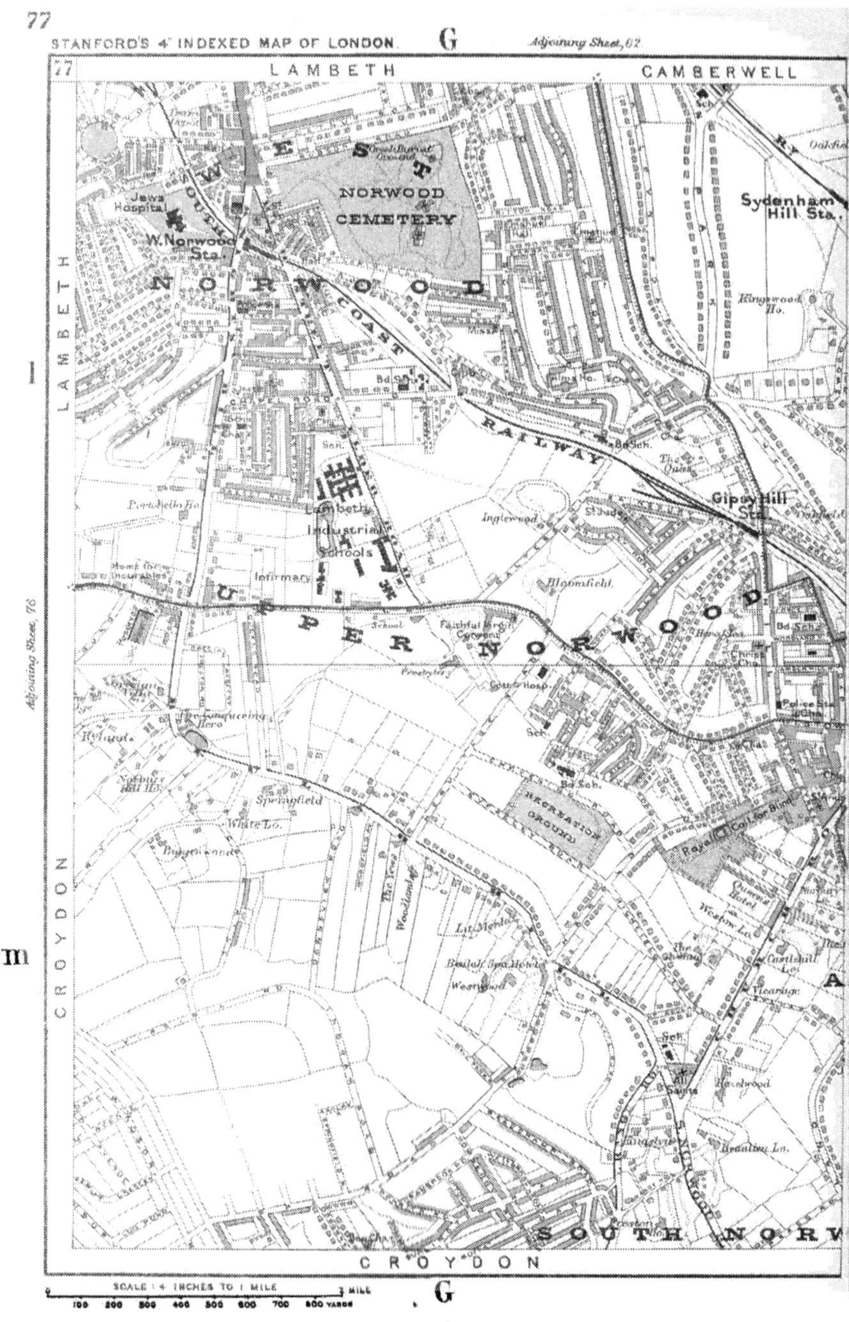
77
STANFORD'S 4" INDEXED MAP OF LONDON.
G
Adjoining Sheet, 62.
LAMBETH
CAMBERWELL
LAMBETH
CROYDON
CROYDON
l
m
Adjoining Sheet, 76
WEST NORWOOD
NORWOOD CEMETERY
Jews Hospital
W. Norwood Sta.
COAST RAILWAY
Sydenham Hill Sta.
Gipsy Hill Sta.
Lambeth Industrial Schools
Infirmary
Home for Incurables
UPPER NORWOOD
Inglewood
Bloomfield
Recreation Ground
Royal Coll. for Blind
Springfield
White Lo.
Rylands
Woodlands
SOUTH NORWOOD
Bd. Sch.
SCALE: 4 INCHES TO 1 MILE
100 200 300 400 500 600 700 800 YARDS
G

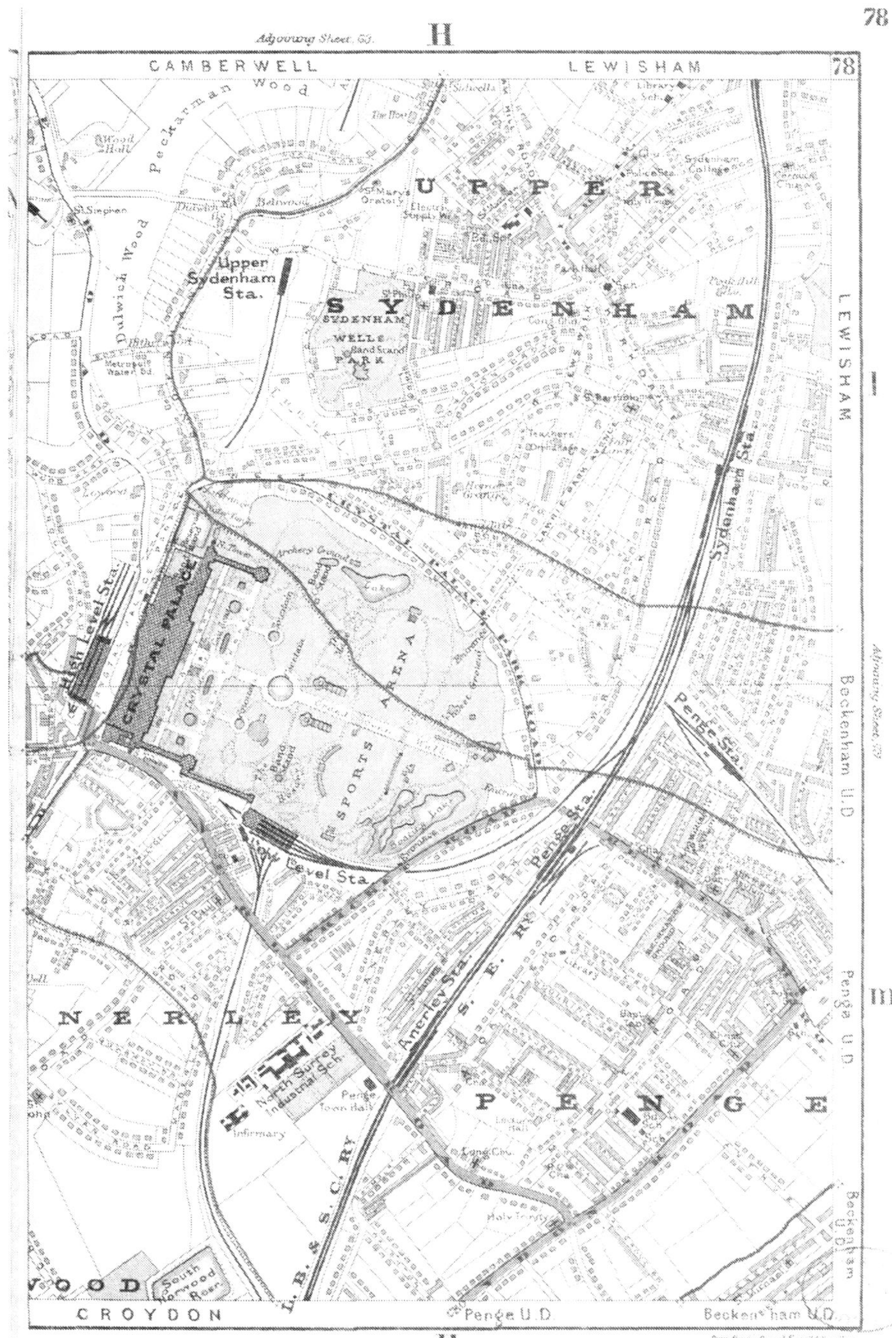

Adjoining Sheet. 63.
H
CAMBERWELL
LEWISHAM
78
UPPER
SYDENHAM
Upper Sydenham Sta.
Dulwich Wood
Peckarman Wood
SYDENHAM WELLS PARK
Sydenham College
Sydenham Sta.
LEWISHAM
I
CRYSTAL PALACE
High Level Sta.
Low Level Sta.
Archery Ground
SPORTS ARENA
CRYSTAL PALACE PARK ROAD
Adjoining Sheet. 79
Beckenham U.D.
Penge Sta.
Penge U.D.
III
NERLEY
Anerley Sta.
S. E. RY
North Surrey Industrial Sch.
Infirmary
Penge Town Hall
PENGE
Library
Holy Trinity
L. B. & S. C. RY
Beckenham U.D.
VOOD
South Norwood Resr.
CROYDON
Penge U.D.
Beckenham U.D.
II

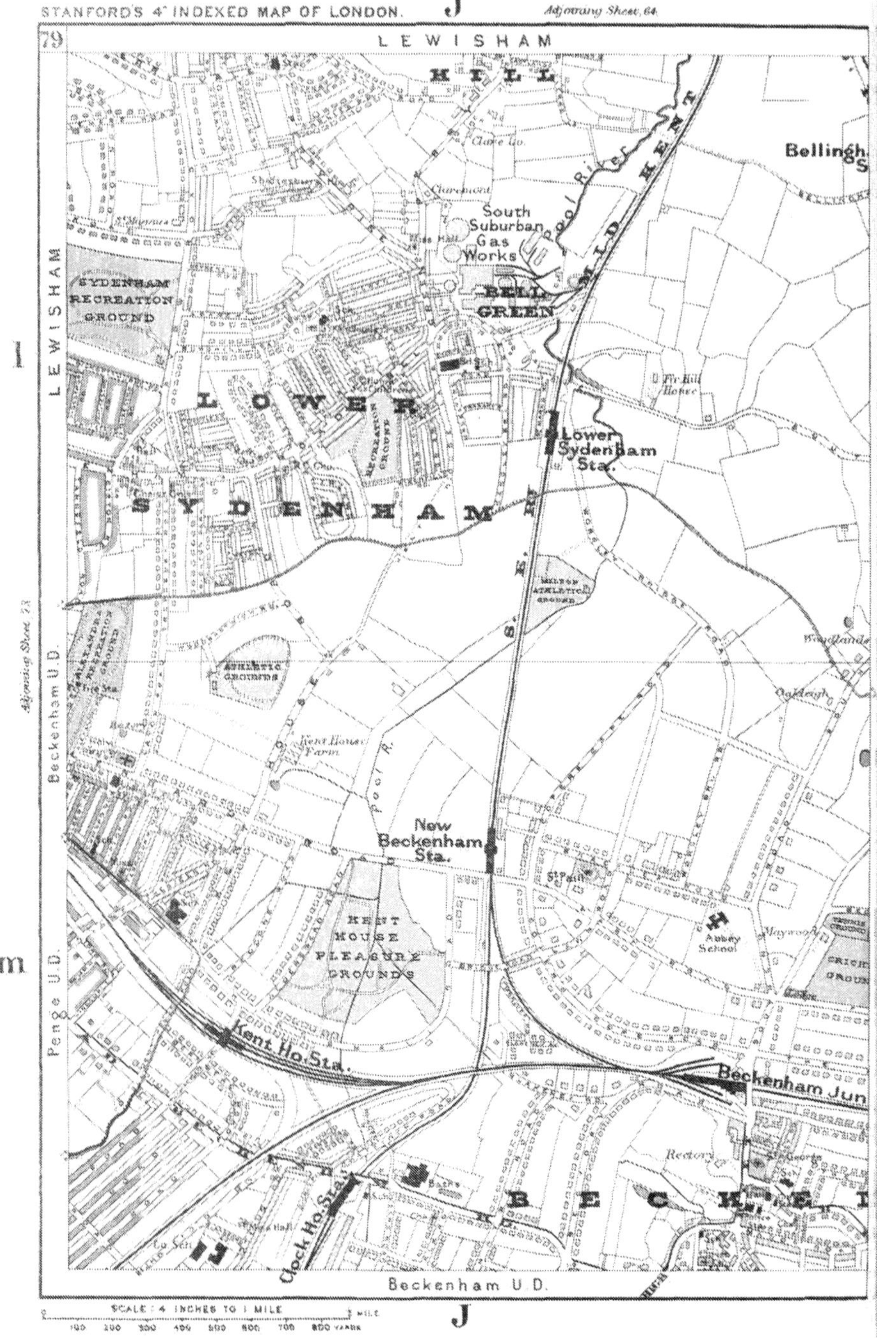

STANFORD'S 4" INDEXED MAP OF LONDON.
J
Adjoining Sheet, 64.
79
LEWISHAM
HILL
Bellingh
South Suburban Gas Works
BELL GREEN
Pool River
MID KENT
SYDENHAM RECREATION GROUND
LEWISHAM
I
LOWER
RECREATION GROUND
Lower Sydenham Sta.
SYDENHAM
Fir Hill House
ATHLETIC GROUNDS
Adjoining Sheet, 73.
Beckenham U.D.
Kent House Farm
Pool R.
New Beckenham Sta.
St. Paul
Abbey School
KENT HOUSE PLEASURE GROUNDS
m
Penge U.D.
Kent Ho. Sta.
Beckenham Jun
Rectory
Clock Ho. Sta.
BECKE
Beckenham U.D.
SCALE: 4 INCHES TO 1 MILE
J

Adjoining Sheet 65

K

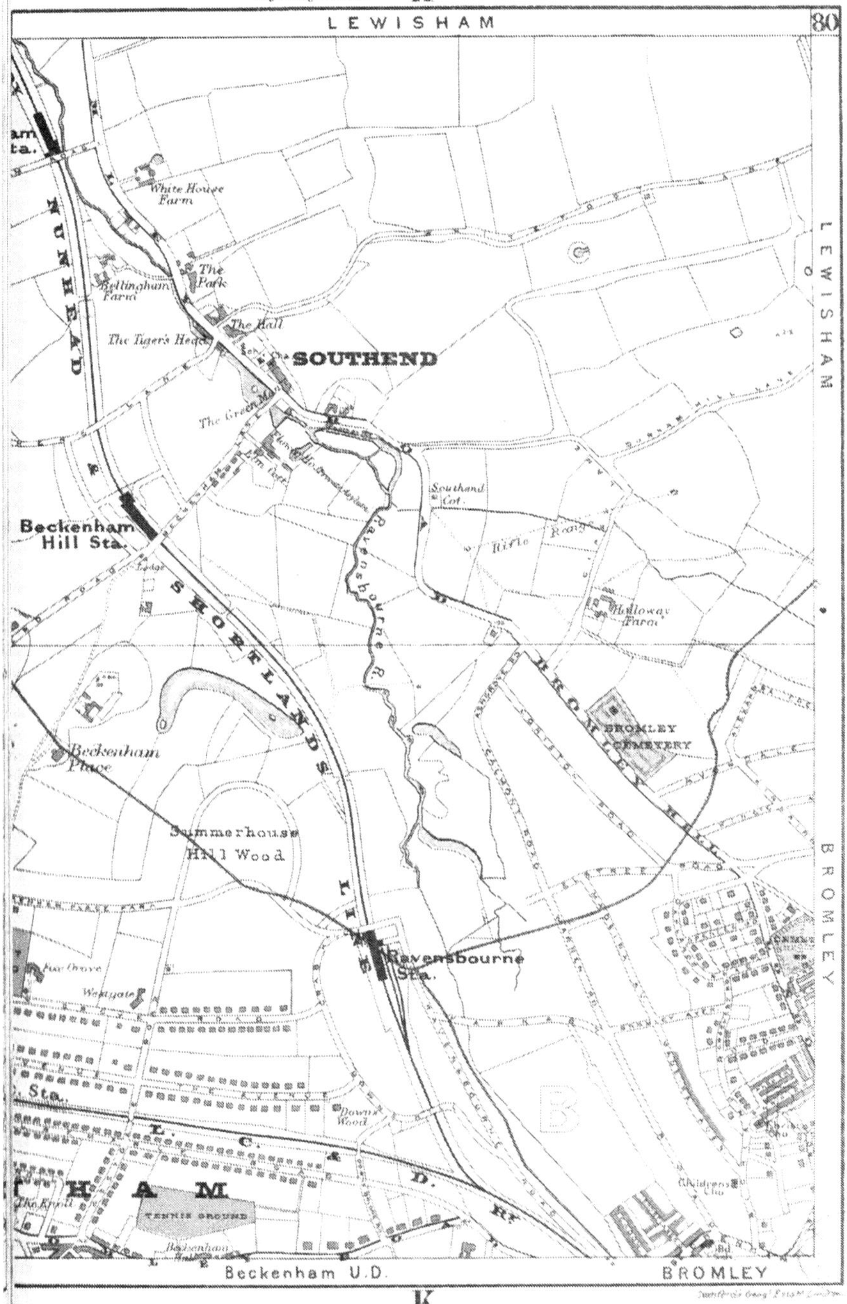

LEWISHAM

80

White House Farm

The Park

Bellingham Farm

The Hall

The Tiger's Head

SOUTHEND

The Green Man

NUNHEAD

Beckenham Hill Sta.

SHORTLANDS

Southend Cot.

Rifle Range

Holloway Farm

Ravensbourne R.

BROMLEY CEMETERY

Beckenham Place

Summerhouse Hill Wood

LINE

Ravensbourne Sta.

Fox Grove

Westgate

Sta.

L. C. & D. RY.

HAM

TENNIS GROUND

Beckenham

Beckenham U.D.

BROMLEY

K

LEWISHAM

BROMLEY

l

m

Adjoining Sheet

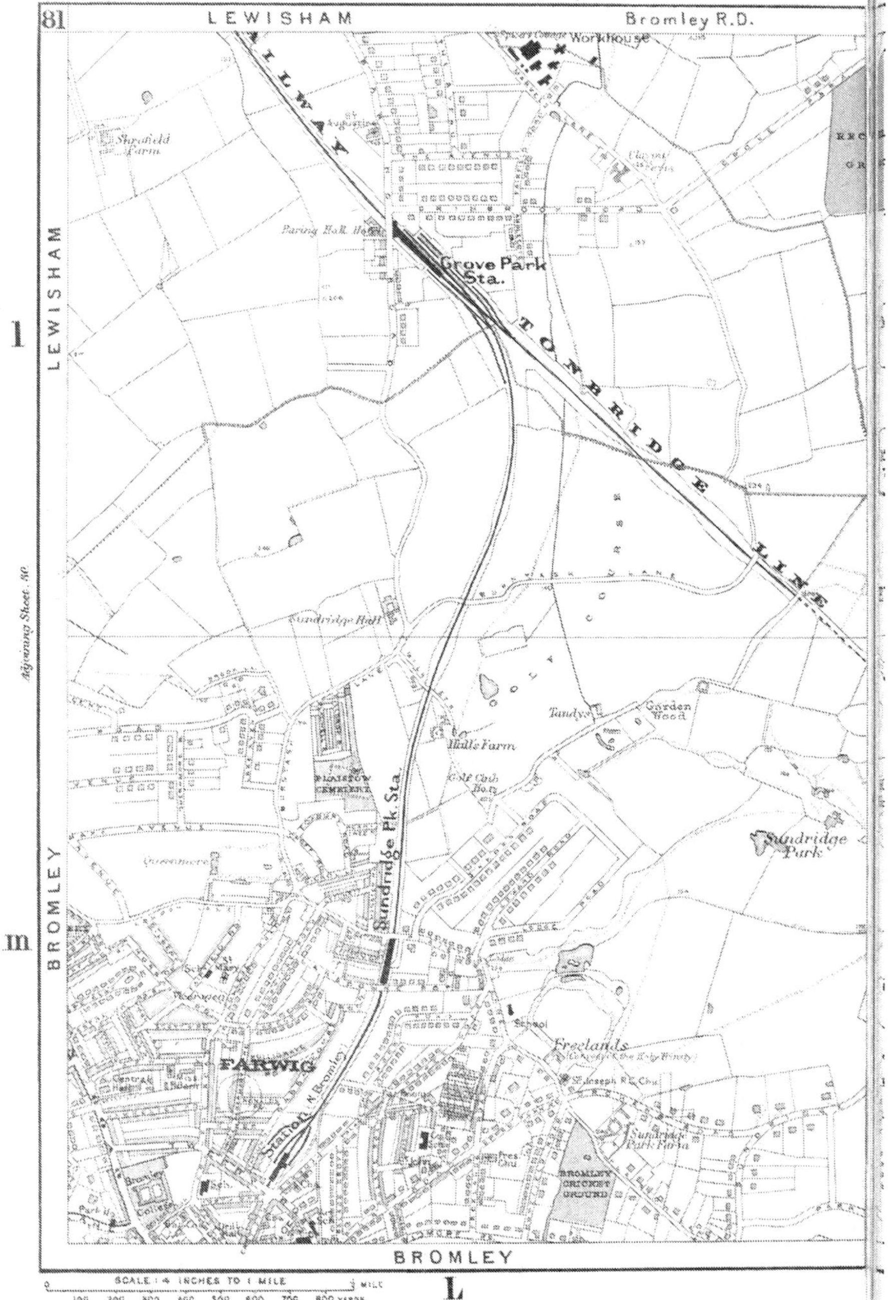

STANFORD'S 4" INDEXED MAP OF LONDON. **L** *Adjoining Sheet. 66.*

81 LEWISHAM Bromley R.D.

Workhouse

Shroffield Farm

Baring Hall Hotel

Grove Park Sta.

RAILWAY

TONBRIDGE LINE

LEWISHAM

l

Adjoining Sheet. 80

Sundridge Hall

GOLF COURSE

Tandys

Garden Wood

Hall's Farm

Golf Club Ho.

PLAISTOW CEMETERY

Sundridge Pk. Sta.

Sundridge Park

BROMLEY

m

School

Freelands

St. Joseph R.C. Chu.

FARWIG

Station

Pres. Chu.

Sundridge Park Farm

BROMLEY CRICKET GROUND

BROMLEY

SCALE: 4 INCHES TO 1 MILE

0 100 200 300 400 500 600 700 800 YARDS ½ MILE

L

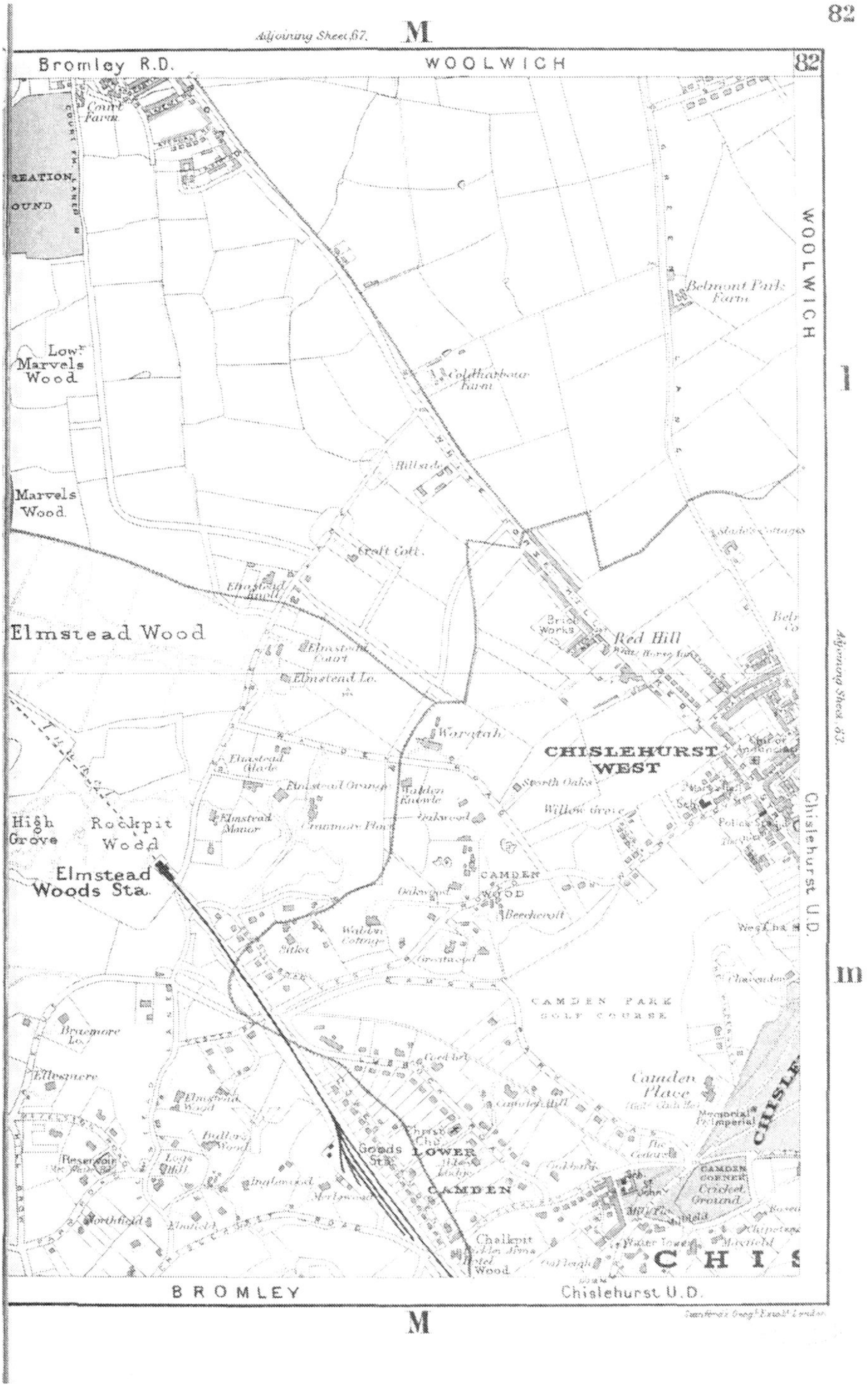

Adjoining Sheet, 67.

M

Bromley R.D.

WOOLWICH

82

Court Farm

RECREATION GROUND

Low[r] Marvels Wood

Marvels Wood

Elmstead Wood

Belmont Park Farm

Coldharbour Farm

Hillside

Croft Cott.

Elmstead Knoll

Brick Works

Red Hill

Elmstead Court

Elmstead Lo.

Waratah

CHISLEHURST WEST

Elmstead Glade

Elmstead Grange

Walden Knowle

Oakwood

High Grove

Rockpit Wood

Elmstead Manor

Elmstead Woods Sta.

CAMDEN WOOD

Beechcroft

Walden Cottage

Greatwood

CAMDEN PARK GOLF COURSE

Braemore Lo.

Ellesmere

Camden Place

Memorial Imperial

Reservoir

Goods Sta.

LOWER CAMDEN

CAMDEN CORNER Cricket Ground

Chalkpit Wood

CHIS

Chislehurst U.D.

WOOLWICH

Adjoining Sheet, 83.

l

m

BROMLEY

Chislehurst U.D.

M

Stanford's Geog[l] Estab[t] London

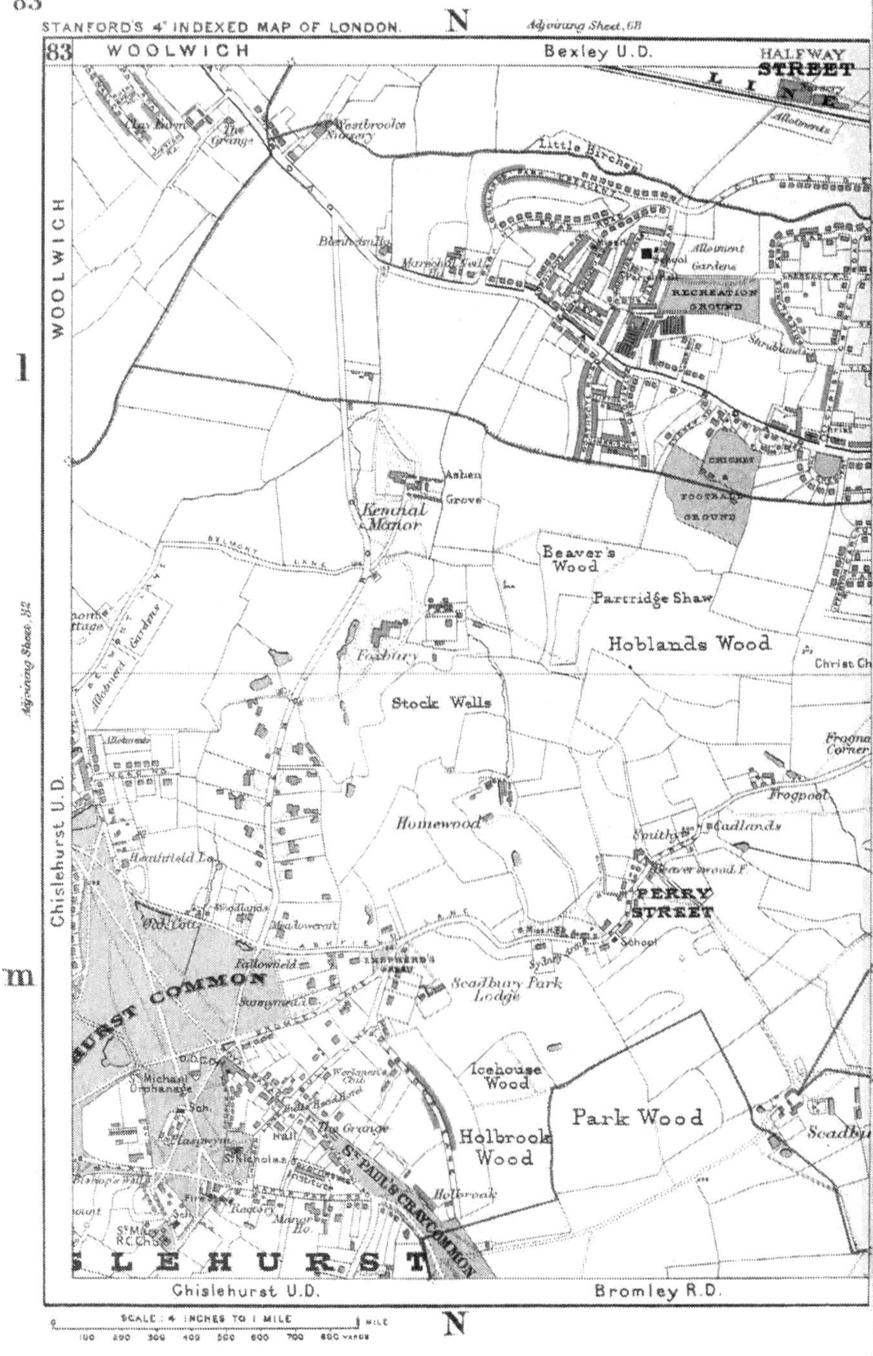
STANFORD'S 4" INDEXED MAP OF LONDON.
N
Adjoining Sheet, 68
83
WOOLWICH
Bexley U.D.
HALFWAY STREET
Westbrooke Nursery
The Grange
Little Birches
Allotment Gardens
School
RECREATION GROUND
WOOLWICH
Ashen Grove
Kennal Manor
BELMONT LANE
Beaver's Wood
Partridge Shaw
Hoblands Wood
Foxbury
Stock Wells
Adjoining Sheet, 82
Chislehurst U.D.
Homewood
Frogpool
Cadlands
PERRY STREET
School
Meadowcroft
Fallowfield
Scadbury Park Lodge
HURST COMMON
St. Michael Orphanage
Icehouse Wood
Park Wood
The Grange
Holbrook Wood
ST PAUL'S CRAY COMMON
Holbrook
Rectory
SLEHURST
Chislehurst U.D.
Bromley R.D.
SCALE: 4 INCHES TO 1 MILE
N
1
m

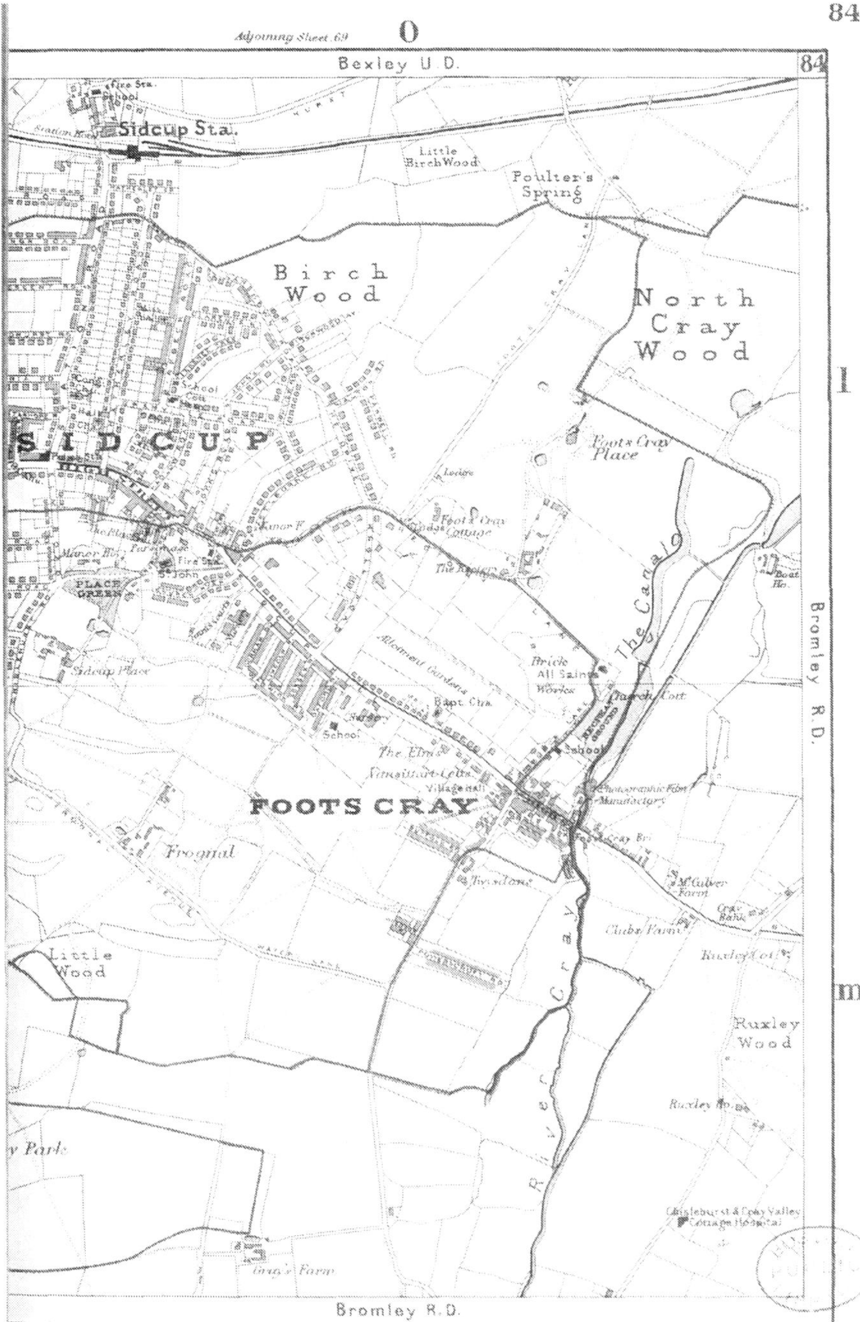
Adjoining Sheet 69
O
Bexley U.D.
84
Sidcup Sta.
Little
Birch Wood
Poulter's
Spring
Birch
Wood
North
Cray
Wood
SIDCUP
Foots Cray
Place
The Cadalo
Foots Cray
Cottage
Allotment Gardens
Brick
All Saints
Works
Church Cott.
Bapt. Ch.
School
The Elms
Vansittart Cotts.
Village Hall
FOOTS CRAY
Photographic Film Manufactory
Foots Cray Bri.
Frognal
Little
Wood
Clubs Farm
Ruxley Cot.
Ruxley
Wood
River Cray
Ruxley Ho.
Chislehurst & Cray Valley
Cottage Hospital
Gray's Farm
Bromley R.D.
O
Stanford's Geog.l Estab.t London
l
m
Bromley R.D.

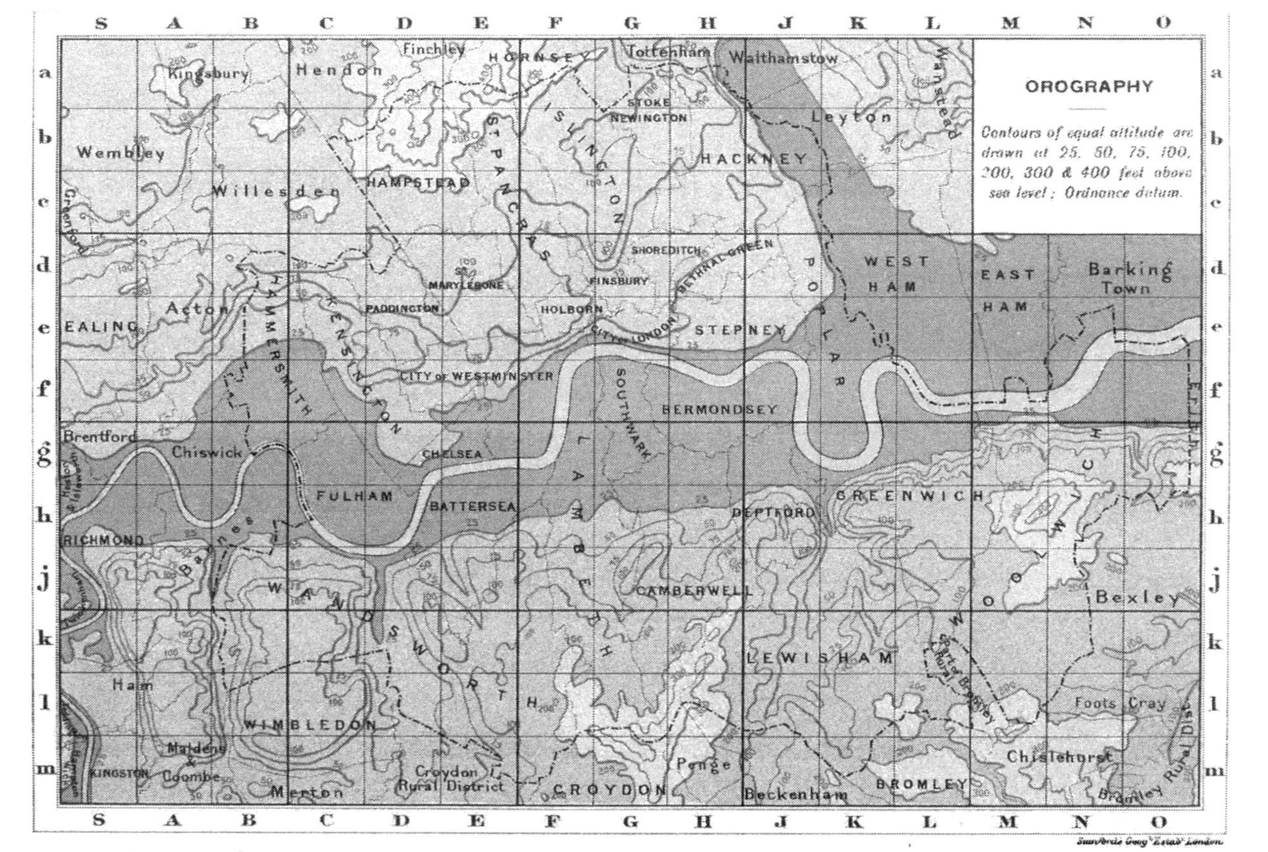
OROGRAPHY
Contours of equal altitude are drawn at 25, 50, 75, 100, 200, 300 & 400 feet above sea level; Ordnance datum.
S A B C D E F G H J K L M N O
a b c d e f g h j k l m
Kingsbury
Hendon
Finchley
HORNSEY
Tottenham
Walthamstow
Wanstead
Wembley
STOKE NEWINGTON
Leyton
ISLINGTON
HACKNEY
Willesden
HAMPSTEAD
ST PANCRAS
SHOREDITCH
BETHNAL GREEN
POPLAR
WEST HAM
EAST HAM
Barking Town
MARYLEBONE
FINSBURY
PADDINGTON
HOLBORN
CITY OF LONDON
STEPNEY
EALING
Acton
HAMMERSMITH
KENSINGTON
CITY of WESTMINSTER
SOUTHWARK
BERMONDSEY
Brentford
Chiswick
CHELSEA
LAMBETH
FULHAM
BATTERSEA
GREENWICH
DEPTFORD
RICHMOND
Barnes
WANDSWORTH
CAMBERWELL
WOOLWICH
Bexley
LEWISHAM
Part of Bromley Rural
Ham
Foots Cray
WIMBLEDON
Maldens & Coombe
KINGSTON
Merton
Croydon Rural District
CROYDON
Penge
Beckenham
BROMLEY
Chislehurst
Bromley Rural Dist.
Sanford's Geog' Estab' London.

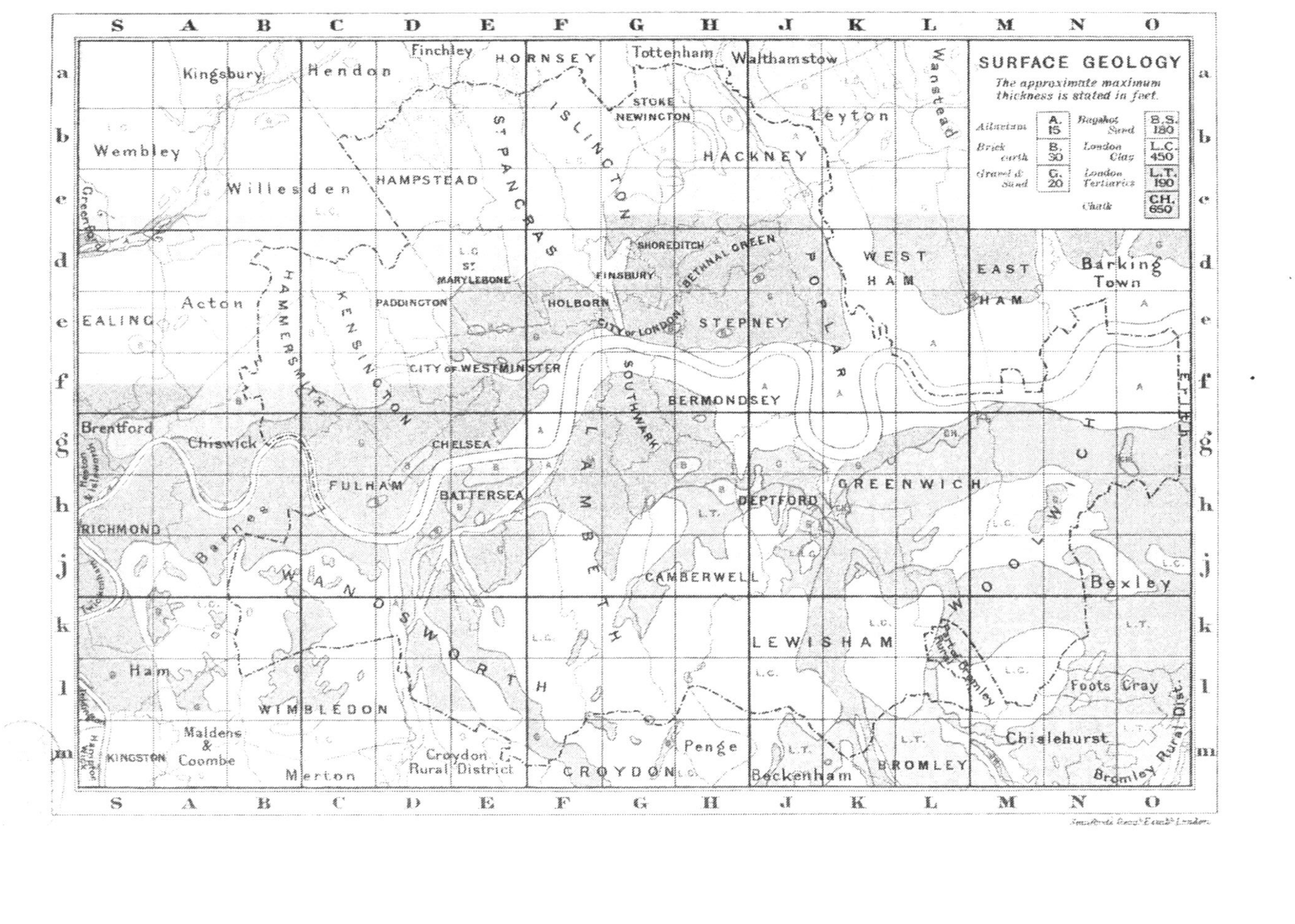
SURFACE GEOLOGY
The approximate maximum thickness is stated in feet.
Alluvium A. 15
Brick earth B. 30
Gravel & Sand G. 20
Bagshot Sand B.S. 180
London Clay L.C. 450
London Tertiaries L.T. 190
Chalk CH. 650
Kingsbury
Hendon
Finchley
HORNSEY
Tottenham
Walthamstow
Wanstead
STOKE NEWINGTON
Leyton
Wembley
Willesden
HAMPSTEAD
ST PANCRAS
ISLINGTON
HACKNEY
Greenford
SHOREDITCH
BETHNAL GREEN
POPLAR
WEST HAM
EAST HAM
Barking Town
ST MARYLEBONE
FINSBURY
EALING
Acton
HAMMERSMITH
KENSINGTON
PADDINGTON
HOLBORN
CITY OF LONDON
STEPNEY
CITY OF WESTMINSTER
SOUTHWARK
BERMONDSEY
Brentford
Chiswick
CHELSEA
LAMBETH
FULHAM
BATTERSEA
DEPTFORD
GREENWICH
WOOLWICH
RICHMOND
Barnes
WANDSWORTH
CAMBERWELL
Bexley
LEWISHAM
Part of Bromley Rural
Ham
WIMBLEDON
Foots Cray
Maldens & Coombe
KINGSTON
Hampton Wick
Merton
Croydon Rural District
CROYDON
Penge
Beckenham
BROMLEY
Chislehurst
Bromley Rural Dist.

AN INDEX

TO THE

Streets, Squares, Roads, Terraces, Places, Lanes, Railway Stations, Hospitals, L.C.C. Fire Stations, the chief Public Buildings, &c., that fall within the limits of the map.

Containing upwards of 12,000 names.

STREET OR PLACE.	BOROUGH.	P.D.	MAP.
Abbeville Road	Hornsey	N.	Fa
Abbeville Road	Wandsworth	S.W.	Fj
Abbey Bldgs.	Westminster	S.W.	Ff
Abbey Creek	West Ham	E.	Kd
Abbey Gardens	Westminster	S.W.	Ff
Abbey Gardens	Marylebone	N.W.	Dd
Abbey Gardens	Wandsworth	S.W.	El
Abbey Grove	Woolwich		Og
Abbey Lane	Hampstead	N.W	Dc
Abbey Lane	West Ham	E.	Kd
Abbey Lodge	Chislehurst, U.D.		Mm
Abbey Mews	Marylebone	N.W.	El
Abbey Place	Lambeth	S.E.	Ff
Abbey Road	Merton U.D.	S.W.	Dm
Abbey Road (part)	Hampstead	N.W.	Dc
Abbey Road	Kensington	W.	Ce
Abbey Road (part)	Marylebone	N.W.	Dd
Abbey Road	West Ham	E.	Kd
Abbey Road	Willesden, U.D.	N.W.	Ad
Abbey Street	Bermondsey	S.E.	Hf
Abbey Street	Bethnal Green	E.	Hd
Abbey Street	West Ham	E.	Ld
Abbey Works	Barking, U.D.	E.	Nd
Abbey Arms	Woolwich		Og
Abbey Mills Pumping Station	West Ham	E.	Kd
Abbeyfield Road	Bermondsey	S.E.	Hg
Abbeyhill	Bexley U.D.		Ok
Abbeywood Road	Woolwich		Og
Abbeywood Station	Woolwich		Og
Abbot Street	Hackney	E.	Hc
Abbotsbury Road	Kensington	W.	Cf
Abbotsford Avenue	Tottenham, U.D.	N.	Ga
Abbotshall Road	Lewisham	S.E.	Kk
Abbotsleigh Road	Wandsworth	S.W.	El
Abbotstone Road	Wandsworth	S.W.	Ch
Abbotswell Street	Lewisham	S.E.	Jj
Abbott Road	Hampstead	N.W.	Dc
Abbott Road	Poplar	E.	He
Abbott's Park Road	Leyton, U.D.	E.	Ka
Abbott's Road	East Ham	E.	Md
Abbott's Wharf	Bermondsey	S.E.	Hf
Abchurch Lane	City of London	E.C.	He
Abchurch Yard	City of London	E.C.	Ge
Abdale Road	Hammersmith	W.	Ce
Abel's Bldgs.	Stepney	E.	He
Abercorn	Richmond		Sj
Abercorn Mews	Marylebone	N.W.	Dd
Abercorn Mews South	Marylebone	N.W.	Dd
Abercorn Place	Marylebone	N.W.	Dd
Abercrombie Place	Holborn	E.C.	Ge
Abercrombie Street	Battersea	S.W.	Eh
Aberdare Gardens	Hampstead	N.W.	Dc
Aberdeen Place	Marylebone	N.W.	Dd
Aberdeen Park	Islington	N.	Gb
Aberdeen Mews	Islington	N.	Gc
Aberdeen Road	Croydon, R.D.	S.W.	Dm
Aberdeen Road	Islington	N.	Gb
Aberdeen Road	Willesden, U.D.	N.W.	Bb
Aberdeen Terrace	Lewisham	S.E.	Kh
Aberdeen Steam Wharf	Stepney	E.	Hf
Aberdeen Steam Navigation Wharf	Poplar	E.	Jf
Aberdour Street	Bermondsey and Southwark	S.E.	Gf
Aberfeldy Avenue	Croydon		Fm
Aberfeldy Street	Poplar	E.	He
Abernethy Road	Lewisham	S.E.	Kj
Abernethy Terrace	Lewisham	S.E.	Kj
Abersham Road	Hackney	E.	Hc
Abery Street	Woolwich		Ng
Abingdon Road	Kensington	W.	Df
Abingdon Street	Bethnal Green	E.	Hd
Abingdon Street	Westminster	S.W.	Ff
Abingdon Villas	Kensington	W.	Df
Abinger Road	Chiswick, U.D.	W.	Bf
Abinger Road	Deptford	S.E.	Jg
Ablett Street	Camberwell	S.E.	Hg
Abney Park Cemetery	Stoke Newington	N.	Hb
Abney Park Terrace	Hackney	N.	Hb
Aboukir Street	Camberwell	S.E.	Hg
Aboyne Road	Wandsworth	S.W.	Dk
Absolom Place	Camberwell	S.E.	Gh
Abyssinia Road	Battersea	S.W.	Ej
Acacia Gardens	Marylebone	N.W.	Dd
Acacia Grove	Camberwell	S.E.	Gk
Acacia Grove	The Maldens & Coombe, U.D.		Am
Acacia Parade	Woolwich		Ng
Acacia Place	Marylebone	N.W.	Dd
Acacia Road	Acton, U.D.	W.	Ae
Acacia Road	Beckenham, U.D.		Jm
Acacia Road	Lewisham	S.E.	Jl
Acacia Road	Leyton, U.D.	E.	Lb
Acacia Road	Marylebone	N.W.	Dd
Acacia Road	Walthamstow, U.D.	E.	Ja
Acanthus Road	Battersea	S.W.	Eh
Acfold Road	Fulham	S.W.	Dh
Achilles Road	Hampstead	N.W.	Db
Achilles Street	Deptford	S.E.	Jh
Acklam Road	Kensington	W.	Ce
Ackland Road	Croydon	S.E.	Hm
Ackmar Road	Fulham	S.W.	Dh
Ackroyd Road	Lewisham	S.E.	Jk
Acland Road	Willesden, U.D.	N.W.	Cc
Acland Street	Stepney	E.	Je
Acol Road	Hampstead	N.W.	Dc
Acorn Bldgs.	Camberwell	S.E.	Hh
Acorn Place	Camberwell	S.E.	Hh
Acorn Pond	Bermondsey	S.E.	Jf
Acorn Street	Camberwell	S.E.	Gh
Acorn Street	City of London	E.C.	He
Acorn Street	Woolwich		Mg
Acorn Terrace	Greenwich	S.E.	Lg
Acorn Wharf	Stepney	E.	Hf
Acorn Yard	Bermondsey	S.E.	Jf
Acre Lane	Lambeth	S.W.	Fj
Acre Road	Kingston		Sm
Acre Square	Wandsworth	S.W.	Fj
Acre Street	Battersea	S.W.	Fh
Acris Street	Wandsworth	S.W.	Dj
Acton Cemetery	Acton, U.D.	W.	Ad
Acton Green	Acton, U.D. and Chiswick, U.D.	W.	Af
Acton Green Road (part)	Acton, U.D.	W.	Af
Acton Green Road (part)	Chiswick, U.D.	W.	Ag
Acton Lane	Acton, U.D. and Willesden, U.D.	W.	Ad
Acton Lane (part)	Acton, U.D.	W.	Af
Acton Lane (part)	Chiswick, U.D.	W.	Af
Acton Lane (part)	Willesden, U.D	N.W.	Ad
Acton Mews	Shoreditch	E.	Hc
Acton Park	Acton, U.D.	W.	Be
Acton Sewage Works	Acton, U.D.	W.	Af
Acton Station	**Acton, U.D.**	W.	Ae
Acton Street	Hackney and Shoreditch	E.	Hc
Acton Street	St. Pancras	W.C.	Fd
Acton Vale	Acton, U.D.	W.	Bf
Acton Isolation Hospital	Acton, U.D.	W.	Be
Acuba Street	Wandsworth	S.W.	Dk
Ada Place	Bethnal Green	E.	Hd
Ada Road	Camberwell	S.E.	Gh
Ada Road	Wembley, U.D.		Sb
Ada Street	Hackney	E.	Hd
Adair Road	Kensington	W.	Cd
Adam Street	Westminster	W.C.	Fe
Adam Street	Marylebone	W.	Ee
Adam Street	Southwark	S.E.	Gf
Adam and Eve Lane	Islington	N.	Gc
Adam and Eve Mews	Kensington	W.	Df
Adam's Mews	Chelsea	S.W.	Dg
Adam's Mews	Westminster	W.	Ee
Adam's Place	Islington	N.	Fc
Adam's Pond	Barnes, U.D.	S.W.	Aj
Adamson Road	Hampstead	N.W.	Dc
Adamson Road	West Ham	E.	Le
Adamsrill Road	Lewisham	S.E.	Jl
Addey Street	Deptford	S.E.	Jh
Addington Crescent	Lambeth	S.E.	Ff
Addington Grove	Lewisham	S.E.	Jl
Addington Road	Hornsey	N.	Ga
Addington Road	Poplar	E.	Jd
Addington Road	West Ham	E.	Ke
Addington Square	Camberwell	S.E.	Gg

STREET OR PLACE.	BOROUGH.	P.D.	MAP.
Addington Street	Lambeth	S.E.	Ff
Addison Court Gard'ns	Hammersmith	W.	Cf
Addison Crescent	Kensington	W.	Cf
Addison Gardens	Hammersmith and Kensington	W.	Cf
Addison Gardens Mans.	Kensington	W.	Cf
Addison Place	Lambeth	S.W.	Gh
Addison Road	Chiswick, U.D.	W.	Bf
Addison Road	Kensington	W.	Cf
Addison Road	Walthamstow, U.D.	E.	Ka
Addison Road	Wanstead, U.D.	E.	La
Addison Road	West Ham	E.	Le
Addison Road North	Kensington	W.	Cf
Addison Road Station	**Kensington**	W.	Cf
Addle Hill	City of London	E.C.	Ge
Addle Street	City of London	E.C.	Ge
Adela Street	Kensington	W.	Cd
Adelaide Place	Camberwell	S.E.	Hh
Adelaide Place	City of London	E.C.	Gf
Adelaide Place	Westminster	W.C.	Fe
Adelaide Road	Hammersmith	W.	Bf
Adelaide Road	Hampstead	N.W.	Ec
Adelaide Road	Lewisham	S.E.	Jj
Adelaide Road	Leyton, U.D.	E.	Kb
Adelaide Road	Richmond		Sj
Adelaide Square	Islington	N.	Gc
Adelaide Street	Westminster	W.C.	Fe
Adelaide Tavern	Hampstead	N.W.	Ec
Adelaide Terrace	Islington	N.	Gd
Adelaide Villas	Richmond		Sj
Adelina Grove	Stepney	E.	He
Adelphi Theatre	**Westminster**	W.C.	Fe
Adelphi Terrace	Westminster	W.C.	Fe
Aden Grove	Stoke Newington	N.	Gb
Aden Terrace	Stoke Newington	N.	Gb
Adeney Road	Fulham	W.	Cg
Adie Road	Hammersmith	W.	Cf
Adine Road	West Ham	E.	Le
Adlard's Wharf	Bermondsey	S.E.	Hf
Adley Street	Hackney	E.	Jc
Admaston Road	Woolwich		Nh
Admiral Mews	Kensington	W.	Cd
Admiral Place	Kensington	W.	Cd
Admiral Terrace	Kensington	W.	Cd
Admiralty	**Westminster**	S.W.	Ff
Ado Place	Lambeth	S.E.	Fg
Adolphus Road	Stoke Newington	N.	Ga
Adolphus Street	Deptford	S.E.	Jh
Adon Mount	Camberwell	S.E.	Hk
Adpar Mews	Paddington	W.	Dd
Adpar Street	Paddington	W.	Dd
Adrian Mews	Kensington	S.W.	Dg
Adrian Terrace	Kensington	S.W.	Dg
Adys Road	Camberwell	S.E.	Hj
Ægis Grove	Battersea	S.W.	Fh
Affleck Street	Finsbury	N.	Fd
Afghan Road	Battersea	S.W.	Eh
Agamemnon Road	Hampstead	N.W.	Db
Agar Street	Westminster	W.C.	Fe
Agate Road	Hammersmith	W.	Cf
Agate Street	West Ham	E.	Le
Agave Road	Willesden, U.D.	N.W.	Cb
Agincourt Road	Hampstead	N.W.	Eb
Agnes Street	Lambeth	S.E.	Ff
Agnes Street	Stepney	E.	Je
Agnes Street	West Ham	E.	Mf
Agnew Road	Lewisham	S.E.	Jk
Agra Place	Stepney	E.	He
Agricultural Hall	**Islington**	N.	Gd
Ailsa Street	Poplar	E.	Ke
Ainger Road	Hampstead and St. Pancras	N.W.	Ec
Ainsley Street	Bethnal Green	E.	Hd
Ainsty Street	Bermondsey	S.E.	Jf
Ainsworth Street	Bethnal Green	E.	Hd
Aintree Avenue	East Ham	E.	Md
Aintree Street	Fulham	S.W.	Cg
Air Street	Westminster	W.	Fe
Airdale Avenue	Chiswick, U.D.	W.	Bg
Airedale Road	Wandsworth	S.W.	Ek
Airlie Gardens	Kensington	W.	Df
Aislibie Road	Lewisham	S.E.	Kj
Aitken Road	Lewisham	S.E.	Kk
Ajax Road	Hampstead	N.W.	Db
Akehurst Street	Wandsworth	S.W.	Bj
Akenside Road	Hampstead	N.W.	Dc
Akerman Road	Lambeth	S.W.	Gh
Alabama Street	Woolwich		Nh
Alacross Road	Ealing	W.	Sf
Alan Road	Wimbledon	S.W.	Cl
Alaska Street	Lambeth	S.E.	Ff
Albacore Crescent	Lewisham	S.E.	Kj
Albany, The	Camberwell	S.E.	Gg
Albany, The	Westminster	W.	Fe
Albany Club	**Kingston**		Sl
Albany Mews	Camberwell	S.E.	Gg
Albany Mews	St. Pancras	N.W.	Ee
Albany Park Road	Ham, U.D., and Kingston		Sl
Albany Place	Islington	N.	Gb
Albany Road	Brentford, U.D.		Sg
Albany Road	Camberwell and Southwark	S.E.	Gg
Albany Road	Chislehurst, U.D.		Mm
Albany Road	Ealing	W.	Se
Albany Road	Hornsey	N.	Ga
Albany Road	Leyton, U.D.	E.	Ka
Albany Road	Walthamstow, U.D.	E.	Ja
Albany Road	West Ham	E.	Le
Albany Road	Wimbledon	S.W.	Dl
Albany Row	Camberwell	S.E.	Gg
Albany Street	Camberwell	S.E.	Gg
Albany Street	St. Pancras	N.W.	Ed
Albany Street	Stepney	E.	Je
Albany Terrace	St. Pancras	N.W.	Ee
Albatross Street	Woolwich		Nh
Albemarle	Wandsworth	S.W.	Ck
Albemarle College	Lewisham	S.E.	Jl
Albemarle Road	Beckenham, U.D.		Km
Albemarle Street	Westminster	W.	Ee
Albemarle Street	Finsbury	E.C.	Gd
Albert Bridge	Battersea and Chelsea	S.W.	Eg
Albert Bridge Road	Battersea	S.W.	Eh
Albert Buildings	Bethnal Green	E.	Hd
Albert Buildings	City of London	E.C.	Ge
Albert Buildings	Shoreditch	E.C.	Gd
Albert Embankment	Lambeth	S.E.	Fg
Albert Gardens	Willesden, U.D.	N.W.	Cd
Albert Gate	Westminster	S.W.	Ef
Albert Grove	Acton, U.D.	W.	Af
Albert Grove	Hackney	E.	Jd
Albert Hall	**Westminster**	S.W.	Df
Albert Memorial	Westminster	S.W.	Df
Albert Place	Islington	N.	Gb
Albert Place	Kensington	W.	Df
Albert Road	Beckenham, U.D.	S.E.	Jm
Albert Road	Bromley, R.D.		Ml
Albert Road	Camberwell	S.E.	Hh
Albert Road	Croydon, R.D.		Em
Albert Road	Deptford	S.E.	Jh
Albert Road	East Ham, West Ham, and Woolwich	E.	Mf
Albert Road	Hackney	E.	Hc
Albert Road	Hornsey	N.	Fa
Albert Road	Kingston		Sm
Albert Road	Leyton, U.D.	E.	Kb
Albert Road	Marylebone and St. Pancras	N.W.	Ed
Albert Road	Richmond		Sj
Albert Road	Tottenham, U.D.	N.	Ga
Albert Road	Walthamstow, U.D.	E.	Ja
Albert Road	Wandsworth	S.W	Ck
Albert Road	West Ham	E.	Lc
Albert Road	West Ham	E.	Ld
Albert Road	Willesdon U.D.	N.W.	Cd
Albert Row	Hendon, U.D.	N.W.	Ca
Albert Square	Lambeth	S.W.	Fh
Albert Square	Stepney	E.	Je
Albert Square	West Ham	E.	Lc
Albert Square Mews	Lambeth	S.W.	Fh
Albert Street	Westminster	S.W	Ef

STREET OR PLACE.	BOROUGH.	P.D.	MAP.
Allenby Road	Lewisham	S.E.	Jl
Allendale Road	Camberwell	S.E.	Gh
Allen's School	Camberwell	S.E.	Gj
Allenswood	Wandsworth	S.W.	Ck
Allerton Road	Stoke Newington	N.	Gb
Allerton Street	Shoreditch	N.	Gd
Allestree Road	Fulham	S.W.	Cg
Alleyn Crescent	Camberwell	S.E.	Gl
Alleyn Park	Camberwell and Lambeth	S.E.	Gk
Alleyn Road	Camberwell	S.E.	Gl
Alleyn School	Camberwell	S.E.	Gj
Allfarthing Lane	Wandsworth	S.W.	Dj
Allhallows Pier	City of London	E.C.	Ge
Allington	Willesden, U.D.	W.	Cd
Allington Place	Westminster	S.W.	Ef
Allington Street	Westminster	S.W.	Ef
Allington Street	Lambeth	S.W.	Fj
Allison Gardens	Camberwell	S.E.	Gk
Allison Grove	Camberwell	S.E.	Gk
Allison Road	Acton, U.D.	W.	Ae
Allison Road	Hornsey and Tottenham, U.D.	N.	Ga
Alloa Road	Deptford	S.E.	Jg
All Saints Orphanage	Lewisham	S.E.	Kh
All Saints Road	Acton U.D.	W.	Af
All Saints Road	Kensington	W.	Ce
All Saints Road	Lewisham	S.E.	Kh
All Saints Street	Islington	N.	Fd
Allsop Place	Marylebone	N.W.	Ed
Allsop Street	Marylebone	N.W.	Ed
All Soul's Avenue	Willesden U.D.	N.W.	Bd
Alma Grove	Bromley		Lm
Alma Grove	Islington	N.	Fc
Alma Place	Hammersmith	W.	Cd
Alma Road	Bermondsey	S.E.	Hg
Alma Road	Bethnal Green	E.	Jd
Alma Road	Bexley U.D. and Foot Cray U.D.		Ol
Alma Road	Islington	N.	Gc
Alma Road	Stepney	E.	Je
Alma Road	Wandsworth	S.W.	Dj
Alma Square	Marylebone	N.W.	Dd
Alma Street	St. Pancras	N.W.	Ec
Alma Street	Shoreditch	N.	Gd
Alma Street	West Ham	E.	Kc
Alma Terrace	Wandsworth	S.W.	Dk
Alma Tavern	Woolwich		Oh
Almack Road	Hackney	E.	Jb
Almeida Street	Islington	N.	Gc
Almeric Road	Battersea	S.W.	Ej
Almington Street	Islington	N.	Fa
Almorah Road	Islington	N.	Gc
Alnwick Road	West Ham	E.	Le
Alperton	Wembley U.D.		Sc
Alperton Hall	Wembley U.D.		Sc
Alperton Lodge	Wembley U.D.		Sc
Alperton Park Road	Wembley U.D.		Sc
Alperton Road	Wembley U.D.		Sc
Alperton Street	Paddington	W.	Cd
Alpha Cottages	Wandsworth	S.W.	Dj
Alpha Place	Chelsea	S.W.	Eg
Alpha Place	St. Pancras	N.W.	Ec
Alpha Place	Willesden U.D.	N.W.	Dd
Alpha Place North	Willesden U.D.	N.W.	Dd
Alpha Road	Deptford	S.E.	Jh
Alpha Road	Poplar	E.	Jf
Alpha Square	Southwark	S.E.	Gg
Alpha Street	Camberwell	S.E.	Hh
Alpha Street	Southwark	S.E.	Gg
Alpha Villas	Islington	N.	Fa
Alpine Road	Deptford	N.W.	Jg
Alric Avenue	Willesden U.D.	S.E.	Bc
Alroy Road	Hornsey	N.	Ga
Alsace Mews	Southwark	S.E.	Gg
Alsace Street	Southwark	S.E.	Gg
Alscot Road	Bermondsey	S.E.	Hf
Alsen Road	Islington	N.	Fb
Alston Road	Wandsworth	S.W.	Dl
Alt Grove	Wimbledon	S.W.	Cm
Altenberg Avenue	Ealing	W.	Sf
Altenburg Gardens	Battersea	S.W.	Ej
Althea Street	Fulham	S.W.	Dh
Althorp Road	Battersea	S.W.	Ek
Althorpe Grove	Battersea	S.W.	Dh
Altmore Avenue	East Ham	E.	Md
Alton House	Wandsworth	S.W.	Bk
Alton Lodge	Wandsworth	S.W.	Bk
Alton Road	Richmond		Sj
Alton Road	Wandsworth	S.W.	Bk
Alton Street	Poplar	E.	Ke
Alva Road	Wandsworth	S.W.	Cj
Alvar Street	Deptford	S.E.	Jh
Alverstone Road	Willesden U.D.	N.W.	Cc
Alverton Street	Deptford	S.E.	Jg
Alveston Mews	Kensington	S.W.	Dg
Alvey Street	Southwark	S.E.	Gg
Alvington Crescent	Hackney	E.	Hc
Alwyne Lane	Islington	N.	Gc
Alwyne Place	Islington	N.	Gc
Alwyne Road	Islington	N.	Gc
Alwyne Road	Wimbledon	S.W.	Cl
Alwyne Square	Islington	N.	Gc
Alwyne Villas	Islington	N.	Gc
Amber Place	Stepney	E.	He
Amberley Mews	Paddington	W.	Dd
Amberley Road	Lewisham	S.E.	Hl
Amberley Road	Leyton U.D.	E.	Ka
Amberley Road	Paddington	W.	Dd
Amberley Wharves	Paddington	W.	Dd
Amblecote Road	Lewisham	S.E.	Ll
Amblers Road	Islington	N.	Gb
Ambleside Avenue	Wandsworth	S.W.	Fl
Ambleside Road	Willesden U.D.	N.W.	Bc
Ambrookhill Wood	Camberwell	S.E.	Hk
Ambrosden Avenue	Westminster	S.W.	Ff
Ambrose Street	Bermondsey	S.E.	Hg
Amelia Street	Southwark	S.E.	Gg
Amelia Street	West Ham	E.	Lf
Amen Corner	City of London	E.C.	Ge
Amen Corner	Wandsworth	S.W.	El
America Square	City of London	E.C.	He
America Street	Southwark	S.E.	Gf
Amerland Road	Wandsworth	S.W.	Cj
Amersham Grove	Deptford	S.E.	Jg
Amersham Road	Deptford	S.E.	Jh
Amersham Road	Wandsworth	S.W.	Cj
Amersham Vale	Deptford	S.E.	Jh
Amery's Place	Southwark	S.E.	Hg
Ames Street	Bethnal Green	E.	Jd
Amesbury Avenue	Wandsworth	S.W.	Fk
Amesbury Road	Bromley		Lm
Amhurst Avenue	Ealing	W.	Se
Amhurst Park	Hackney & Stoke Newington	N.	Ga
Amhurst Passage	Hackney	N.	Hb
Amhurst Road	Ealing	W.	Se
Amhurst Road	Hackney	E.&N.	Hb
Amhurst Street	Wandsworth	S.W.	Bj
Amhurst Terrace	Hackney	E.	Hb
Amies Street	Battersea	S.W.	Eh
Amity Grove	Wimbledon	S.W.	Bm
Amity Road	West Ham	E.	Lc
Amner Road	Battersea	S.W.	Ej
Amor Road	Hammersmith	W.	Cf
Amott Road	Camberwell	S.E.	Hj
Amoy Place	Poplar & Stepney	E.	Je
Amphitheatre Row	Lambeth	S.E.	Ff
Ampney Road	Wandsworth	S.W.	Fl
Ampthill Square	St. Pancras	N.W.	Fd
Ampton Street	St. Pancras	N.W.	Fd
Amwell Street	Finsbury	E.C.	Fd
Amyruth Street	Lewisham	S.E.	Jj
Anatola Road	Islington	N.	Fb
Ancaster Street	Woolwich		Nh
Anchor Place	Shoreditch	E.	Hd
Anchor Street	Bermondsey	S.E.	Hg
Anchor Street	Stepney	E.	Je
Anchor Wharf	City of London	E.C.	Ge
Anchor and Hope Lane	Greenwich		Lg
Anchor in Hope, The	Bexley, U.D.		Nh
Ancill Street	Fulham	W.	Cg
Ancona Road	Islington	N.	Gb
Ancona Road	Woolwich		Ng
Andalus Road	Lambeth	S.W.	Fh
Anderson Street	Chelsea	S.W.	Eg

STREET OR PLACE.	BOROUGH.	P.D.	MAP.
Anderson's Road	Hackney	E.	Jc
Anderson's Walk	Lambeth	S.E.	Fg
Andoe Road	Battersea	S.W.	Dh
Andover Gardens	Islington	N.	Fb
Andover Grove	Islington	N.	Fb
Andover Place	Paddington	N.W.	Dd
Andover Road	Islington	N.	Fb
Andover Row	Islington	N.	Fb
Andover Street	Islington	N.	Fb
Andover Yard	Islington	N.	Fb
André Street	Hackney	E.	Hb
Andrew Street	Poplar	E.	Hc
Andrew Street	West Ham	E.	Mf
Andrews Road	Hackney	E.	Hd
Anerley	Croydon and Penge, U.D.		Hm
Anerley Grove	Penge, U.D.	S.E.	Hm
Anerley Park	Penge, U.D.	S.E.	Hm
Anerley Park Road	Penge, U.D.	S.E.	Hm
Anerley Road	Penge, U.D.	S.E.	Hm
Anerley Station	**Penge, U.D.**	**S.E.**	**Hm**
Anerley Street	Battersea	S.W.	Eh
Anerley Vale	Penge, U.D.	S.E.	Hm
Angel, The	Islington	N.	Gd
Angel Lane	Stepney	E.	He
Angel Lane	West Ham	E.	Kc
Angel Place	West Ham	E.	Kc
Angel Road	Hammersmith	W.	Cg
Angel Station	**Islington**	**E.C.**	**Gd**
Angel Street	City of London	E.C.	Ge
Angell Park Gardens	Lambeth	S.W.	Gh
Angell Road	Lambeth	S.W.	Gh
Angerstein's Wharf	Greenwich		Lf
Angle Street	Leyton, U.D.	E.	Kb
Angle Street	Southwark	S.E.	Gg
Anglers Lane	St. Pancras	N.W.	Ec
Angles, The	Barnes, U.D.	S.W.	Aj
Angles Road	Wandsworth	S.W.	Fl
Anglesea Avenue	Woolwich		Mg
Anglesea House	Woolwich		Mh
Anglesea Road	Woolwich		Mg
Anglesea Street	Bethnal Green	E.	Hd
Angola Mews	Kensington	W.	Ce
Angrave Street	Shoreditch	E.	Hd
Angus Street	Deptford	S.E.	Jh
Anhalt Road	Battersea	S.W.	Eg
Anley Road	Hammersmith	W.	Cf
Ann Street	Finsbury	W.C.	Fd
Ann Street	Islington	N.	Gd
Ann Street	Lambeth	S.E.	Ff
Ann Street	Poplar	E.	Je
Ann Street	Stepney	E.	Hd
Ann Street	Stepney	E.	Jc
Ann Street	Woolwich		Mg
Ann Street	Woolwich		Ng
Annable Street	Poplar	E.	Ke
Annendale	Hendon, U.D.	N.W.	Da
Annandale Road	Chiswick U.D.	W.	Bg
Annandale Road	Greenwich	S.E.	Lg
Anne Street	Stepney	E.	Je
Anne Street	West Ham	E.	Le
Anne Boleyn's Tower	East Ham	E.	Ld
Annesley Road	Islington	N.	Fb
Annette Road	Islington	N.	Fb
Annett's Crescent	Islington	N.	Gc
Anning Street	Shoreditch	N.	Hd
Annis Road	Hackney	E.	Jc
Ann's Avenue	Camberwell	S.E.	Hg
Ann's Buildings	Southwark	S.E.	Gg
Ann's Place	Bermondsey	S.E.	Hf
Ann's Place	Woolwich	E.	Mf
Ann's Row	Southwark	S.E.	Gg
Ansdell Road	Camberwell	S.E.	Hh
Anselm Road	Fulham	S.W.	Dg
Anson Road	Islington	N.	Fb
Anson Road	Willesden, U.D.	N.W.	Cb
Anstey Road	Camberwell	S.E.	Hh
Anstey Road	Lambeth	S.E.	Gj
Antcliffe Street	Stepney	E.	He
Anthony Street	Bermondsey	S.E.	Hf
Anthony Street	Stepney	E.	He
Antill Place	Stepney	E.	Je
Antill Road	Tottenham, U.D	N.	Ha
Antill Road	Poplar and Stepney	E.	Jd
Anton Street	Hackney	E.	Hc
Antrim Mews	Hampstead	N.W.	Ec
Antrim Street	Hampstead	N.W.	Ec
Antrobus Road	Acton, U.D.	W	Af
Antwerp Street	Hackney	E.	Hd
Apollo Theatre	**Westminster**	**W.**	**Fe**
Apothecaries Hall	City of London	E.C.	Ge
Appach Road	Lambeth	S.W.	Fj
Appian Road	Poplar	E.	Jd
Apple Market	Kingston		Sm
Appleby Road	Hackney	E.	Hc
Appleby Street	Shoreditch	E.	Hd
Appleford Road	Kensington	W.	Cd
Applegarth Road	Hammersmith	W.	Cf
Apple Tree Yard	Westminster	S.W.	Ff
Appold Street	Shoreditch	E.C.	Ge
Approach Road	Bethnal Green	E.	Hd
Approach Road	Lambeth	S.E.	Gk
Approach Road	Westminster	W.C.	Fe
April Street	Hackney	E.	Hb
Apsley Road	Walthamstow U.D.	E.	Ja
Apsley Road	Wandsworth	S.W.	Dj
Apsley Street	Stepney	E.	Je
Aquila Street	Marylebone	N.W.	Dd
Aquinas Street	Lambeth	S.E.	Gf
Arab Boy, The	Wandsworth	S.W.	Cj
Arabin Road	Lewisham	S.E.	Jj
Aragon Road	East Ham	E.	Md
Arbery Road	Bethnal Green and Stepney	E.	Jd
Arbour Square	Stepney	E.	Je
Arbuthnot Road	Deptford	S.E.	Jh
Arcade, The	Westminster	W.	Ee
Arcadia Street	Poplar	F.	Je
Arch Road	Richmond		Ah
Arch Street	Southwark	S.E.	Gf
Arch Street Buildings	Southwark	S.E.	Gf
Archbishop's Place	Lambeth	S.W.	Fk
Archbishop's Pk., The	Lambeth	S.E.	Ff
Archdale Road	Camberwell	S.E.	Hj
Archel Road	Fulham	W.	Cg
Archer Mews	Kensington	W.	Ce
Archer Street	Westminster	W.	Fe
Archer Street	Kensington	W.	Ce
Archer Street	Lambeth	S.W.	Fg
Archer Street	St. Pancras	N.W.	Fc
Archer Villas	Kensington	W.	Ce
Archibald Road	Islington	N.	Fb
Archibald Street	Poplar	E.	Jd
Archway Road	Hornsey and Islington	N.	Ea Fa
Archway Street	Barnes, U.D.	S.W.	Bh
Archway Tavern	Islington	N.	Fb
Archway Villas	Islington	N.	Fa
Arcola Street	Hackney	E.	Hb
Ardbeg Road	Camberwell	S.E.	Gj
Arden Street	Battersea	S.W.	Fg
Ardfern Avenue	Croydon	S.W.	Fm
Ardfillan Road	Lewisham	S.E.	Kk
Ardgowan Road	Lewisham	S.E.	Kk
Ardilawn Road	Islington	N.	Gb
Ardleigh Road	Hackney	N.	Gc
Ardlethen	Hendon U.D.	N.W.	Da
Ardlui Road	Lambeth	S.E.	Gk
Ardmere Road	Lewisham	S.E.	Kj
Ardoch Road	Lewisham	S.E.	Kk
Argent Street	Southwark	S.E.	Gf
Argyle Cottages	Hammersmith	W.	Bg
Argyle Lodge	Wandsworth	S.W.	Ck
Argyle Place	Hammersmith	W.	Bg
Argyle Road	Ealing	W.	Se
Argyle Road	Hammersmith	W.	Bg
Argyle Road	Leyton U.D. and West Ham	E.	Kb
Argyle Road	West Ham	E.	Le
Argyle Square	St. Pancras	W.C.	Fd
Argyle Street	St. Pancras	W.C.	Fd
Argyll Place	Westminster	W.	Fe
Argyll Road	Hendon U.D.	N.W.	Ba
Argyll Road	Kensington	W.	Df

STREET OR PLACE.	BOROUGH.	P.D.	MAP.
Argyll Road	Stepney	E.	Jd
Argyll Street	Westminster	W.	Ee
Argylle Street	Leyton U.D.	E.	Kb
Arica Place	Bermondsey	S.E.	Hf
Arica Road	Deptford	S.E	Jj
Ariel Road	Hampstead	N.W.	Dc
Aristotle Road	Wandsworth	S.W.	Fj
Arkley Crescent	Walthamstow U.D.	E.	Ja
Arkley Road	Walthamstow U.D.	E.	Ja
Arklow Road	Deptford	S.E.	Jg
Arkwright Road	Hampstead	N.W.	Db
Arkwright Street	West Ham	E.	Le
Arlborough Road	Wandsworth	S.W.	Ej
Arlesford Road	Lambeth	S.W.	Fg
Arloe Street	Bethnal Green	E.	Hd
Arlingford Road	Lambeth	S.W.	Fj
Arlington Park Gdns.	Chiswick U.D.	W.	Ag
Arlington Park Gdns. North	Chiswick U.D.	W.	Ag
Arlington Road	Ealing	W.	Se
Arlington Road	St. Pancras	N.W.	Ec
Arlington Road	Twickenham U.D.		Sj
Arlington Square	Islington	N.	Gd
Arlington Street	Westminster	S.W.	Ef
Arlington Street	Finsbury	E.C.	Gd
Arlington Street	Islington and Shoreditch	N.	Gd
Arlington Terrace	St. Pancras	N.W.	Ec
Arliss Road	Battersea	S.W.	Eh
Armada Street	Greenwich	S.E.	Jg
Armadale Road	Fulham	S.W.	Dg
Armagh Road	Poplar	E.	Jd
Arminger Road	Hammersmith	W.	Bf
Armitage Cottages	Greenwich	S.E.	Lg
Armitage Road	Greenwich	S.E.	Lg
Armstrong Place	Woolwich		Ng
Armstrong Street	Woolwich		Ng
Army Street	Wandsworth	S.W.	Fh
Army Clothing Depôt	Westminster	S.W.	Fg
Army and Navy Stores	Westminster	S.W.	Ff
Army Service Barracks	Woolwich		Mg
Arneway Street	Westminster	S.W.	Ff
Arngask Road	Lewisham	S.E.	Kk
Arnold Circus	Bethnal Green	E.	Hd
Arnold Road	Camberwell	S.E.	Hj
Arnold Road	Croydon R.D.	S.W.	Em
Arnold Road	Poplar	E.	Jd
Arnott Street	Southwark	S.E.	Gf
Arnside Street	Southwark	S.E.	Gg
Arodene Road	Lambeth	S.W.	Fj
Arpley Road	Penge U.D.	S.E.	Hm
Arragon Gardens	Wandsworth	S.W.	Fm
Arrol Road	Beckenham U.D.		Hm
Arrow Street	West Ham	E.	Kc
Arsenal Football Ground	Woolwich		Ng
Arsenal Station	**Woolwich**		Ng
Arterberry Road	Wimbledon	S.W.	Cm
Artesian Road	Paddington	W.	De
Arthingworth Street	West Ham	E.	Kd
Arthur Mews	Islington	N.	Fc
Arthur Road	Beckenham U.D.		Jm
Arthur Road	East Ham	E.	Md
Arthur Road	Islington	N.	Fb
Arthur Road	Kingston		Am
Arthur Road	Lambeth	S.W.	Gh
Arthur Road	Stoke Newington	N.	Hb
Arthur Road	Wimbledon	S.W.	Cl
Arthur Street	Battersea	S.W.	Eh
Arthur Street	Bethnal Green	E.	Hd
Arthur Street	Camberwell	S.E.	Hg
Arthur Street	Chelsea	S.W.	Dg
Arthur Street	Westminster	S.W.	Ef
Arthur Street	Hackney	E.	Jc
Arthur Street	Holborn	W.C.	Fe
Arthur Street	Poplar	E.	Ke
Arthur Street	Southwark	S.E.	Gg
Arthur Street	West Ham	E.	Kd
Arthur Street	West Ham	E.	Le
Arthur Street	West Ham	E.	Lf
Arthur Street	Woolwich		Ng
Arthur Street East	City of London	E.C.	Ge
Arthur Street West	City of London	E.C.	Ge
Arthur Street Yard	Holborn	W.C.	Fe
Arthur Villas	Southwark	S.E.	Gg
Arthurdon Street	Lewisham	S.E.	Jj
Artichoke Mews	Camberwell	S.E.	Gh
Artichoke Place	Camberwell	S.E.	Gh
Artillery Court	Bermondsey	S.E.	Hf
Artillery Lane	Bermondsey	S.E.	Hf
Artillery Lane	City of London and Stepney	E.	He
Artillery Passage	Stepney	E.	He
Artillery Place	Bermondsey	S.E.	Hf
Artillery Place	Woolwich		Mg
Artillery Practice Gr'nd	Woolwich		Of
Artillery Row	Westminster	S.W.	Ff
Artillery Street	Bermondsey	S.E.	Hf
Artillery Street	Bethnal Green	E.	Hd
Artillery Yard	Shoreditch	E.C.	Ge
Arundel Gardens	Kensington	W.	Ce
Arundel Grove	Islington	N.	Hc
Arundel Place	Westminster	S.W.	Fe
Arundel Place	Islington	N.	Gc
Arundel Road	Islington	N.	Hc
Arundel Road	Leyton, U.D. and West Ham	E.	Kb
Arundel Square	Islington	N.	Fc
Arundel Street	Westminster	W.C.	Fe
Arundel Street	Islington	N.	Hc
Arundel Terrace	Barnes, U.D.	S.W.	Bg
Arundell Street	Westminster	W.	Fe
Arvon Road	Islington	N.	Gb
Ascalon Street	Battersea	S.W.	Eh
Ascham Street	St. Pancras	N.W.	Fb
Ascot Avenue	Ealing	W.	Sf
Ascot Road	Tottenham, U.D.	N.	Ga
Ascot Street	West Ham	E.	Le
Asgill House	Richmond		Sj
Asgill Road	Richmond		Sj
Ash Grove	Hackney	E.	Hd
Ash Grove	Hendon, U.D.	N.W.	Cb
Ash Street	Southwark	S.E.	Gg
Ashborne Road	Croydon, R.D.		Em
Ashbourne Grove	Camberwell	S.E.	Hj
Ashbourne Grove	Chiswick, U.D.	W.	Bg
Ashbourne Terrace	Wimbledon	S.W.	Cm
Ashbridge Road	Leyton, U.D.	E.	La
Ashbrook Road	Islington	N.	Fa
Ashburn Mews	Kensington	S.W.	Df
Ashburn Place	Kensington	S.W.	Df
Ashburnham Grove	Greenwich	S.E.	Kh
Ashburnham Road	Chelsea	S.W.	Dg
Ashburnham Road	Greenwich	S.E.	Kh
Ashburnham Road	Willesden, U.D.	N.W.	Cd
Ashburton Grove	Islington	N.	Gb
Ashburton House	Fulham	S.W.	Cj
Ashburton Road	West Ham	E.	Le
Ashburton Terrace	West Ham	E.	Ld
Ashbury Road	Battersea	S.W.	Eh
Ashby Road	Deptford	S.E.	Jh
Ashby Road	Islington	N.	Gc
Ashchurch Grove	Hammersmith	W.	Bf
Ashchurch Pk. Avenue	Hammersmith	W.	Bf
Ashchurch Terrace	Hammersmith	W.	Bf
Ashcombe Road	Wimbledon	S.W.	Dl
Ashcombe Street	Fulham	S.W.	Dh
Ashcroft Road	Stepney	E.	Jd
Ashdon Road	Willesden, U.D.	N.W.	Bc
Ashdown Road	Kingston		Sm
Ashdown Street	St. Pancras	N.W.	Ec
Ashen Grove	Chislehurst, U.D.		Nl
Ashen Grove	Wandsworth	S.W.	Dj
Ashenden Road	Hackney	E.	Jb
Ashen Tree Court	City of London	E.C.	Ge
Ashfield Lane	Chislehurst, U.D.		Nm
Ashfield Place	Stepney	E.	Je
Ashfield Road	Tottenham, U.D.	N.	Ga
Ashford Road	Walthamstow, U.D.	E.	Ja
Ashford Road	Willesden, U.D.	N.W.	Cb
Ashford Street	Shoreditch	N.	Gd
Ashford Terrace	Camberwell	S.E.	Hj
Ash Grove Road	Lewisham	S.E.	Hk

STREET OR PLACE.	BOROUGH.	P.D.	MAP.
Ashington Road	Fulham	S.W.	Ch
Ashlake Road	Wandsworth	S.W.	Fl
Ashland Place	Marylebone	W.	Ee
Ashleigh	Wimbledon	S.W.	Cl
Ashleigh Road	Barnes, U.D.	S.W.	Bh
Ashley Cottages	Kensington	W.	Cg
Ashley Gardens	Westminster	S.W.	Ff
Ashley Place	Westminster	S.W.	Ef
Ashley Road	Deptford	S.E.	Jh
Ashley Road	Islington	N.	Fa
Ashley Road	Richmond		Sh
Ashley Road	Wimbledon	S.W.	Dl
Ashlin Road	Leyton, U.D. and West Ham	E.	Kb
Ashlone Road	Wandsworth	S.W.	Ch
Ashmead Road	Deptford	S.E.	Jh
Ashmere Grove	Lambeth	S.W.	Fj
Ashmole Place	Lambeth	S.W.	Fg
Ashmore Road	Paddington	W.	Cd
Ashmount Road	Islington	N.	Fa
Ashmount Road	Tottenham, U.D.	N.	Ha
Ashness Road	Battersea	S.W.	Ej
Ashton Place	Hackney	N.	Ha
Ashton Road	Ealing	W.	Se
Ashton Street	Poplar	E.	Ke
Ashton Street	West Ham	E.	Kc
Ashton Terrace	Fulham	S.W.	Cg
Ashurst Street	Battersea	S.W.	Eh
Ashvillle Road	Leyton, U.D.	E.	Kb
Ashwell Road	Bethnal Green	E.	Jd
Ashwin Street	Hackney	E.	Hc
Ashworth Road	Paddington	W.	Dd
Aske Street	Shoreditch	N.	Gd
Aske's School	Deptford	S.E.	Jh
Askew Crescent	Hammersmith	W.	Bf
Askew Road	Hammersmith	W.	Bf
Askew Street	Hackney	E.	Hc
Aslett Street	Wandsworth	S.W.	Dk
Asmara Street	Hampstead	N.W.	Cb
Aspath Road	Wembley, U.D.		Sb
Aspen Place	Hammersmith	W.	Bg
Aspenlea Road	Fulham	W.	Cg
Aspinall Road	Deptford	S.E.	Jj
Aspinden Road	Bermondsey	S.E.	Hg
Aspland Grove	Hackney	E.	Hc
Aspley Road	Wandsworth	S.W.	Dj
Assam Street	Stepney	E.	He
Assaye Street	Lambeth	S.W.	Fg
Assembly Place	Bermondsey	S.E.	Hf
Assembly Place	Stepney	E.	He
Astbury Road	Camberwell	S.E.	Hh
Aste Street	Poplar	E.	Kg
Aster Place	Southwark	S.E.	Gf
Astey's Row	Islington	N.	Gc
Astlett Street	Wandsworth	S.W.	Dk
Astley Avenue	Willesden, U.D	N.W.	Cb
Astley Street	Camberwell	S.E.	Hg
Aston Road	Merton U.D.	S.W.	Cm
Aston Road	Kensington	W.	Ce
Aston Street	Stepney	E.	Je
Astonville Street	Wandsworth	S.W.	Dk
Astoria Place	Lewisham	S.E.	Kh
Astrop Mews	Hammersmith	W.	Cf
Astwood Road	Kensington	S.W.	Df
Asylum Road	Camberwell	S.E.	Hh
Asylum Road	Kingston		Am
Asylum Road	Wandsworth	S.W.	Dk
Atalanta Road	Fulham	S.W.	Ch
Atbara Road	Teddington, U.D.		Sl
Atheldene Road	Wandsworth	S.W.	Dk
Athelstane Mews	Islington	N.	Fa
Athelstane Road	Islington	N.	Gb
Asbelstane Road	Poplar	E.	Jd
Athenæum (Camden)	Islington	N.	Fb
Athenlay Road	Camberwell	S.E.	Jj
Atherden Road	Hackney	E.	Hb
Atherfold Road	Lambeth	S.W.	Fh
Atherstone Mews	Kensington	S.W.	Df
Atherton Grange	Wimbledon	S.W.	Cl
Atherton Road	West Ham	E.	Lc
Atherton Street	Battersea	S.W.	Eh
Athol Place	Islington	N.	Fb
Athol Street	Poplar	E.	He

STREET OR PLACE.	BOROUGH.	P.D.	MAP.
Athol Terrace	Wandsworth	S.W.	Fl
Atkins Road	Leyton, U.D.	E.	Ka
Atkins Road	Wandsworth	S.W.	Fk
Atkinson Street	Deptford	S.E.	Jg
Atkinson Morley's Convalescent Hospital	Wimbledon	S.W.	Bm
Atlantic Road	Lambeth	S.W.	Fj
Atlas Street	Greenwich	S.E.	Kh
Atlas Brick & Tile Works	Acton, U.D.	N.W.	Bd
Atley Street	Poplar	E.	Jd
Atney Road	Wandsworth	S.W.	Cj
Atterbury Court	City of London	E.C.	Ge
Atterbury Road	Hornsey	N.	Ga
Atterbury Street	Westminster	S.W.	Fg
Attneave Street	Finsbury	W.C.	Fd
Atwell Road	Camberwell	S.E.	Hh
Atwell Street	Camberwell	S.E.	Hh
Atworth Street	Poplar	E.	Kf
Auberon Street	Woolwich	E.	Mf
Aubert Park	Islington	N.	Gb
Aubrey House	Kensington	W.	Cf
Aubrey Road	Hornsey	N.	Fa
Aubrey Road	Kensington	W.	Cf
Aubrey Road	West Ham	E.	Lb
Aubrey Villas	Kensington	W.	Cf
Aubrey Walk	Kensington	W.	Cf
Aubyn Street	Lambeth	S.E.	Ff
Auckland Hill	Lambeth	S.E.	Gl
Auckland Road	Battersea	S.W.	Ej
Auckland Road	Bethnal Green	E.	Jd
Auckland Road	Leyton, U.D.	E.	Kb
Auckland Street	Lambeth	S.E.	Fg
Audley Road	Richmond		Sj
Audley Square	Westminster	W.	Ee
Audrey Street	Shoreditch	E.	H[illegible]
Augusta Court	Chelsea	S.W.	Dg
Augusta Street	Poplar	E.	K[illegible]
Augustine Road	Hammersmith	W.	Cf
Augustus Road	Hammersmith	W.	Bf
Augustus Road	Wandsworth	S.W.	Ck
Augustus Street	St. Pancras	N.W.	Ed
Augustus Square	St. Pancras	N.W.	Ed
Aulay Street	Camberwell	S.E.	Hg
Aulton Place	Lambeth	S.E.	Gg
Aumerle Villa	Wembley, U.D.		Sa
Aurelius Road	Wandsworth	S.W.	Fj
Auriol Mansions	Fulham	W.	Cf
Auriol Road	Fulham	W.	Cf
Austin Court	Bethnal Green	E.	Hd
Austin Friars	City of London	E.C.	Ge
Austin Friars Passage	City of London	E.C.	Ge
Austin Friars Square	City of London	E.C.	Ge
Austin Road	Battersea	S.W.	Eh
Austin Street	Bethnal Green and Shoreditch	E.	Hd
Austral Street	Southwark	S.E.	Gf
Australian Avenue	City of London	E.C.	Ge
Autumn Street	Poplar	E.	Jd
Avalon Road	Fulham	S.W.	Dh
Avebury Road	Wimbledon	S.W.	Cm
Avebury Street	Shoreditch	N.	Gd
Aveley Road	Hackney	E.	Hb
Ave Maria Lane	City of London	E.C.	Ge
Avenell Road	Islington	N.	Gb
Avenons Road	West Ham	E.	Le
Avenue, The	Acton U.D.	W.	Bf
Avenue, The	Beckenham U.D.		Km
Avenue, The	Camberwell	S.E.	Gl
Avenue, The	Croydon	S.E.	Gm
Avenue, The	Merton U.D.	S.W	Bm
Avenue, The	Ealing	W.	Se
Avenue, The	Greenwich	S.E.	Kh
Avenue, The	Hackney	E.	Ha
Avenue, The	Hackney	E.	Hb
Avenue, The	Heston and Isleworth U.D. and Twickenham U.D.		Sj
Avenue, The	Kensington	S.W.	Dg
Avenue, The	Lambeth	S.W	Fj
Avenue, The	Lewisham	S.E.	Lh
Avenue, The	Lewisham	S.E.	Ll
Avenue, The	Richmond		Ah

STREET OR PLACE.	BOROUGH.	P.D.	MAP.
Balfern Street	Battersea	S.W.	Eh
Balfour Mews	Westminster	W.	Ee
Balfour Road	Ealing	W.	Sf
Balfour Road	Islington	N.	Gb
Balfour Road	Wimbledon	S.W.	Dm
Balgowan Street	Woolwich		Ng
Balham Empire	**Wandsworth**	S.W.	Ek
Balham Grove	Wandsworth	S.W.	Ek
Balham High Road	Wandsworth	S.W.	Ek
Balham Hill	Wandsworth	S.W.	Ek
Balham New Road	Wandsworth	S.W.	Ek
Balham Park Road	Battersea and Wandsworth	S.W.	Ek
Balham Station	**Wandsworth**	S.W.	Ek
Balin Place	Southwark	S.E.	Gf
Ball Alley	City of London	E.C.	Ge
Ball Court	City of London	E.C.	Ge
Ball Street	Kensington	W.	Df
Ballance Road	Hackney	E.	Jc
Ballantine Street	Wandsworth	S.W.	Dj
Ballard	The Maldens and Coombe, U.D.		Am
Ballater Road	Lambeth	S.W.	Fj
Ballina Street	Lewisham	S.E.	Jk
Ballingdon Road	Battersea	S.W.	Ej
Balloch Road	Lewisham	S.E.	Kk
Ball's Pond Place	Islington	N.	Gc
Ball's Pond Road	Hackney and Islington	N.	Gc
Balmer Road	Poplar	E.	Jd
Balmes Road	Hackney	N.	Gc
Balmoral Avenue	Croydon		Fm
Balmoral Grove	Islington	N.	Fc
Balmoral Road	East Ham and West Ham	E.	Lc
Balmoral Road	Leyton, U.D.	E.	Kb
Balmoral Road	Willesden, U.D.	N.W.	Bc
Balmoral Terrace	Lewisham	S.E.	Kk
Balmuir Gardens	Wandsworth	S.W.	Cj
Baltic Court	Finsbury	E.C.	Gd
Baltic Place	Finsbury	E.C.	Gd
Baltic Street	Finsbury	E.C.	Gd
Baltic Wharf	Stepney	E.	Hf
Balvernie Grove	Wandsworth	S.W.	Dj
Balzac Street	Wandsworth	S.W.	Fj
Bampton Road	Lewisham	S.E.	Jl
Banbury Court	Westminster	W.C.	Fe
Banbury Road	Hackney	E.	Jc
Banbury Street	Battersea	S.W.	Eh
Banchory Road	Greenwich	S.E.	Lh
Bancroft Road	Stepney	E.	Jd
Bandon Road	Bethnal Green	E.	Jd
Bangalore Street	Wandsworth	S.W.	Ch
Bangor Court	Southwark	S.E.	Gf
Bangor Road	Brentford, U.D.		Ag
Bangore Street	Kensington	W.	Ce
Banim Street	Hammersmith	W.	Bf
Banim Terrace	Hammersmith	W.	Bf
Bank Buildings	City of London	E.C.	Ge
Bank Buildings	Islington	N.	Fc
Bank Grove	Ham, U.D.		Sl
Bank Lane	Kingston		Sm
Bank Mansions	Lambeth	S.E.	Gj
Bank Station	**City of London**	E.C.	Ge
Bank Street	Wandsworth	S.W.	Dj
Bank of England	**City of London**	E.C.	Ge
Banks Cricket Ground	Lewisham	S.E.	Jk
Bankside	Southwark	S.E.	Gf
Bankton Road	Lambeth	S.W.	Fj
Banner Street	Finsbury	E.C.	Gd
Banning Street	Greenwich	S.E.	Kg
Bannockburn Road	Woolwich		Ng
Banstead Street	Camberwell	S.E.	Hh
Bantry Place	Bermondsey	S.E.	Hf
Banville Grove	Bermondsey	S.E.	Hf
Banyard Road	Bermondsey	S.E.	Hf
Barandon Street	Kensington	W.	Ce
Barbaco Road	Bromley		Lm
Barbara Street	Islington	N.	Fc
Barbel Street	Southwark	S.E.	Gf
Barberton Road	Lewisham	S.E.	Jl
Barbican	City of London	E.C.	Ge
Barbican Court	City of London	E.C.	Ge

STREET OR PLACE.	BOROUGH.	P.D.	MAP.
Barchester Street	Poplar	E.	Ke
Barclay Road	Fulham	S.W.	Dg
Barclay Road	Leyton, U.D.	E.	La
Barclay Road	Walthamstow, U.D.	E.	Ka
Barclay Street	St. Pancras	N.W.	Fd
Barclay's Brewery	Southwark	S.E.	Gf
Barcombe Avenue	Wandsworth	S.W.	Fk
Barden Street	Woolwich		Nh
Bardolph Road	Islington	N.	Fb
Bardsley Street	Greenwich	S.E.	Kg
Barfett Street	Paddington	W.	Cd
Barfield Road	Leyton, U.D.	E.	La
Barford Street	Islington	N.	Gd
Barforth Road	Camberwell	S.E.	Hj
Barge Lock	Stepney	E.	Je
Barge Yard	City of London	E.C.	Ge
Bargery Road	Lewisham	S.E.	Kk
Baring Road	Lewisham	S.E.	Lk
Baring Street	Islington and Shoreditch	N.	Ga
Bark Place	Paddington	W.	De
Barker Street	Kensington	S.W.	Dg
Barking Creek	Barking, U.D.		Nd
Barking Level	Barking, U.D.		Od
Barking Point	Barking, U.D.		Oe
Barking Reach	Barking, U.D.		Oe
Barking Road	East Ham	E.	Md
Barking Road	West Ham	E.	Le
Barking Sewage Works	Barking, U.D.		Nd
Barkston Gardens	Kensington	S.W.	Dg
Barkston Mansions	Kensington	S.W.	Dg
Barkston Mansions West	Kensington	S.W.	Dg
Barkworth Road	Bermondsey	S.E.	Gj
Barkworth Road	Camberwell	S.E.	Gj
Barlborough Street	Deptford	S.E.	Jh
Barlby Road	Kensington	W.	Ce
Barlow Mews	Westminster	W.	Ee
Barlow Road	Acton, U.D.	W.	Ae
Barlow Street	Southwark	S.E.	Gg
Barmeston Road	Lewisham	S.E.	Jk
Barmore Street	Battersea	S.W.	Dh
Barmouth Road	Wandsworth	S.W.	Dj
Barn Hill	Wembley, U.D.		Aa
Barn Street	Stoke Newington	N.	Gb
Barnard Road	Battersea	S.W.	Ej
Barnard's Wharf	Bermondsey	S.E.	Jf
Barnby Street	St. Pancras	N.W.	Fd
Barn Elms Park	Barnes, U.D.	S.W.	Bh
Barnes Cemetery	Barnes U.D.	S.W.	Bh
Barnas Common	Barnes U.D.	S.W.	Bh
Barnes Green	Barnes U.D.	S.W.	Bh
Barnes Pikle	Ealing	W.	Se
Barnes Station	**Barnes, U.D.**	S.W.	Bh
Barnes Street	Stepney	E.	Je
Barnes Terrace	Greenwich	S.E.	Jg
Barnet Grove	Bethnal Green	E.	Hd
Barnett Street	Stepney	E.	He
Barney Street	West Ham	E.	Kd
Barnfield Road	Lambeth	S.E.	Gl
Barnfield Road	Woolwich		Ng
Barnham Street	Bermondsey	S.E.	Hf
Barnham St. Dwellings	Bermondsey	S.E.	Hf
Barnmead Road	Beckenham U.D.		Jm
Barnsbury Grove	Islington	N.	Fc
Barnsbury Mews	Islington	N.	Gc
Barnsbury Park	Islington	N.	Gc
Barnsbury Road	Islington	N.	Fd
Barnsbury Square	Islington	N.	Fc
Barnsbury Street	Islington	N.	Gc
Barnsbury Terrace	Islington	N.	Fc
Barnsdale Road	Paddington	W.	Cd
Barnsley Street	Bethnal Green	E.	Hd
Barnwell Road	Lambeth	S.W.	Gj
Barnwood Road	West Ham	E.	Lf
Baron Road	West Ham	E.	Le
Baron Street	Finsbury	N.	Gd
Baroness Road	Bethnal Green	E.	Hd
Barons, The	Twickenham U.D.		Sj
Barons Place	Southwark	S.E.	Gf
Barons Court Station	**Hammersmith**	W.	Cg

STREET OR PLACE.	BOROUGH.	P.D.	MAP.
Barons Court Road	Fulham	W.	Cg
Baronsfield Road	Twickenham U.D.		Sj
Baronsmeade Road	Barnes U.D.	S.W.	Bh
Barque Street	Poplar	E.	Kg
Barrack Field	Woolwich		Mg
Barrel Wharf	Poplar	E.	Kg
Barrett Grove	Stoke Newington	N.	Hb
Barrett Street	Marylebone	W.	Ee
Barrett's Green Road	Acton U.D.	N.W.	Ad
Barrington Mansions	Lambeth	S.W.	Gh
Barrington Road	Hornsey	N.	Fa
Barrington Road	Lambeth	S.W.	Gh
Barron's Place	Southwark	S.E.	Gf
Barrow Road	Wandsworth	S.W.	Fm
Barrowgate Road	Chiswick U.D.	W.	Ag
Barrow Hill Road	Marylebone	N.W.	Ed
Barrow's Place	Lewisham	S.E.	Kj
Barry Road	Camberwell	S.E.	Hj
Barry Road	Willesden U.D.	N.W.	Ac
Barset Road	Camberwell	S.E.	Hh
Barston Road	Lambeth	S.E.	Gk
Barter Street	Holborn	W.C.	Fe
Barth Road	Woolwich		Ng
Bartholomew Close	City of London	E.C.	Ge
Bartholomew Lane	City of London	E.C.	Ge
Bartholomew Place	City of London	E.C.	Ge
Bartholomew Road	East Ham	E.	Md
Bartholomew Road	St. Pancras	N.W	Fc
Bartholomew Road	Wandsworth	S.W.	El
Bartholomew Square	Finsbury	E.C.	Gd
Bartholomew Villas	St. Pancras	N.W.	Fc
Bartle Avenue	East Ham	E.	Md
Bartlett's Buildings	City of London	E.C.	Ge
Bartlett's Buildings	Shoreditch	E.C.	Ge
Bartlett's Street	Poplar	E.	Ke
Bartley Street	Lambeth	S.W.	Fj
Barton Court	Shoreditch	N.	Hd
Barton Street	Westminster	S.W.	Ff
Barton Street	Fulham	W.	Cg
Bartram Road	Lewisham	S.E.	Jj
Barwick Road	West Ham	E.	Lb
Basil Mansions	Chelsea	S.W.	Ef
Basil Street	Chelsea and Kensington	S.W.	Ef
Basildon Road	Woolwich		Og
Basin, The	Westminster	W.	Df
Basin, The	Wanstead U.D.	E.	La
Basing Place	Shoreditch	E.	Hd
Basing Road	Camberwell	S.E.	Hh
Basing Road	Kensington	W.	Ce
Basing Square	Shoreditch	E.	Hd
Basinghall Avenue	City of London	E.C.	Ge
Basinghall Street	City of London	E.C.	Ge
Baskerville Road	Wandsworth	S.W.	Ek
Basnett Road	Battersea	S.W.	Eh
Bassano Street	Camberwell	S.E.	Hj
Bassant Road	Woolwich		Nh
Bassein Park Road	Hammersmith	W.	Bf
Bassett Road	Kensington	W.	Ce
Bassett Street	St. Pancras	N.W.	Ec
Bassingham Road	Wandsworth	S.W.	Dk
Basterfield Street	Finsbury	E.C.	Gd
Bastion Road	Woolwich		Og
Bastwick Street	Finsbury	E.C.	Gd
Basuto Road	Fulham	S.W.	Dh
Batavia Road	Deptford	S.E.	Jh
Bateman's Row	Shoreditch	N.	Hd
Bateman's Street	Westminster	W.	Fe
Bateson Street	Woolwich		Ng
Bath Cottages	Poplar	E.	Ke
Bath Court	Shoreditch	E.C.	Gd
Bath Grove	Bethnal Green	E.	Hd
Bath Place	Camberwell	S.E.	Gg
Bath Place	Shoreditch	E.C.	Gd
Bath Place	Stepney	E.	He
Bath Road	Chiswick U.D.	W.	Bf
Bath Road	Croydon R.D.		Dm
Bath Row	Finsbury	E.C.	Gd
Bath Row	Hackney	E.	Hc
Bath Street	Bethnal Green and Stepney	E.	He
Bath Street	Finsbury	E.C.	Gd
Bath Street	Poplar	E.	Ke
Bath Street	Southwark	S.E.	Gf
Bath Terrace	Southwark	S.E.	Gf
Bathmore Street	Woolwich		Lg
Bathurst Gardens	Willesden U.D.	N.W.	Bd
Bathurst Mews	Paddington	W.	De
Bathurst Street	Paddington	W.	De
Batoum Gardens	Hammersmith	W.	Cf
Batsford Road	Deptford	S.E.	Jh
Batson Street	Hammersmith	W.	Bf
Batson Street	Stepney	E.	Je
Batsworth Road	Croydon R.D.		Dm
Battenberg Road	Richmond		Ah
Battersea Boro' Wharf	Battersea	S.W.	Dh
Battersea Bridge	Battersea and Chelsea	S.W.	Dg
Battersea Bridge Road	Battersea	S.W.	Eh
Battersea Electric light Station	Battersea	S.W.	Dh
Battersea High Street	Battersea	S.W.	Dh
Battersea Park	Battersea	S.W.	Eg
Battersea Park Pier	Battersea	S.W.	Eg
Battersea Park Road	Battersea	S.W.	Eh
Battersea Park Road Station	**Battersea**	S.W.	Eh
Battersea Park Station	**Battersea**	S.W.	Eh
Battersea Polytechnic	Battersea	S.W.	Eh
Battersea Reach	Battersea	S.W.	Dh
Battersea Rise	Battersea	S.W.	Ej
Battersea Square	Battersea	S.W.	Dh
Battersea Square Pier	Battersea	S.W.	Dh
Battersea Station	**Battersea**	S.W.	Dh
Battle Bridge Lane	Bermondsey	S.E.	Gf
Battle Bridge Road	St. Pancras	N. & N.W.	Fd
Battle Bridge Stairs	Bermondsey	S.E.	Gf
Battledean Road	Islington	N.	Gc
Batty Court	Stepney	E.	He
Batty Place	Stepney	E.	He
Batty Street	Stepney	E.	He
Batty's Gardens	Stepney	E.	He
Baulk, The	Wandsworth	S.W.	Dj
Bavent Road	Lambeth	S.E.	Gh
Bawdale Road	Camberwell	S.E.	Hj
Bawick Road	West Ham	E.	Le
Bawtree Road	Deptford	S.E.	Jg
Baxendale Street	Bethnal Green	E.	Hd
Baxter Grove	Islington	N.	Gc
Baxter Road	Islington	N.	Gc
Baxter Road	West Ham	E.	Le
Baxter's Court	Hackney	N.E.	Hc
Bay Street	Hackney	E.	Hc
Bayer Street	Finsbury	E.C.	Gd
Bayford Road	Willesden U.D.	N.W.	Cd
Bayford Street	Hackney	E.	Hc
Bayham Road	Acton U.D.	W.	Af
Bayham Place	St. Pancras	N.W.	Fc
Bayham Street	St. Pancras	N.W.	Fc
Bayley Street	Holborn	W.C.	Fe
Bayonne Road	Fulham	W.	Cg
Bayston Road	Hackney	N.	Hb
Bayswater Hill	Paddington	W.	De
Bayswater Road	Kensington and Paddington	W.	De
Bayswater Station	**Paddington**	W.	De
Bayswater Terrace	Paddington	W.	De
Bayswater Walk	Kensington	W.	Df
Baythorne Street	Stepney	E.	Je
Bazon Street	Lambeth	S.E.	Ff
Beech Road	Croydon	S.W.	Fm
Beachcroft Buildings	Stepney	E.	Je
Beachcroft Road	Leyton U.D.	E.	Kb
Beachy Road	Poplar		Jc
Beacondale Road	Lambeth	S.E.	Gl
Beacon Hill	Islington	N.	Fb
Beacon Road	Lewisham	S.E.	Kj
Beacon Street	Southwark	S.E.	Gg
Beaconsfield Road	Ealing	W.	Sf
Beaconsfield Road	Fulham	S.W.	Ch
Beaconsfield Road	Greenwich	S.E.	Lg
Beaconsfield Road	Leyton U.D.	E.	Kb
Beaconsfield Road	The Maldens and Coombe U.D.		Am

STREET OR PLACE.	BOROUGH.	P.D.	MAP.
Belgrave Mews	Westminster	S.W.	Ef
Belgrave Mews East	Westminster	S.W.	Ef
Belgrave Mews North South and West	Westminster	S.W.	Ef
Belgrave Place	Westminster	S.W.	Ef
Belgrave Road	Westminster	S.W.	Eg
Belgrave Road	Croydon R.D.		Dm
Belgrave Road	Leyton U.D.	E.	Ka
Belgrave Road	Marylebone	N.W.	Dd
Belgrave Road	Walthamstow U.D.	E.	Ka
Belgrave Square	Westminster	S.W.	Ef
Belgrave Street	Westminster	S.W.	Ef
Belgrave Street	St. Pancras	W.C.	Fd
Belgrave Street	Shoreditch	E.	Hc
Belgrave Street	Stepney	E.	Je
Belgrave Terrace	Lambeth	S.W.	Fh
Belham Street	Camberwell	S.E.	Gh
Belhaven Street	Bethnal Green and Stepney	E.	Jd
Belinda Road	Lambeth	S.W.	Gh
Belitha Villas	Islington	N.	Fc
Bell, The	Ealing	W.	Se
Bell Court	Bermondsey	S.E.	Gf
Bell Court	Lambeth	S.E.	Ff
Bell Court	Stepney	E.	He
Bell Green	Lewisham	S.E.	Jl
Bell Green Cottages	Lewisham	S.E.	Jl
Bell Green Terrace	Lewisham	S.E.	Jl
Bell Grove	Bexley U.D.		Oj
Bell Hotel	Bromley		Lm
Bell Lane	Stepney	E.	He
Bell Lane	Wandsworth	S.W.	Dj
Bell Road	Poplar	E.	Ke
Bell Street	Greenwich	S.E.	Kg
Bell Street	Marylebone	N.W.	Ee
Bell Street Cottages	Marylebone	N.W.	Ee
Bell Yard	City of London	E.C.	Ga
Bell Yard	City of London and Westminster	W.C.	Fe
Bell Yard	Finsbury	E.C.	Gd
Bellamy Street	Wandsworth	S.W.	Ek
Bellamy's Wharf	Bermondsey	S.E.	Jf
Bellasis Avenue	Wandsworth	S.W.	Fk
Bellefield Road	Lambeth	S.W.	Fj
Belle Isle	Islington	N.	Fc
Bellenden Road	Camberwell	S.E.	Hh
Belle Sauvage Yard	City of London	E.C.	Ge
Belleville Road	Battersea	S.W.	Ej
Bellevue Crescent	Hampstead	N.W.	Db
Bellevue Place	Stepney	E.	He
Bellevue Road	Battersea and Wandsworth	S.W.	Ek
Bellevue Road	Kingston		Sm
Bellevue Terrace	Greenwich	S.E.	Lg
Bellew Street	Wandsworth	S.W.	Dl
Bell's Garden Road	Camberwell	S.E.	Hh
Bellgrove Park	Bexley		Oj
Bellingham Station	**Lewisham**	S.E.	Jl
Bell Moor	Hampstead	N.W.	Db
Bellot Street	Greenwich	S.E.	Kg
Bellow Street	Stepney	E.	He
Bell's Alley	Fulham	S.W.	Dh
Bell Wharf Lane	City of London	E.C.	Ge
Bellwood Road	Camberwell	S.E.	Jj
Belmont	Wandsworth	S.W.	Ck
Belmont Cottages	Chislehurst U.D.		Ml
Belmont Grove	Chiswick U.D.	W.	Af
Belmont Grove	Lewisham	S.E.	Kj
Belmont Hill	Lewisham	S.E.	Kj
Belmont Lane	Chislehurst U.D.		Nl
Belmont Park	Lewisham	S.E.	Kj
Belmont Park Road	Leyton U.D.	E.	Ka
Belmont Place	Woolwich		Mg
Belmont Road	Beckenham U.D.		Jm
Belmont Road	Chislehurst U.D.		Mm
Belmont Road	Chiswick U.D.	W.	Af
Belmont Road	Lewisham	S.E.	Kj
Belmont Road	Wandsworth	S.W.	Fh
Belmont Street	St. Pancras	N.W.	Ec
Belmont Candle Works	Battersea	S.W.	Dh
Belmore Street	Lambeth	S.W.	Fh

STREET OR PLACE.	BOROUGH.	P.D.	MAP.
Belper Street	Islington	N.	Fc
Belsham Street	Hackney	E.	Hc
Belshaw Street	Hackney	E.	Jc
Belsize Avenue	Ealing	W.	Sf
Belsize Avenue	Hampstead	N.W.	Ec
Belsize Crescent	Hampstead	N.W.	Dc
Belsize Crescent Stables	Hampstead	N.W.	Ec
Belsize Grove	Hampstead	N.W.	Ec
Belsize Lane	Hampstead	N.W.	Ec
Belsize Mews	Hampstead	N.W.	Ec
Belsize Park	Hampstead	N.W.	Dc
Belsize Park Gardens	Hampstead	N.W.	Ec
Belsize Park Mews	Hampstead	N.W.	Ec
Belsize Park Terrace	Hampstead	N.W.	Ec
Belsize Place	Hampstead	N.W.	Ec
Belsize Road	Hampstead	N.W.	Dc
Belsize Road Mews	Hampstead	N.W.	Dc
Belsize Square	Hampstead	N.W.	Dc
Belsize Station	**Hampstead**	N.W.	Ec
Belsize Terrace	Hampstead	N.W.	Ec
Beltham Road	Fulham	S.W.	Dh
Belton Road	Leyton U.D.	E.	Kb
Belton Road	Willesden U.D.	N.W.	Bc
Beltwood	Camberwell	S.E.	Hl
Belvedere Avenue	Wimbledon	S.W.	Cl
Belvedere Buildings	Southwark	S.E.	Gf
Belvedere Crescent	Lambeth	S.E.	Ff
Belvedere Drive	Wimbledon	S.W.	Cl
Belvedere Place	Southwark	S.E.	Gf
Belvedere Road	Lambeth	S.E.	Ff
Belvedere Road	Penge U.D.	S.E.	Hm
Belvedere Grove	Wimbledon	S.W.	Cl
Belvernie Road	Wandsworth	S.W.	Ck
Belvoir Road	Camberwell	S.E.	Hk
Belward Avenue	Finsbury	E.C.	Gd
Bemerton Street	Islington	N.	Fc
Bemish Road	Wandsworth	S.W.	Ch
Benares Road	Woolwich		Ng
Benbow Road	Hammersmith	W.	Bf
Benbow Street	Greenwich	S.E.	Jg
Bendall Street	Woolwich		Nh
Bendmere (Bendmeer) Road	Wandsworth	S.W.	Ch
Bendon Valley	Wandsworth	S.W.	Dk
Benedict Road	Croydon R.D.		Dm
Benedict Road	Lambeth	S.W.	Fh
Benfield Street	Battersea	S.W.	Dh
Bengeo Street	West Ham	E.	Le
Bengeworth Road	Lambeth	S.E.	Gh
Benham Place	Hampstead	N.W.	Db
Benham Street	Battersea	S.W.	Dh
Benhill Road	Camberwell	S.E.	Gh
Benin Street	Lewisham	S.E.	Kk
Benjamin Street	Finsbury	E.C.	Ge
Ben Jonson Road	Stepney	E.	Je
Ben Jonson Square	Stepney	E.	Je
Benledi Street	Poplar	E.	He
Benmore Street	Deptford	S.E.	Kg
Benn Street	Hackney	E.	Jc
Bennerley Road	Battersea	S.W.	Ej
Bennett Park	Greenwich	S.E.	Lh
Bennett Street	Chiswick U.D.	W.	Bg
Bennett Street	Westminster	S.W.	Ff
Bennett Street	Greenwich	S.E.	Kh
Bennett Street	Southwark	S.E.	Gf
Bennett's Hill	City of London	E.C.	Ge
Bennett's Mews	Southwark	S.E.	Gf
Bennett's Place	Stepney	E.	He
Bennett's Terrace	Kensington	W.	Cd
Bensham Grove	Croydon		Sm
Benson Avenue	East Ham	E.	Ld
Benson Road	Lewisham	S.E.	Hk
Benthall Road	Hackney	N.	Hb
Bentham Road	Hackney	E.	Jc
Bentinck Court	Marylebone	N.W.	Ed
Bentinck Mews	Marylebone	N.W.	Ee
Bentinck Street	Westminster	S.W.	Fg
Bentinck Street	Marylebone	W.	Ee
Bentley Road	Hackney	N.	Hc
Bentley Terrace	Poplar	E.	Ke
Benton's Lane	Lambeth	S.E.	Gl
Benwell Road	Islington	N.	Gb
Benworth Street	Poplar	E.	Jd

STREET OR PLACE.	BOROUGH	P.D.	MAP.
Blackwall Point	Greenwich	S.E.	Kf
Blackwall Reach	Poplar	E.	Kf
Blackwall Station	**Poplar**	E.	Kf
Blackwall Tunnel	Poplar	E.	Kf
Blackwater Street	Camberwell	S.E.	Hj
Blackwood Street	Southwark	S.E.	Gg
Bladon Terrace	Wandsworth	S.W.	Fl
Blagdon Street	Lewisham	S.E.	Kj
Blagrove Road	Kensington	W.	Ce
Blair Street	Poplar	E.	He
Blake Road	Fulham	S.W.	Dh
Blake Street	Deptford	S.E.	Jg
Blake Hall	Wanstead U.D.	E.	La
Blake Hall Road	Wanstead U.D.	E.	La
Blakeley's Buildings	Greenwich	S.E.	Kf
Blakeley's Cottages	Greenwich	S.E.	Kf
Blakeney Avenue	Beckenham U.D.		Jm
Blakeney Road	Beckenham U.D.		Jm
Blakenham Road	Wandsworth	S.W.	El
Blake's Road	Camberwell	S.E.	Gg
Blakesley Avenue	Ealing	W.	Se
Blakesley Street	Stepney	E.	He
Blanchard Road	Hackney	E.	Hc
Blanchard Street	Hackney	E.	Hc
Blanche Street	West Ham	E.	Le
Bland Avenue	Camberwell	S.E.	Hh
Blanfields Street	Wandsworth	S.W.	Ej
Blandford Avenue	Beckenham U.D.		Jm
Blandford Mews	Marylebone	W.	Ee
Blandford Road	Acton U.D.	W.	Bf
Blandford Road	Beckenham U.D.		Hm
Blandford Road	Ealing	W.	Sf
Blandford Square	Marylebone	N.W.	Ed
Blandford Street	Marylebone	W.	Ee
Blanmerle Road	Woolwich		Mk
Blantyre Street	Chelsea	S.W.	Dg
Blashford Street	Lewisham	S.E.	Kk
Blechynden Street	Hammersmith & Kensington	W.	Ce
Blegborough Road	Wandsworth	S.W.	El
Blendon Hall	Bexley		Ok
Blendon Road	Bexley		Ok
Blendon Road	Woolwich		Ng
Blendon Row	Southwark	S.E.	Gg
Blendon Wood	Bexley		Ok
Blenheim Crescent	Kensington	W.	Ce
Blenheim Gardens	Willesden U.D.	N.W.	Cc
Blenheim Grove	Camberwell	S.E.	Hh
Blenheim House	Foots Cray U.D.		Nl
Blenheim Mansions	Westminster	S.W.	Ff
Blenheim Place	Marylebone	N.W.	Dd
Blenheim Road	Acton U.D. and Chiswick U.D.	W.	Bf
Blenheim Road	Chiswick U.D.	W.	Ag
Blenheim Road	Merton U.D.	S.W.	Bm
Blenheim Road	Islington	N.	Fb
Blenheim Road	Leyton U.D.	E.	Kb
Blenheim Road	Marylebone	N.W.	Dd
Blenheim Street	Chelsea	S.W.	Eg
Blenheim Street	Westminster	W.	Ee
Blenheim Terrace	Marylebone	N.W.	Dd
Blenkarne Road	Battersea	S.W.	Ej
Blessington Road	Lewisham	S.E.	Kj
Blewett Street	Southwark	S.E.	Gg
Blicke Street	Lambeth	S.W.	Fg
Blind Lane	Barnes U.D.	S.W.	Aj
Blind Lane	Bexley U.D. and Woolwich		Nh
Blind Lane	Wembley U.D.		Sb
Blind Manor Way	Woolwich		Of
Blisset Street	Greenwich	S.E.	Kh
Blithdale Road	Woolwich		Og
Blithfield Street	Kensington	W.	Df
Blockhouse Street	Deptford	S.E.	Hg
Bloemfontein Avenue	Hammersmith	W.	Bf
Bloemfontein Road	Hammersmith	W.	Be
Blomfield Gardens	Hammersmith	W.	Ce
Blomfield Mews	Hammersmith	W.	Cf
Blomfield Place	Paddington	W.	De
Blomfield Place	Westminster	W.	Ee
Blomfield Road	Hammersmith	W.	Ce
Blomfield Road	Paddington	W.	De
Blomfield Street	City of London	E.C.	Ge

STREET OR PLACE.	BOROUGH.	P.D.	MAP.
Blomfield Street	Hackney	E.	Hc
Blomfield Street	Paddington	W.	De
Blomfield Terrace	Hammersmith	W.	Cf
Blomfield Villas	Hammersmith	W.	Ce
Blondel Street	Battersea	S.W.	Eh
Blondin Street	Poplar	E.	Jd
Bloomburg Street	Westminster	S.W.	Fg
Bloomfield	Lambeth	S.E.	Gl
Bloomfield Place	Chelsea & Westminster	S.W.	Eg
Bloomfield Place	Westminster	S.W.	Ee
Bloomfield Yard	Kensington	W.	Ce
Bloomfield Road	Ealing	W.	Se
Bloomfield Road	Hornsey	N.	Ea
Bloomfield Road	Kingston		Sm
Bloomfield Road	Stepney	E.	Je
Bloomfield Road	Woolwich		Ng
Bloomfield Street	Poplar	E.	Ke
Bloomfield Terrace	Westminster	S.W.	Eg
Bloom-park Road	Fulham	S.W.	Ch
Bloomsbury Court	Holborn	W.C.	Fe
Bloomsbury Place	Holborn	W.C.	Fe
Bloomsbury Square	Holborn	W.C.	Fe
Bloomsbury Street	Holborn	W.C.	Fe
Bloomsbury Street	Poplar	E.	Ke
Bloomwood Road	Lambeth	S.E.	Gl
Blossom Place	Stepney	E.	He
Blossom Street	Stepney	E.	He
Blount Street	Stepney	E.	Je
Blows Place	Lewisham	S.E.	Kj
Blucher Road	Camberwell	S.E.	Gh
Blucher Street	Southwark	S.E.	Gg
Blue Anchor Alley	Finsbury	E.C.	Gd
Blue Anchor Lane	Bermondsey	S.E.	Hg
Blue Anchor Yard	Stepney	E.	He
Blue Ball Yard	Westminster	S.W.	Ff
Blue Boar Court	City of London	E.C.	Ge
Bluecoat Girl's School	Greenwich	S.E.	Kh
Blue Cross Street	Westminster	W.C.	Fe
Blundell Street	Islington	N.	Fc
Blunts Road	Woolwich		Mj
Blurton Road	Hackney	E.	Jb
Blyth Road	Bromley		Km
Blyth Road	West Ham	E.	Kc
Blythe Hill	Lewisham	S.E.	Jk
Blythe Hill Lane	Lewisham	S.E.	Jk
Blythe Road	Hammersmith	W.	Cf
Blythe Street	Bethnal Green	E.	Hd
Blythe Vale	Lewisham	S.E.	Jk
Blythe's Yard	Finsbury	E.C.	Gd
Blythwood Road	Islington	N.	Fa
Board of Trade	**Westminster**	S.W.	Ff
Boar's Head Court	City of London	E.C.	Fe
Boar's Head Yard	Stepney	E.	He
Boat Street	Hackney	E.	Jb
Boden Mews	Camberwell	S.E.	Gh
Bodley Street	Southwark	S.E.	Gg
Bodney Road	Hackney	E.	Hc
Bohemia Avenue	Chiswick U.D.	W.	Bf
Bohemia Place	Hackney	E.	He
Bohn Street	Stepney	E.	Je
Bolan Street	Battersea	S.W.	Eh
Bold Grove	Bromley R.D.		Om
Bolden Street	Deptford	S.E.	Jh
Bolena Road	Deptford	S.E.	Hg
Boleyn Castle	East Ham	E.	Ld
Boleyn Road	East Ham	E.	Md
Boleyn Road	Hackney, Islington and Stoke Newington	N.	Hc
Boleyn Road	West Ham	E.	Lc
Bolingbroke Grove	Battersea	S.W.	Ej
Bolingbroke Road	Battersea	S.W.	Dh
Bolingbroke Road	Hammersmith	W.	Cf
Bolivia Road	Deptford	S.E.	Hg
Bollo Bridge	Acton U.D.	W.	Af
Bollo Lane	Acton U.D. and Brentford U.D.	W.	Af
Bollo Bridge Road	Acton U.D.	W.	Af
Bolney Street	Lambeth	S.W.	Fg
Bolsover Street	Marylebone	W.	Ee
Bolt Court	City of London	E.C.	Fe
Bolt-in-Tun Court	City of London	E.C.	Fe

STREET OR PLACE.	BOROUGH.	P.D.	MAP.
Bolton Gardens	Chiswick U.D.	W.	Bg
Bolton Gardens	Holborn	W.C.	Fe
Bolton Gardens	Kensington	S.W.	Dg
Bolton Gardens	Willesden U.D.	N.W.	Cd
Bolton Gardens Mews	Kensington	S.W.	Dg
Bolton Gardens South	Kensington	S.W.	Dg
Bolton Gardens West	Kensington	S.W.	Dg
Bolton Mews	Kensington	S.W.	Dg
Bolton Mews	Kensington	W.	Ce
Bolton Place	Finsbury	W.C.	Fd
Bolton Road	Acton U.D.	W.	Af
Bolton Road	Chiswick U.D.	W.	Ag
Bolton Road	Hampstead	N.W.	Dc
Bolton Road	Kensington	W.	Ce
Bolton Road	West Ham	E.	Lc
Bolton Road	Willesden U.D.	N.W.	Bc
Bolton Street	Westminster	W.	Ef
Bolton Street	Lambeth	S.E.	Gg
Boltons, The	Kensington	S.W.	Dg
Bolwell Street	Lambeth	S.E.	Fg
Bombay Street	Bermondsey	S.E.	Hg
Bomore Road	Kensington	W.	Ce
Bonamy Street	Camberwell	S.E.	Gj
Bonar Road	Camberwell	S.E.	Hg
Bonar Terrace	Camberwell	S.E.	Hg
Bonchurch Road	Kensington	W.	Ce
Bond Court	City of London	E.C.	Ge
Bond Place	Lambeth	S.E.	Ff
Bond Road	Croydon R.D.		Em
Bond Street	Chiswick U.D.	W.	Bg
Bond Street	Finsbury	W.C.	Fd
Bond Street	Lambeth	S.W.	Fg
Bond Street	West Ham	E.	Kc
Bond Street Station	**Westminster**	W.	Ee
Bonfield Road	Lewisham	S.E.	Kj
Bonham Road	Lambeth	S.W.	Fj
Bonheur Road	Acton U.D.	W.	Bf
Boniface Street	Lambeth	S.E.	Ff
Bonlay Road	Lambeth	S.W.	Fj
Bonnar Road	Camberwell	S.E.	Hh
Bonner Road	Bethnal Green	E.	Hd
Bonner Street	Bethnal Green	E.	Jd
Bonneville Road	Wandsworth	S.W.	Fj
Bonnington Square	Lambeth	S.W.	Fg
Bonny Street	St. Pancras	N.W.	Fc
Bonny Downs Road	East Ham	E.	Md
Bonsor Street	Camberwell	S.E.	Gh
Bonwell Street	Bethnal Green	E.	Jd
Bookham Street	Shoreditch	N.	Gd
Boone Place	Lewisham	S.E.	Kj
Boone Street	Lewisham	S.E.	Kj
Boones Road	Lewisham	S.E.	Kj
Boord Street	Greenwich	S.E.	Kf
Boot Street	Shoreditch	N.	Gd
Booth Street	Stepney	E.	He
Booth Street Bldgs.	Stepney	E.	He
Boothby Road	Islington	N.	Fa
Bootmakers' Institute	Barnes U.D.	S.W.	Ah
Border Crescent	Beckenham U.D.	S.E.	Hl
Border Road	Beckenham U.D. and Lewisham	S.E.	Hl
Boreham Street	Bethnal Green	E.	Hd
Borland Road	Camberwell	S.E.	Hj
Borneo Street	Wandsworth	S.W.	Ch
Borough High Street	Bermondsey and Southwark	S.E.	Gf
Borough Market	Southwark	S.E.	Gf
Borough Road	Kingston		Am
Borough Road	Southwark	S.E.	Gf
Borough Station	**Southwark**	S.E.	Gf
Borough Theatre	**West Ham**	E.	Kc
Borrett Road	Southwark	S.E.	Gg
Borrodaile Road	Wandsworth	S.W.	Dj
Borthwick Road	Hendon U.D.	N.W.	Ba
Borthwick Road	Leyton U.D.	E.	Kb
Boscastle Road	St. Pancras	N.W.	Eb
Boscawen Street	Deptford	S.E.	Jg
Boscomb Avenue	Leyton U.D.	E.	Ka
Boscombe Road	Hammersmith	W.	Bf
Boss Court	City of London	E.C.	Ge
Boss Street	Bermondsey	S.E.	Hf
Bostall Heath	Woolwich		Og
Bostall Hill	Woolwich		Og

STREET OR PLACE.	BOROUGH.	P.D.	MAP.
Bostall House	Woolwich		Og
Bostall Lane	Woolwich		Og
Bostall Lodge	Woolwich		Og
Bostall Wood	Woolwich		Og
Bostock Street	Stepney	E.	Hf
Boston House	Brentford U.D.		Sg
Boston Place	Marylebone	N.W.	Ed
Boston Road	Brentford U.D.		Sg
Boston Road	Walthamstow U.D.	E.	Ja
Boston Street	Marylebone	N.W.	Ed
Boston Street	Shoreditch	E.	Hd
Boston Yard	Islington	N.	Fd
Boston Park Road	Brentford U.D.		Sg
Boswell Court	Holborn	W.C.	Fe
Bosworth Road	Kensington	W.	Cd
Botanic Gardens	Marylebone	N.W.	Ed
Botanic Gardens	Chelsea	S.W.	Eg
Botany Bay Lane	Chislehurst U.D.		Nm
Bothwell Street	Fulham	S.W.	Cg
Botolph Alley	City of London	E.C.	Ge
Botolph Lane	City of London	E.C.	Ge
Botolph Road	Poplar	E.	Kd
Bott's Mews	Paddington	W.	De
Boucher Place	Southwark	S.E.	Gf
Boulton Road	West Ham	E.	Le
Boundaries Road	Battersea and Wandsworth	S.W.	Ek
Boundary Court	Stepney	E.	He
Boundary Lane	Camberwell and Southwark	S.E.	Gg
Boundary Mews	Hampstead	N.W.	Dc
Boundary Mews	Kensington	W.	De
Boundary Road	Barking U.D.		Nd
Boundary Road	East Ham and West Ham	E.	Ld
Boundary Road	Hammersmith & Kensington	W.	Cf
Boundary Road	Hampstead and Marylebone	N.W.	Dc
Boundary Road	Leyton U.D. and Walthamstow U.D.	E.	Ja
Boundary Road	Paddington	W.	Dd
Boundary Row	Southwark	S.E.	Gf
Boundary Street	Bethnal Green & Shoreditch	E.	Hd
Bourdon Road	Penge U.D.	S.E.	Hm
Bourdon Buildings	Westminster	W.	Ee
Bourdon Street	Westminster	W.	Ee
Bourke Road	Willesden U.D.	N.W.	Bc
Bourn Road	Leyton U.D.	E.	Lb
Bourne Road	Hornsey	N.	Fa
Bourne Terrace	Kensington	W.	Ce
Bournemouth Road	Camberwell	S.E.	Hh
Bournes Place	Battersea	S.W.	Dh
Bournevale Road	Wandsworth	S.W.	Fl
Bournville Road	Lewisham	S.E.	Jk
Bourton Street	Camberwell	S.E.	Hj
Bousfield Road	Deptford	S.E.	Jh
Boutflower Road	Battersea	S.W.	Dj
Bouverie Road	Stoke Newington	N.	Gb
Bouverie Street	City of London	E.C.	Ge
Bouverie Street	Paddington	W.	De
Bovay Place	Islington	N.	Fb
Bovay Street	Islington	N.	Fb
Boveney Road	Lewisham	S.E.	Jk
Bovill Place	Lewisham	S.E.	Jk
Bovill Road	Lewisham	S.E.	Jk
Bovingdon Road	Fulham	S.W.	Dh
Bow Bridge	Poplar and West Ham	E.	Kd
Bow Common Lane	Poplar and Stepney	E.	Jd
Bow Creek	Poplar and West Ham	E.	Ke
Bow Lane	City of London	E.C.	Ge
Bow Lane	Poplar	E.	Ke
Bow Road	Poplar	E.	Jd
Bow Road Station	**Poplar**	E.	Jd
Bow Station	**Poplar**	E.	Kd
Bow Street	Westminster	W.C.	Fe
Bow Street	West Ham	E.	Kc

STREET OR PLACE.	BOROUGH.	P.D.	MAP.
Bowater Place	Greenwich	S.E.	Lh
Bowater Road	Woolwich		Mf
Bowdale Road	Camberwell	S.E.	Hj
Bowden Street	Lambeth	S.E.	Gg
Bowdon Road	Leyton U.D.	E.	Ka
Bowen Street	Poplar	E.	Ke
Bower Avenue	Greenwich	S.E.	Kh
Bower Road	Hackney	E.	Jc
Bower Street	Stepney	E.	Je
Bowerdean Street	Fulham	S.W.	Dh
Bowering Place	Camberwell	S.E.	Hh
Bowerman Road	Lewisham	S.E.	Jj
Bowles Road	Camberwell	S.E.	Hg
Bowles' Wharf	Stepney	E.	Je
Bowley Street	Poplar	E.	Jf
Bowling Green	Southwark	S.E.	Gf
Bowling Green Lane	Finsbury	E.C.	Gd
Bowling Green Lane	Southwark	S.E.	Gf
Bowling Green Row	Woolwich		Mg
Bowling Green Street	Lambeth	S.E.	Fg
Bowling Green Street	Stoke Newington	N.	Hb
Bowling Green Walk	Shoreditch	N.	Gd
Bowness Road	Lewisham	S.E.	Kk
Bowstead Road	Wandsworth	S.W.	Ch
Box Street	Poplar	E.	Je
Boxall Road	Camberwell	S.E.	Gk
Boxley Street	West Ham	E.	Lf
Boxmoor Street	Hammersmith	W.	Cf
Boxworth Grove	Islington	N.	Fc
Boy Court	City of London	E.C.	Ge
Boyd Road	West Ham	E.	Le
Boyd Street	Stepney	E.	He
Boyer Road	Willesden U.D.	N.W.	Dd
Boyle Street	Westminster	W.	Ee
Boyne Road	Lewisham	S.E.	Kj
Boyson Road	Southwark	S.E.	Gg
Brabant Court	City of London	E.C.	Ge
Brabazon Street	Poplar	E.	Ke
Brabourn Grove	Camberwell	S.E.	Hh
Bracewell Road	Hammersmith	W.	Ce
Brackenbury Road	Hammersmith	W.	Bf
Brackley Road	Beckenham U.D.		Jm
Brackley Road	Chiswick U.D.	W.	Bg
Brackley Street	City of London	E.C.	Ge
Brackley Terrace	Chiswick U.D.	W.	Bg
Bracklyn Street	Shoreditch	N.	Gd
Bracknell Gardens	Hampstead	N.W.	Db
Bracondale Road	Woolwich		Of
Bracton Road	Bermondsey	S.E.	Jg
Bracy Street	Islington	N.	Fa
Brad Street	Lambeth	S.E.	Gf
Bradbourne Street	Fulham	S.W.	Dh
Bradbury Street	Hackney	N.	Hc
Braddon Street	Bermondsey	S.E.	Hf
Braddyll Street	Greenwich	S.E.	Kg
Braden Street	Paddington	W.	Dd
Bradfield Road	West Ham	E.	Lf
Bradford Avenue	City of London	E.C.	Ge
Bradford Road	Lewisham	S.E.	Hl
Bradgate Road	Lewisham	S.E.	Jk
Brading Road	Lambeth	S.W.	Fk
Bradiston Road	Paddington	W.	Cd
Bradley Gardens	Ealing	W.	Se
Bradley Street	Lambeth	S.W.	Fg
Bradley Street	West Ham	E.	Le
Bradley's Buildings	Finsbury	N.	Gd
Bradmore Chambers	Hammersmith	W.	Cf
Bradmore Grove	Hammersmith	W.	Cf
Bradmore Lane	Hammersmith	W.	Cf
Bradmore-Park Road	Hammersmith	W.	Bf
Bradshaw Street	Camberwell	S.E.	Hg
Bradstock Road	Hackney	E.	Jc
Bradwell Street	Stepney	E.	Jd
Brady Street	Bethnal Green & Stepney	E.	He
Brady Street Bldgs.	Bethnal Green & Stepney	E.	He
Brady's Buildings	Bethnal Green	E.	Hd
Braemar Road	Brentford U.D.		Sg
Braemar Road	Tottenham U.D.	N.	Ha
Braemar Road	West Ham	E.	Le
Braemar Street	Bethnal Green	E.	Jd
Braemar Villas	Camberwell	S.E.	Jj
Braidwood Road	Lewisham	S.E.	Kk
Brailsford Road	Lambeth	S.W.	Fj
Braime's Buildings	City of London	E.C.	Ge
Braithwaite Place	Paddington	W.	Dd
Bramah Road	Lambeth	S.W.	Gh
Bramber Road	Fulham	W.	Cg
Bramblebury Road	Woolwich		Ng
Bramcote Road	Bermondsey and Camberwell	S.E.	Hg
Bramcote Road	Wandsworth	S.W.	Bj
Bramerton Street	Chelsea	S.W.	Dg
Bramfield Road	Battersea	S.W.	Ej
Bramford Road	Wandsworth	S.W.	Dj
Bramham Gardens	Kensington	S.W.	Dg
Bramhope	Woolwich		Lg
Bramley Mews	Kensington	W.	Ce
Bramley Road	Hammersmith & Kensington	W.	Ce
Bramley Street	Kensington	W.	Ce
Brampton Road	Bexley U.D. and Erith U.D.		Oh
Brampton Road	Hackney	E.	Jc
Brampton Road	Tottenham U.D.	N.	Ga
Bramshill Road	Lambeth	S.E.	Gl
Bramshill Gardens	St. Pancras	N.W.	Eb
Bramshill Road	Willesden U.D.	N.W.	Bd
Bramshot Avenue	Woolwich		Lg
Bramwell Street	Battersea and Wandsworth	S.W.	Eh
Brancaster Road	Wandsworth	S.W.	Fl
Branch Hill	Hampstead	N.W.	Db
Branch Hill House	Hampstead	N.W.	Db
Branch Hill Park	Hampstead	N.W.	Db
Branch Hill Side	Hampstead	N.W.	Db
Branch Hill Side Mans.	Hampstead	N.	Db
Branch Place	Shoreditch	N.	Gc
Brand Place	Islington	N.	Fb
Brand Street	Greenwich	S.E.	Kh
Brand Street	Islington	N.	Fb
Brand Street	Marylebone	N.W.	Ed
Brandenburgh Road	Fulham	W.	Cg
Brandenburgh Road	Chiswick U.D.	W.	Ag
Brandlehow Road	Wandsworth	S.W.	Cj
Brandon Mansions	Fulham	W.	Cg
Brandon Road	Islington	N.	Fc
Brandon Road	Lambeth	S.W.	Fj
Brandon Street	Southwark	S.E.	Gg
Brandram Road	Lewisham	S.E.	Kj
Brandreth Road	Wandsworth	S.W.	Ek
Branksea Road	Fulham	S.W.	Ch
Branksome Road	Lambeth	S.W.	Fj
Branscombe Street	Lewisham	S.E.	Kj
Bransford Street	Kensington	W.	Cd
Branstone Road	Richmond		Ah
Branston Street	Kensington	W.	Cd
Brantome Place	St. Pancras	W.C.	Fd
Brashiers' Cottages	Wandsworth	S.W.	Fk
Brassey Square	Battersea	S.W.	Eh
Brathway Road	Wandsworth	S.W.	Dk
Bratley Street	Bethnal Green	E.	Hd
Bratton Street	Islington	N.	Fc
Braunton Mansions	Finsbury	E.C.	Gd
Bravington Mews	Paddington	W.	Cd
Bravington Road	Paddington	W.	Cd
Braxfield Road	Lewisham	S.E.	Jj
Braxted Park	Wandsworth	S.W.	Fm
Braxton Road	Wandsworth	S.W.	Fm
Brayard's Road	Camberwell	S.E.	Hh
Brayburne Avenue	Wandsworth	S.W.	Fh
Braydon Road	Hackney	N.	Ha
Bread Street	City of London	E.C.	Ge
Bread Street Hill	City of London	E.C.	Ge
Breakespeare's Road	Deptford	S.E.	Jj
Bream Street	Poplar	E.	Je
Bream's Buildings	City of London & Westminster	E.C.	Fe
Brechin Place	Kensington	S.W.	Dg
Brecknock Road	Islington and St. Pancras	N. & N.W.	Fb
Brecknock Terrace	Islington	N.	Fb
Brecknock Arms	St. Pancras	N.W.	Fc
Brecon Road	Fulham	W.	Cg
Breden Road	Lambeth	S.E.	Gh

STREET OR PLACE.	BOROUGH.	P.D.	MAP.
Breemount Road	Greenwich	S.E.	Lg
Breer Street	Fulham	S.W.	Dh
Breezer's Hill	Stepney	E.	He
Bremen Street	Hackney	E.	Hc
Brenda Road	Wandsworth	S.W.	Ek
Brent Bridge	Hendon U.D.	N.W.	Ca
Brent Bridge House	Hendon U.D.	N.W.	Ca
Brent Cottages	Wembley		Sd
Brent Gasworks	Hendon U.D.	N.W.	Ca
Brent Green	Hendon U.D.	N.W.	Ca
Brent Lodge	Hendon U.D.	N.W.	Ca
Brent, River	Brentford U.D.		Sg
Brent Road	West Ham	E.	Le
Brent Road	Woolwich		Mh
Brent Street	Brentford U.D.		Sg
Brent Street	Hendon U.D.	N.W.	Ca
Brentford-ait	Brentford U.D.		Sg
Brentford Market	Brentford U.D.		Ag
Brentford Road	Acton U.D. Brentford U.D. & Chiswick U.D.	W.	Ag
Brentford Sanatorium	Brentford U.D.	W.	Sf
Brentford Sewage Wks	Brentford U.D.		Sf
Brenthurst Road	Willesden U.D.	N.W.	Bc
Brenton Street	Stepney	E.	Je
Brents' Court	Southwark	S.E.	Gf
Brentview Road	Hendon U.D.	N.W.	Ba
Brett Road	Hackney	E.	Hc
Brettell Street	Southwark	S.E.	Gg
Bretton Street	Greenwich	S.E.	Jg
Brewer Place	Woolwich		Mg
Brewer Street	Chelsea	S.W.	Dg
Brewer Street	Westminster	S.W.	Ef
Brewer Street	Westminster	W.	Fe
Brewer Street	Finsbury	W.C.	Gd
Brewer Street	Woolwich		Mg
Brewer Street North	Finsbury	E.C.	Ge
Brewer's Buildings	Finsbury	E.C.	Gd
Brewer's Green	Westminster	S.W.	Ff
Brewer's Lane	City of London	E.C.	Ge
Brewer's Place	Westminster	S.W.	Fe
Brewer's Quay	City of London	E.C.	Hf
Brewer's Wharf	Stepney	E.	Hf
Brewer's Yard	Islington	N.	Gd
Brewery Road	Islington	N.	Fc
Brewery Road	Woolwich		Ng
Brewery Street	Wandsworth	S.W.	Eh
Brewery Yard	Poplar	E.	Jd
Brewhouse Lane	Greenwich	S.E.	Kg
Brewhouse Lane	Stepney	E.	Hf
Brewhouse Lane	Wandsworth	S.W.	Cj
Brewhouse Yard	Finsbury	E.C.	Ge
Brewster Road	Leyton U.D.	E.	Ka
Brewster Gardens	Hammersmith & Kensington	W.	Ce
Briant Street	Deptford	S.E.	Jh
Briar Road	Willesden U.D.	N.W.	Cb
Briar Walk	Wandsworth	S.W.	Bj
Briardale Gardens	Hampstead	N.W.	Db
Briardale Street	Hampstead	N.W.	Db
Briarwood Road	Wandsworth	S.W.	Fj
Brick Court	Bethnal Green	E.	Jd
Brick Lane	Bethnal Green & Stepney	E.	Hd He
Brick Lock	Hackney	E.	Jb
Brick Street	Westminster	W.	Ef
Brickfield Recreation Ground	Stepney	E.	Je
Brickfield Road	Poplar	E.	Kd
Brickhill Lane	City of London	E.C.	Ge
Bride Court	City of London	E.C.	Ge
Bride Lane	City of London	E.C.	Ge
Bride Street	Islington	N.	Fc
Bridewell Place	City of London	E.C.	Ga
Bridge Avenue	Hammersmith	W.	Cg
Bridge Cottages	Camberwell	S.E.	Gh
Bridge Cottages	Wandsworth	S.W.	El
Bridge Drive	Battersea	S.W.	Eg
Bridge House	Hendon U.D.	N.W.	Ca
Bridge Lane	Hendon U.D.	N.W.	Ca
Bridge Place	Westminster	S.W.	Eg
Bridge Place	Southwark	S.E.	Ge
Bridge Road	Beckenham U.D.		Jm
Bridge Road	Croydon R.D.	S.W.	Dm
Bridge Road	Hammersmith	W.	Cg
Bridge Road	Hampstead and St. Pancras	N.W.	Ec
Bridge Road	Poplar	E.	Jf
Bridge Road	West Ham	E.	Kd
Bridge Road	West Ham	E.	Ld
Bridge Road	Willesden U.D.	N.W.	Bc
Bridge Road West	Battersea	S.W.	Dh
Bridge Street	Greenwich	S.E.	Kg
Bridge Street	Hackney	E.	Jc
Bridge Street	Hammersmith	W.	Cg
Bridge Street	Richmond		Sj
Bridge Street	Stepney	E.	Jd
Bridge Street	Westminster	S.W.	Ff
Bridge Street	Willesden U.D.	N.W.	Dd
Bridge Terrace	Hammersmith	W.	Cg
Bridge Terrace	Stepney	E.	Je
Bridge End Road	Wandsworth	S.W.	Dj
Bridge House Road	Lewisham	S.E.	Jj
Bridges Place	Hackney	E.	Jc
Bridgewater Square	City of London	E.C.	Ge
Bridgewater Court	City of London	E.C.	Ge
Bridgewater Street	City of London	E.C.	Ge
Bridgewater Street	St. Pancras	N.W.	Fd
Bridgford Street	Wandsworth	S.W.	Dk
Bridgman Road	Acton U.D.	W.	Af
Bridle Lane	Westminster	W.	Fe
Bridport Place	Shoreditch	N.	Gd
Bridport Road	Croydon		Fm
Bridport Street	Marylebone	N.W.	Ed
Bridport Terrace	Wandsworth	S.W.	Fh
Bridson Street	Camberwell	S.E.	Hg
Brief Street	Lambeth	S.W.	Gh
Brier Walk	Wandsworth	S.W.	Cj
Brierley Road	Leyton U.D.	E.	Kb
Brierley Street	Bethnal Green	E.	Jd
Brig Street	Poplar	E.	Kg
Brigade Place	Greenwich	S.E.	Kh
Bright Street	Poplar	E.	Ke
Brightfield Road	Lewisham	S.E.	Kj
Brightlingsea Place	Stepney	E.	Je
Brighton Avenue	Walthamstow U.D.	E.	Ja
Brighton Dwellings	Bermondsey	S.E.	Hg
Brighton Grove	Deptford	S.E.	Jh
Brighton Road	East Ham	E.	Md
Brighton Road	Stoke Newington	N.	Hb
Brighton Road	West Ham	E.	Kd
Brighton Terrace	Battersea	S.W.	Eh
Brighton Terrace	Lambeth	S.W.	Fj
Brighton Terrace	Lewisham	S.E.	Jk
Bright's Buildings	Wandsworth	S.W.	Dj
Brightside Road	Lewisham	S.E.	Kj
Brightwell Crescent	Wandsworth	S.W.	El
Brill Street	Fulham	S.W.	Dh
Brilliant Street	Stepney	E.	Je
Brimington Road	Camberwell	S.E.	Hh
Brindle Mews	Greenwich	S.E.	Lj
Brindley Street	Deptford	S.E.	Jh
Brindley Street	Paddington	W.	De
Brisbane Road	Leyton U.D.	E.	Kb
Brisbane Cottages	Camberwell	S.E.	Gh
Brisbane Street	Camberwell	S.E.	Gh
Briscoe Road	Croydon R.D.	S.W.	Dl
Bristol Gardens	Paddington	W.	Dd
Bristol Mews	Paddington	W.	Dd
Bristol Road	East Ham	E.	Lc
Briston Grove	Hornsey	N.	Fa
Bristow Street	Shoreditch	N.	Gd
Britannia, The	St. Pancras	N.W.	Ec
Britannia Dwellings	Shoreditch	N.	Gd
Britannia Gardens	Shoreditch	N.	Gd
Britannia Place	Camberwell	S.E.	Hh
Britannia Road	Fulham	S.W.	Dg
Britannia Row	Islington	N.	Gc
Britannia Street	St. Pancras	W.C.	Fd
Britannia Street	Shoreditch	N.	Gd
Britannia Terrace	Camberwell	S.E.	Hh
Britannia Theatre	Shoreditch	N.	Gd
Britannia Wharf	Fulham	S.W.	Dh
British Grove	Hammersmith	W.	Bg
British Street	Poplar	E.	Jd

STREET OR PLACE.	BOROUGH.	P.D.	MAP.
Brook's Road	West Ham	E.	Ld
Brook's Wharf	City of London	E.C.	Ge
Brook's Wharf Lane	City of London	E.C.	Ge
Brook's Yard	City of London	E.C.	Ge
Brooksby Street	Islington	N.	Gc
Brooksby's Mews	Islington	N.	Gc
Brooksby's Walk	Hackney	E.	Jc
Brookside Road	Islington	N.	Fb
Brooksville Avenue	Willesden U.D.	N.W.	Cc
Brookville Road	Fulham	S.W.	Cg
Brookwood Road	East Ham	E.	Md
Brookwood Road	Wandsworth	S.W.	Ck
Broom Hall	Teddington U.D.		Sl
Broom House	Fulham	S.W.	Dh
Broom Road	Teddington U.D.		Sl
Broom Wood	Bromley R.D.		Om
Broomfield Hill	Ham U.D.		Al
Broomfield Road	Richmond		Ah
Broomhall	Woolwich		Mh
Broomhill Road	Wandsworth	S.W.	Dj
Broomhouse Lane	Fulham	S.W.	Ch
Broomhouse Road	Fulham	S.W.	Dh
Broom's Alley	Fulham	S.W.	Dh
Broomsleigh Street	Hampstead	N.W.	Dc
Broomwood Gardens	Battersea	S.W.	Ej
Broomwood Park	Battersea	S.W.	Ej
Broomwood Road	Battersea	S.W.	Ej
Broseley Grove	Lewisham	S.E.	Jl
Brougham Road	Acton U.D.	W.	Ae
Brougham Road	Hackney and Shoreditch	E.	Hc
Brougham Street	Battersea	S.W.	Eh
Broughton Lodge	Hornsey	N.	Fa
Broughton Road	Ealing	W.	Se
Broughton Road	Fulham	S.W.	Dh
Broughton Road	Stoke Newington	N.	Gb
Broughton Road Approach	Fulham	S.W.	Dh
Broughton Street	Battersea	S.W.	Eh
Brouncher Road	Acton U.D.	W.	Af
Brown Place	Hackney	E.	Hb
Brown Street	Westminster	W.	Ee
Brown Street	Marylebone	W.	Ee
Brownhill Gardens	Lewisham	S.E.	Kk
Brownhill Market	Lewisham	S.E.	Kk
Brownhill Road	Lewisham	S.E.	Kk
Browning Road	Leyton U.D.	E.	La
Browning's Court	Stepney	E.	He
Brownlow Mews	St. Pancras	W.C.	Fd
Brownlow Road	Shoreditch	E.	Hc
Brownlow Road	West Ham	E.	Lb
Brownlow Road	Willesden U.D.	N.W.	Bc
Brownlow Street	Holborn	W.C.	Fe
Brownlow Street	Shoreditch	E.	Hc
Brown's Court	Chelsea	S.W.	Dg
Brown's Court	Marylebone	W.	Ee
Brown's Grounds	Camberwell	S.E.	Jj
Brown's Road	West Ham	E.	Ld
Brown's Terrace	Camberwell	S.E.	Gg
Brownswood Road	Stoke Newington	N.	Gb
Broxash Street	Battersea	S.W.	Ej
Broxbourn Road	Leyton U.D.	E.	Lb
Broxholm Road	Lambeth and Wandsworth	S.E.	Fl
Broxted Road	Lewisham	S.E.	Jk
Bruce Road	Croydon R.D.		Em
Bruce Road	Poplar	E.	Kd
Bruce Road	Willesden U.D.	N.W.	Bc
Brunan Road	Wandsworth	S.W.	Ej
Brune Place	Southwark	S.E.	Gg
Brunel Road	Bermondsey	S.E.	Hf
Brunel Street	West Ham	E.	Le
Brunner Road	Walthamstow U.D.	E.	Ja
Brunswick Close	Finsbury	E.C.	Ge
Brunswick Court	Bermondsey	S.E.	Hf
Brunswick Gardens	Kensington	W.	Df
Brunswick Mews	Kensington	W.	De
Brunswick Place	Marylebone	N.W.	Ed
Brunswick Place	Penge U.D.	S.E	Hm
Brunswick Place	Shoreditch	N.	Gd
Brunswick Place	Southwark	S.E	Gf
Brunswick Place	Stepney	E.	Je
Brunswick Road	Bexley		Oj
Brunswick Road	Camberwell	S.E.	Gh
Brunswick Road	Islington	N.	Fb
Brunswick Road	Kingston		Am
Brunswick Road	Leyton U.D.	E.	Ka
Brunswick Road	Poplar	E.	Ke
Brunswick Road	Tottenham U.D.	N.	Ha
Brunswick Row	Southwark	S.E.	He
Brunswick Square	Camberwell	S.E.	Gh
Brunswick Square	St. Pancras	W.C.	Fd
Brunswick Square	Shoreditch	E.	Hd
Brunswick Street	Hackney	E.	Jc
Brunswick Street	Islington	N.	Fd
Brunswick Street	Poplar	E.	Kf
Brunswick Street	Shoreditch	E.	Hd
Brunswick Street	Southwark	S.E.	Gf
Brunswick Street	Stepney	E.	He
Brunswick Street	Walthamstow U.D.	E.	Ka
Brunswick Terrace	Hammersmith	W.	Cg
Brunswick Terrace	Southwark	S.E.	Gg
Brunton's Cottages	Stepney	E.	Je
Brunton's Place	Stepney	E.	Je
Brushfield Street	City of London and Stepney	E.	He
Brushwood Road	Richmond		Ag
Brussels Road	Battersea	S.W.	Dj
Bruton Mews North	Westminster	W.	Ee
Bruton Mews South	Westminster	W.	Ee
Bruton Street	Westminster	W.	Ee
Bryan Mews	Islington	N.	Fd
Bryan Place	Islington	N.	Fd
Bryan Street	Islington	N.	Fd
Bryan Vale	Islington	N.	Fd
Bryan's Court	Bermondsey	S.E.	Jf
Bryanston Mansions	Marylebone	N.	Ee
Bryanston Mews	Marylebone	W.	Ee
Bryanston Mews East	Marylebone	W.	Ee
Bryanston Mews West	Marylebone	W.	Ee
Bryanston Place	Marylebone	W.	Ee
Bryanston Road	Hornsey	N.	Fa
Bryanston Square	Marylebone	W.	Ee
Bryanston Street	Marylebone	W.	Ee
Bryant Road	Croydon	S.W.	Fm
Bryant Street	West Ham	E.	Kc
Bryantwood Road	Islington	N.	Gb
Bryden Grove	Lewisham	S.E.	Jl
Brydges Road	West Ham	E.	Kc
Bryett Road	Islington	N.	Fb
Bryner Road	Camberwell	S.E.	Hg
Brynmaer Road	Battersea	S.W.	Eh
Buccleuch Cottages	Hackney	E.	Ha
Buccleuch House	Richmond		Sj
Buccleuch Place	Hackney	E.	Ha
Buccleugh Road	Lambeth	S.E.	Gk
Buchan Road	Camberwell	S.E.	Hh
Buchanan Gardens	Willesden U.D.	N.W.	Bd
Bucharest Road	Wandsworth	S.W.	Dk
Buck Hill Walk	Westminster	W.	De
Buck House	Walthamstow U.D.	E.	Ka
Buck Lane	Walthamstow U.D.	E.	Ka
Buck Lane	Kingsbury U.D.		Ba
Buck Street	St. Pancras	N.W.	Ec
Buckenham Place	Southwark	S.E.	Gf
Buckenham Square	Southwark	S.E.	Gf
Buckenham Street	Southwark	S.E.	Hg
Buckeridge Street	Stepney	E.	Jd
Buckhold Road	Wandsworth	S.W.	Dj
Buckhurst Street	Bethnal Green	E.	He
Buckingham Avenue	Croydon		Fm
Buckingham Court	Westminster	S.W.	Ff
Buckingham Gate	Westminster	S.W.	Ff
Buckingham Gate Mansions	Westminster	S.W.	Ef
Buckingham Palace	**Westminster**	S.W.	Ef
Buckingham Palace Gardens	Westminster	S.W.	Fg
Buckingham Palace Mansions	Westminster	S.W.	Fg
Buckingham Palace Road	Westminster	S.W.	Fg

STREET OR PLACE.	BOROUGH.	P.D.	MAP.
Camberwell Road	Camberwell and Southwark	S.E.	Gg
Camberwell Workhouse	Camberwell	S.E.	Gh
Camberwell New Road	Camberwell and Lambeth	S.E.	Gh Gg
Camberwell New Road Station	**Camberwell**	S.E.	Gh
Camberwell Station Road	Camberwell	S.E.	Gh
Camborn Road	Wandsworth	S.W.	Dk
Cambray Road	Wandsworth	S.W.	Fk
Cambria Road	Lambeth	S.E.	Gh
Cambria Street	Fulham	S.W.	Dh
Cambrian Road	Leyton U.D.	E.	Ja
Cambrian Road	Richmond		Sj
Cambridge Avenue	Merton U.D.	S.W.	Cm
Cambridge Avenue	Willesden U.D.	N.W.	Dd
Cambridge Barracks	Woolwich		Mg
Cambridge Buildings	Bethnal Green	E.	He
Cambridge Buildings	Westminster	S.W.	Fg
Cambridge Circus	Bethnal Green	E.	Hd
Cambridge Circus	Holborn and Westminster	W.C.	Fe
Cambridge Cottages	Woolwich		Mg
Cambridge Gardens	Kensington	W.	Ce
Cambridge Gardens	Willesden U.D.	N.W.	Dd
Cambridge Gate	St. Pancras	N.W.	Ed
Cambridge Gate Mews	St. Pancras	N.W.	Ed
Cambridge Grove	Penge U.D.	S.E.	Hm
Cambridge Grove Road	Kingston		Am
Cambridge Heath Station	**Bethnal Green**	E.	Hd
Cambridge House	Hampton Wick U.D.		Sm
Cambridge Lodge	The Maldens and Coombe U.D.		Am
Cambridge Mansions	Marylebone	N.W.	Ee
Cambridge Mews	Paddington	W.	Ee
Cambridge Music Hall	**Stepney**	E.	He
Cambridge Park	Leyton U.D. and Wanstead U.D.	E.	La
Cambridge Park	Twickenham U.D.		Sj
Cambridge Park Gardens	Twickenham U.D.		Sj
Cambridge Place	Bethnal Green	E.	Hd
Cambridge Place	Kensington	W.	Df
Cambridge Place	Paddington	W.	De
Cambridge Place	St. Pancras	N.W.	Ed
Cambridge Place	Willesden U.D.	N.W.	Dd
Cambridge Place	Woolwich		Ng
Cambridge Road	Barnes U.D.	S.W.	Bh
Cambridge Road	Battersea	S.W.	Eh
Cambridge Road	Bethnal Green and Stepney	E.	Hd
Cambridge Road	Bromley		Lm
Cambridge Road	Chiswick U.D.	W.	Ag
Cambridge Road	Croydon and Penge U.D.	S.E.	Hm
Cambridge Road	Foots Cray U.D.		Nl
Cambridge Road	Hammersmith	W.	Bg
Cambridge Road	Kingston and The Maldens and Coombe U.D.		Am
Cambridge Road	Lewisham	S.E.	Lj
Cambridge Road	Leyton U.D. and Wanstead U.D.	E.	La
Cambridge Road	Richmond		Ag
Cambridge Road	Twickenham U.D.		Sj
Cambridge Road	Walthamstow U.D.	E.	Ja
Cambridge Road	Willesden U.D.	N.W.	Dd
Cambridge Road	Wimbledon	S.W.	Bm
Cambridge Row	St. Pancras	N.W.	Fc
Cambridge Square	Paddington	W.	Ee
Cambridge Street	Bethnal Green	E.	Hd
Cambridge Street	Camberwell and Southwark	S.E.	Gg
Cambridge Street	Westminster	S.W.	Eg
Cambridge Street	Hammersmith	W.	Cf
Cambridge Street	Paddington	W.	Ee
Cambridge Street	St. Pancras	N.W.	Fd
Cambridge Terrace	Greenwich		Lg
Cambridge Terrace	Islington	N.	Gd
Cambridge Terrace	Lewisham	S.E.	Jk
Cambridge Terrace	Paddington	W.	De
Cambridge Terrace	St. Pancras	N.W.	Ed
Cambridge Yard	Camberwell	S.E.	Gg
Cambridge Yard	St. Pancras	N.W.	Fd
Cambrook Street	Camberwell	S.E.	Hg
Camrose Street	Woolwich		Og
Camden Avenue	Camberwell	S.E.	Hh
Camden Buildings	Islington	N.	Fc
Camden Corner	Chislehurst U.D.		Mm
Camden Gardens	Hammersmith	W.	Cf
Camden Gardens	St. Pancras	N.W.	Ec
Camden Grove	Camberwell	S.E.	Hh
Camden Grove	Chislehurst U.D.		Mm
Camden Grove North	Camberwell	S.E.	Hg
Camden Hill	Chislehurst U.D.		Mm
Camden Hill Road	Lambeth	S.E.	Gl
Camden Mews	St. Pancras	N.W.	Fc
Camden Park Road	Chislehurst U.D.		Mm
Camden Park Road	St. Pancras	N.W.	Fc
Camden Park Wood	Chislehurst U.D.		Mm
Camden Passage	Islington	N.	Gd
Camden Place	Chislehurst U.D.		Mm
Camden Road	Islington and St. Pancras	N. N.W.	Fc Fc
Camden Road	Walthamstow U.D.	E.	Ja
Camden Road	Wanstead U.D.	E.	La
Camden Road Mews	St. Pancras	N.W	Fc
Camden Road Station	**St. Pancras**	N.W.	Fc
Camden Row	Lewisham	S.E.	Kh
Camden Square	Camberwell	S.E.	Hh
Camden Square	St. Pancras	N.W.	Fc
Camden Street	Bethnal Green	E.	Hd
Camden Street	Camberwell	S.E.	Hh
Camden Street	Islington	N.	Gd
Camden Street	St. Pancras	N.W.	Fc
Camden Theatre	**St. Pancras**	N.W.	Fd
Camden Wood	Chislehurst U.D.		Mm
Camden Town Station (Hampstead Tube)	**St. Pancras**	N.W.	Ec
Camden Town Station (N. L. Ry.)	**St. Pancras**	N.W.	Fc
Camel Place	Shoreditch	N.	Hd
Camel Yard	Shoreditch	N.	Hd
Camelia Street	Lambeth	S.W.	Fh
Camelot Street	Southwark	S.E.	Gf
Camels, The	Wandsworth	S.W.	Ck
Camera Gardens	Chelsea	S.W.	Dg
Camera Square	Chelsea	S.W.	Dg
Cameron Place	Stepney	E.	He
Cameron Street	East Ham	E.	Me
Camilla Road	Bermondsey	S.E.	Hg
Camlet Street	Bethnal Green	E.	Hd
Camomile Street	City of London	E.C.	Ge
Camp View	Wimbledon	S.W.	Bl
Campana Road	Fulham	S.W.	Dh
Campbell Cottages	Woolwich		Mg
Campbell Road	East Ham	E.	Md
Campbell Road	Islington	N.	Gb
Campbell Road	Leyton U.D. and West Ham	E.	Lb
Campbell Road	Poplar	E.	Kd
Campbell Road	West Ham	E.	Mf
Campbell Street	Fulham	W.	Cg
Campbell Street	Paddington	W.	Dd
Campdale Road	Islington	N.	Fb
Campden Buildings	Hampstead	N.W.	Db
Campden Grove	Kensington	W.	Df
Campden Hill	Kensington	W.	Df
Campden Hill Gardens	Kensington	W.	Df
Campden Hill Place	Kensington	W.	Cf
Campden Hill Road	Kensington	W.	Df
Campden Hill Square	Kensington	W.	Cf
Campden Hill Terrace	Kensington	W.	Df
Campden Hill West	Kensington	W.	Df
Campden House	Kensington	W.	Df
Campden House Court	Kensington	W.	Df
Campden House Mews	Kensington	W.	Df
Campden House Road	Kensington	W.	Df
Campden House Ter.	Kensington	W.	Df

STREET OR PLACE.	BOROUGH.	P.D.	MAP.
Carlton Mews	Kensington	W.	Ce
Carlton Mews	Paddington	W.	Cd
Carlton Mews	Paddington	N.W.	Dd
Carlton Mews	Wandsworth	S.W.	Cj
Carlton Place	Bethnal Green	E.	Hd
Carlton Road	Acton U.D.	W.	Af
Carlton Road	Chislehurst U.D.		Nl
Carlton Road	Deptford	S.E.	Jj
Carlton Road	Ealing	W.	Se
Carlton Road	East Ham	E.	Md
Carlton Road	Hampstead	N.W.	Eb
Carlton Road	Hornsey	N.	Ga
Carlton Road	Leyton U.D.	E.	La
Carlton Road	St. Pancras	N.W.	Ec
Carlton Road	Stepney	E.	Jd
Carlton Road	Wandsworth	S.W.	Cj
Carlton Square	Camberwell	S.E.	Hh
Carlton Square	Stepney	E.	Jd
Carlton Street	St. Pancras	N.W.	Ec
Carlton Street	West Ham	E.	Le
Carlton Street	Westminster	S.W.	Fe
Carlton Tavern	St. Pancras	N.W.	Ec
Carlton Terrace	Lewisham	S.E.	Hl
Carlton Terrace	Paddington	W.	Cd
Carlton Vale	Paddington and Willesden U.D.	N.W.	Dd
Carlton Villas	Paddington	W.	Dd
Carlwell Street	Wandsworth	S.W.	Dl
Carlyle Avenue	Willesden U.D.	N.W.	Ac
Carlyle Gardens	Chelsea	S.W.	Eg
Carlyle Mansions	Chelsea	S.W.	Eg
Carlyle Pier	Chelsea	S.W.	Dg
Carlyle Road	Ealing	W.	Sf
Carlyle Square	Chelsea	S.W.	Dg
Carmalt Gardens	Wandsworth	S.W.	Cj
Carmarthen Place	Bermondsey	S.E.	Gf
Carmarthen Street	Islington	N.	Gc
Carmelite Convent	Kensington	W.	Ce
Carmelite Street	City of London	E.C.	Ge
Carmen Street	Poplar	E.	Ke
Carminia Road	Wandsworth	S.W.	Ek
Carnaby Street	Westminster	W.	Fe
Carnac Street	Lambeth	S.E.	Gl
Carnarvon Road	Leyton U.D.	E.	Ka
Carnforth Road	Hammersmith	W.	Cg
Carnwath House	Fulham	S.W.	Dh
Carnwath Road	Fulham	S.W.	Dh
Caroline Cottages	Hackney	E.	Hb
Caroline Court	Islington	N.	Fb
Caroline Mews	Paddington	W.	De
Caroline Place	Chelsea	S.W.	Ef
Caroline Place	Finsbury	E.C.	Fd
Caroline Place	Fulham	W.	Cg
Caroline Place	Hammersmith	W.	Cf
Caroline Place	Holborn	W.C.	Fd
Caroline Place	Islington	N.	Gc
Caroline Place	Paddington	W.	De
Caroline Place	Shoreditch	E.	Hd
Caroline Road	West Ham	E.	Ld
Caroline Road	Wimbledon	S.W.	Cm
Caroline Street	Camberwell	S.E.	Hg
Caroline Street	Westminster	S.W.	Eg
Caroline Street	Hackney	E.	Hb
Caroline Street	Holborn	W.C.	Fe
Caroline Street	St. Pancras	N.W.	Fc
Caroline Street	Stepney	E.	Je
Carpenter Street	Battersea	S.W.	Eh
Carpenter Street	Westminster	S.W.	Ff
Carpenter Street	Westminster	W.	Ee
Carpenter St. Bldgs.	Westminster	W.	Ee
Carpenter's Place	Camberwell	S.E.	Gj
Carpenter's Road	Poplar	E.	Jc
Carpenter's Road	West Ham	E.	Kc
Carr Street	Stepney	E.	Je
Carrier Street	Holborn	W.C.	Fe
Carrington Mews	Westminster	W.	Ef
Carrol Place	St. Pancras	N.W.	Eb
Carron Wharf	Stepney	E.	Hf
Carroun Road	Lambeth	S.W.	Fg
Carson Road	Lambeth	S.E.	Gk
Carson Road	West Ham	E.	Le
Carswell Road	Lewisham	S.E.	Kk
Carter Lane	City of London	E.C.	Ge
Carter Street	Bethnal Green & Stepney	E.	He
Carter Street	City of London	E.	He
Carter Street	Southwark	S.E.	Gg
Carter Street	Stepney	E.	Jd
Carteret Mews	Westminster	S.W.	Ff
Carteret Street	Westminster	S.W.	Ff
Carthew Road	Hammersmith	W.	Bf
Carthew Terrace	Hammersmith	W.	Bf
Carthew Villas	Hammersmith	W.	Bf
Carthusian Street	City of London and Finsbury	E.C.	Ge
Carting Lane	Westminster	W.C.	Fe
Carton Street	Wandsworth	S.W.	Dj
Cartwright Street	Stepney	E.	He
Carvary Road	West Ham	E.	Le
Carville Hall	Brentford U.D.		Sg
Carwell Road	Lewisham	S.E.	Kk
Cary Road	Leyton U.D.	E.	Lb
Carysfort Road	Hornsey	N.	Fa
Carysfort Road	Stoke Newington	N.	Gb
Casella Road	Deptford	S.E.	Jh
Casella Road East	Deptford	S.E.	Jh
Casewick Road	Lambeth	S.E.	Gl
Casilon Road	Woolwich		Og
Casket Street	Bethnal Green	E.	Hd
Caspian Street	Camberwell	S.E.	Gg
Cassala Road	Battersea	S.W.	Eh
Casselden Road	Willesden U.D.	N.W.	Bc
Cassidy Road	Fulham	S.W.	Dh
Cassilda Road	Woolwich		Og
Cassillis Road	Twickenham U.D.		Sj
Cassimer Terrace	Hackney	E.	Hb
Cassiobury Road	Walthamstow U.D.	E.	Ja
Cassland Crescent	Hackney	E.	Jc
Cassland Road	Hackney	E.	Jc
Cason Street	Stepney	E.	He
Castalia Street	Poplar	E.	Kf
Castellain Road	Paddington	W.	Dd
Castello Avenue	Wandsworth	S.W.	Cj
Casterton Street	Hackney	E.	Hc
Casterton Terrace	Hackney	E.	Hc
Castigny Place	Finsbury	E.C.	Gd
Castlands Road	Lewisham	S.E.	Jl
Castle, The	Hendon U.D.	N.W.	Db
Castle Court	City of London	E.C.	Ge
Castle Hill House	Ealing	W.	Se
Castle Hill Park Road	Ealing	W.	Sd
Castle House	Woolwich		Mh
Castle Lane	Westminster	S.W.	Ff
Castle Mews	Marylebone	W.	Fe
Castle Place	Finsbury	E.C.	Gd
Castle Place	St. Pancras	N.W.	Ec
Castle Road	St. Pancras	N.W.	Ec
Castle Square	Camberwell	S.E.	Hh
Castle Street	Battersea	S.W.	Dh
Castle Street	City of London	E.C.	Ge
Castle Street	East Ham	E.	Ld
Castle Street	Finsbury	E.C.	Gd
Castle Street	Hackney & Stoke Newington	N.	Hc
Castle Street	Holborn	E.C.	Ge
Castle Street	Holborn and Westminster	W.C.	Fe
Castle Street	Southwark	S.E.	Gf
Castle Street	West Ham	E.	Kc
Castle Street East	Marylebone	W.	Fe
Castle Wood	Woolwich		Mh
Castle Yard	City of London	E.C.	Ge
Castlebar	Lewisham	S.E.	Hk
Castlebar Court	Ealing	Sd	W.
Castlebar Crescent	Ealing	Se	W.
Castlebar Hill	Ealing	Se	W.
Castlebar House	Ealing	Se	W.
Castlebar Mews	Ealing	Sd	W.
Castlebar Park	Ealing	Sd	W.
Castlebar Road	Ealing	Se	W.
Castle Baynard Wharf	City of London	Ge	E.C.
Castledine Road	Penge U.D.	Hm	S.E.
Castlegate	Richmond	Sh	
Castlehill Lodge	Croydon	Gm	S.E.
Castlemain Road	Camberwell	Hg	S.E.

STREET OR PLACE.	BOROUGH.	P.D.	MAP.
Challis Court	Stepney	E.	He
Chalmers Street	Battersea and Wandsworth	S.W.	Eh
Chaloner Road	Fulham	W.	Cg
Chalsey Road	Lewisham	S.E.	Jj
Chalton Street	St. Pancras	N.W.	Fd
Chamber Court	Stepney	E.	He
Chamber Street	Stepney	E.	He
Chamberlain's Wharf	Bermondsey	S.E.	Gf
Chamberlayne Wood Road	Willesden U.D.	N.W.	Cd
Chamber's Lane	Willesden U.D.	N.W.	Cc
Chambord Street	Bethnal Green	E.	Hd
Champion Crescent	Lewisham	S.E.	Jl
Champion Grove	Camberwell	S.E.	Gh
Champion Hill	Camberwell	S.E.	Gh
Champion Park	Camberwell	S.E.	Gh
Champion Park	Lewisham	S.E.	Jl
Champion Park Road	Lewisham	S.E.	Jl
Champion Terrace	Greenwich		Lg
Champion Yard	Shoreditch	E.	Hd
Champion's Alley	Westminster	S.W.	Ff
Chance Street	Bethnal Green	E.	Hd
Chancel Street	Southwark	S.E.	Gf
Chancellor Road	Lambeth	S.E.	Gk
Chancellor Street	Hammersmith	W.	Cg
Chancellor's Road	Fulham and Hammersmith	W.	Cg
Chancelot Road	Woolwich		Og
Chancery Lane	Beckenham U.D.		Km
Chancery Lane	City of London, Holborn and Westminster	W.C.	Fe
Chancery Lane Court	Westminster	W.C.	Fe
Chancery Lane Station	**Holborn**	W.C.	Fe
Chandler Street	Stepney	E.	Hf
Chandos Avenue	Ealing	W.	Sf
Chandos Metal Works	Acton U.D.	N.W.	Bd
Chandos Road	Willesden U.D.	N.W.	Cb
Chandos Road East	West Ham	E.	Kc
Chandos Road West	West Ham	E.	Kc
Chandos Street	Westminster	W.C.	Fe
Chandos Street	Marylebone	W.	Ee
Change Alley	City of London	E.C.	Ge
Change Alley	Lambeth	S.E.	Gl
Channelsea River	West Ham	E.	Kd
Channelsea Road	West Ham	E.	Kd
Channelsea Road South	West Ham	E.	Kd
Channelsea Street	West Ham	E.	Kd
Chant Square	West Ham	E.	Kc
Chant Street	West Ham	E.	Kc
Chantry Road	Lambeth	S.W.	Fh
Chapel Alley	Brentford U.D.		Sg
Chapel Court	Westminster	W.	Ee
Chapel Grove	St. Pancras	N.W.	Fd
Chapel House	Bexley		Ok
Chapel House	Ealing	W.	Se
Chapel House Street	Poplar	E.	Kg
Chapel Lane	Woolwich		Mk
Chapel Mews	Westminster	S.W.	Ef
Chapel Passage	Bethnal Green	E.	Hd
Chapel Place	Kensington and Westminster	S.W.	Ef
Chapel Place	Marylebone	W.	Ee
Chapel Place	Southwark	S.E.	Gf
Chapel Place Mews	Westminster	S.W.	Ef
Chapel Place North	Westminster	W.	Ee
Chapel Place South	Westminster	W.	Ee
Chapel Road	Bexley U.D.		Oh
Chapel Road	Croydon R.D.		Dm
Chapel Road	Ealing	W.	Se
Chapel Road	Hackney	N.	Ha
Chapel Road	Kensington	W.	Ce
Chapel Road	Lambeth	S.E.	Gl
Chapel Road	Twickenham U.D.		Sj
Chapel Row	Bexley U.D.		Nk
Chapel Row	Finsbury	E.C.	Gd
Chapel Street	Bethnal Green	E.	Hd
Chapel Street	Chelsea	S.W.	Dg
Chapel Street	City of London	E.C.	Ge
Chapel Street	Westminster	S.W.	Ef
Chapel Street	Finsbury	N.	Gd
Chapel Street	Hammersmith	W.	Bg
Chapel Street	Holborn	W.C.	Fe
Chapel Street	Lambeth	S.W.	Fh
Chapel Street	Marylebone	N.W.	Ee
Chapel Street	St. Pancras	N.W.	Fd
Chapel Street	Stepney	E.	He
Chapel Street	West Ham	E.	Kc
Chapel Street	West Ham	E.	Lc
Chapel Street	West Ham	E.	Lf
Chapel Street	Woolwich		Mg
Chapel Yard	Stepney	E.	Je
Chaplin Road	Lewisham	S.E.	Hk
Chaplin Road	Wembley U.D.		Sc
Chaplin Road	West Ham	E.	Ld
Chaplin Road	Willesden U.D.	N.W.	Bc
Chapman Place	Stepney	E.	He
Chapman Road	Hackney	E.	Jc
Chapman Road	West Ham	E.	Ld
Chapter Chambers	Westminster	S.W.	Fg
Chapter Road	Southwark	S.E.	Gg
Chapter Road	Willesden U.D.	N.W.	Bc
Chapter Road	Willesden U.D.	N.W.	Cc
Chapter Street	Westminster	S.W.	Fg
Chapter Terrace	Southwark	S.E.	Gg
Chardmore Road	Hackney	N.	Ha
Charford Road	Leyton U.D.	E.	Kb
Chargeable Lane	West Ham	E.	Le
Chargeable Street	West Ham	E.	Le
Charlow Brewery	Stepney	E.	Je
Charing Cross	Westminster	S.W.	Ff
Charing Cross Mans.	Westminster	W.C.	Ff
Charing Cross Pier	Westminster	W.C.	Ff
Charing Cross Road	Holborn & Westminster	W.C.	Fe
Charing Cross Station	**Westminster**	W.C.	Ff
Chariot Mews	Paddington	W.	Ee
Charlbert Street	Marylebone	N.W.	Ed
Charlbury Grove	Ealing	W.	Se
Charlcroft Road	Lewisham	S.E.	Kj
Charlemont Road	East Ham	E.	Md
Charles Buildings	Westminster	W.C.	Fe
Charles Court	Finsbury	E.C.	Gd
Charles Lane	Marylebone	N.W.	Ed
Charles Mews	Paddington	W.	De
Charles Place	Deptford	S.E.	Jh
Charles Place	Hackney	N.	Hc
Charles Place	Stepney	E.	Je
Charles Square	Shoreditch	N.	Gd
Charles Street	Barnes U.D.	S.W.	Eh
Charles Street	Chelsea	S.W.	Ef
Charles Street	City of London	E.C.	Ge
Charles Street	Ealing	W.	Se
Charles Street	Finsbury and Holborn	E.C.	Ge
Charles Street	Hackney	E.	Hc
Charles Street	Islington	N.	Fc
Charles Street	Islington	N.	Gc
Charles Street	Kensington	W.	Cf
Charles Street	Kensington	W.	Df
Charles Street	Lewisham	S.E.	Hl
Charles Street	Marylebone	N.W.	Dd
Charles Street	Poplar	E.	Ke
Charles Street	St. Pancras	N.W.	Fd
Charles Street	Shoreditch	N.	Gd
Charles Street	Southwark	S.E.	Gf
Charles Street	Stepney	E.	He
Charles Street	West Ham	E.	Ld
Charles Street	West Ham	E.	Lf
Charles Street	Westminster	W.	Ef
Charles Street	Westminster	S.W.	Ef
Charles Street	Westminster	W.C.	Fe
Charles Street	Westminster	S.W.	Ff
Charles Street	Woolwich		Mg
Charles Street	Woolwich		Ng
Charles Street Mews	Deptford	S.E.	Jh
Charleston Street	Southwark	S.E.	Gg
Charlesworth Street	Islington	N.	Fc
Charleville Circus	Beckenham U.D. and Lewisham	S.E.	Hl
Charleville Mansions	Fulham	W.	Cg
Charleville Road	Fulham	W.	Cg
Charlewood Road	Wandsworth	S.W.	Cj
Charlmont Road	Wandsworth	S.W.	El
Charlotte Cottages	Lambeth	S.E.	Gl

STREET OR PLACE.	BOROUGH.	P.D.	MAP.
Charlotte Court	City of London	E.C.	Ge
Charlotte Mews	Holborn	W.C.	Fe
Charlotte Mews West	Marylebone	W.	Fe
Charlotte Place	Islington	N.	Dj
Charlotte Place	Wandsworth	S.W.	Fc
Charlotte Place	Paddington	W.	Dd
Charlotte Place	Woolwich		Mg
Charlotte Street	Bermondsey and Camberwell	S.E.	Hg
Charlotte Street	Bethnal Green	E.	Hd
Charlotte Street	Chelsea	S.W.	Ef
Charlotte Street	Holborn	W.C.	Fe
Charlotte Street	Islington	N.	Fd
Charlotte Street	Marylebone and St. Pancras	W.	Fe
Charlotte Street	Shoreditch	N.	Ga
Charlotte Street	Southwark	S.E.	Gf
Charlotte Street	West Ham	E.	Ke
Charlotte Street	West Ham	E.	Le
Charlotte Street	Woolwich		Ng
Charlotte Terrace	Islington	N.	Fd
Charlotte Terrace	Kensington	W.	Ce
Charlton Cemetery	Greenwich		Mg
Charlton Crescent	Islington	N.	Gd
Charlton House	Woolwich		Lg
Charlton Lane	Greenwich		Lg
Charlton Lane	Greenwich		Kh
Charlton Place	Islington	N.	Gd
Charlton Park Terrace	Greenwich		Lg
Charlton Road	Greenwich	S.E.	Lg
Charlton Road	West Ham	E.	Ld
Charlton Road	Willesden U.D	N.W.	Bc
Charlton Station	**Woolwich**		Lg
Charlton Street	Marylebone	W.	Ee
Charlton Terrace	Greenwich	S.E.	Kg
Charlton Ballast Wharf	Greenwich		Lf
Charlton King's Road	St. Pancras	N.W.	Fb
Charlwood Place	Westminster	S.W.	Fg
Charlwood Road	Wandsworth	S.W.	Cj
Charlwood Street	Westminster	S.W.	Fg
Charlwood Terrace	Wandsworth	S.W.	Cj
Charnock Road	Hackney	E.	Hb
Charrington Street	St. Pancras	N.W.	Fd
Charsley Road	Lewisham	S.E.	Kk
Charterhouse	**Finsbury**	E.C.	Gd
Charterhouse Bldgs.	Finsbury	E.C.	Gd
Charterhouse Mews	Finsbury	E.C.	Ge
Charterhouse Square	Finsbury	E.C.	Ge
Charterhouse Street	City of London, Finsbury and Holborn	E.C.	Ge
Charteris Road	Islington	N.	Ga
Charteris Road	Willesden. U.D.	N.W.	Cc
Chartfield Avenue	Wandsworth	S.W.	Cj
Chartham Street	Lambeth	S.E.	Gl
Chartley Place	Lambeth	S.E.	Ff
Chase, The	Wandsworth	S.W.	Ej
Chasefield Road	Wandsworth	S.W.	El
Chaseley Street	Stepney	E.	Je
Chaston Street	St. Pancras	N.W.	Ec
Chatham Avenue	Shoreditch	N.	Gd
Chatham Buildings	City of London	E.C.	Ge
Chatham Place	Hackney	E.	Hc
Chatham Place	Southwark	S.E.	Gg
Chatham Road	Battersea	S.W.	Ej
Chatham Road	Kingston		Am
Chatham Street	Battersea	S.W.	Eh
Chatham Street	Southwark	S.E.	Gg
Chatsworth Avenue	Merton U.D.	S.W.	Cm
Chatsworth Gardens	Acton U.D.	W.	As
Chatsworth Road	Hackney	E.	Jb
Chatsworth Road	Lambeth	S.E.	Gk
Chatsworth Road	West Ham	E.	Lc
Chatsworth Road	Willesden U.D.	N.W.	Cc
Chatteris Road	Southwark	S.E.	Gf
Chatteris Square	Southwark	S.E.	Gf
Chatterton Road	Islington	N.	Gb
Chatto Road	Battersea	S.W.	Ej
Chaucer Road	Acton U.D.	W.	Ae
Chaucer Road	Lambeth	S.E.	Gj
Chaucer Road	Wanstead U.D.	E.	La
Chaucer Road	West Ham	E.	Lc
Chauntler Road	West Ham	E.	Le

STREET OR PLACE.	BOROUGH.	P.D.	MAP.
Cheam Place	Camberwell	S.E.	Gg
Cheapside	City of London	E.C.	Ge
Chelmer Road	Hackney	E.	Jc
Chelmer Terrace	Hackney	E.	Jc
Chelmsford Road	Leyton U.D.	E.	Ka
Chelmsford Road	Walthamstow U.D.	E.	Ja
Chelmsford Street	Fulham	W.	Cg
Chelsea, Borough of, Wharf	Chelsea	S.W.	Dh
Chelsea Barracks	Westminster	S.W.	Eg
Chelsea Bridge Road	Chelsea and Westminster	S.W.	Eg
Chelsea Creek	Chelsea	S.W.	Dh
Chelsea Court	Chelsea	S.W.	Eg
Chelsea Embankment	Chelsea	S.W.	Eg
Chelsea Embankment Gardens	Chelsea	S.W.	Eg
Chelsea Gardens	Chelsea	S.W.	Eg
Chelsea Grove	Kensington	S.W.	Dg
Chelsea Pier	Chelsea	S.W.	Eg
Chelsea Reach	Chelsea	S.W.	Eg
Chelsea Station	**Chelsea**	S.W.	Dg
Chelsea Workhouse	Chelsea	S.W.	Dg
Chelsfield Road	Wandsworth	S.W.	Fk
Chelsham Road	Wandsworth	S.W.	Fh
Cheltenham Road	Leyton U.D.	E.	Ka
Cheltenham Terrace	Chelsea	S.W.	Eg
Chilvers Street	Greenwich	S.E.	Lg
Chelverton Road	Fulham	S.W.	Cj
Cheney Street	St. Pancras	N.W.	Fd
Cheneys Road	Leyton U.D.	E.	Lb
Chenies Mews	St. Pancras	W.C.	Fd
Chenies Place	St. Pancras	N.W.	Fd
Chenies Street	Holborn	W.C.	Fe
Chenies St. Chambers.	Holborn	W.C.	Fe
Cheniston Gardens	Kensington	W.	Df
Chepstow Place	Kensington and Paddington	W.	De
Chepstow Villas	Camberwell	S.E.	Hj
Chepstow Villas	Kensington	W.	De
Chepstow Yard	Paddington	W.	De
Chepstow Tavern Mews	Paddington	W.	De
Chequers, The	Wembley U.D.		Sc
Chequers Street	Finsbury	E.C.	Gd
Cherlton Square	Wandsworth	S.W.	Ek
Cherry Garden Place	Bermondsey	S.E.	Hf
Cherry Garden Stairs	Bermondsey	S.E.	Hf
Cherry Garden Street	Bermondsey	S.E.	Hf
Cherry Gardens Pier	Bermondsey	S.E.	Hf
Cherry Orchard	Greenwich		Lh
Cherrytree Alley	Finsbury	E.C.	Ge
Cherrytree Court	City of London	E.C.	Ge
Chertsey Road	Chiswick U.D.	W.	Ag
Chertsey Road	Leyton U.D.	E.	Kb
Chertsey Street	Wandsworth	S.W.	El
Cherwell Street	Battersea	S.W.	Fj
Chesfield	Teddington U.D.		Sm
Chesham Mansions	Chelsea	S.W.	Ef
Chesham Place	Chelsea and Westminster	S.W.	Ef
Chesham Mews	Westminster	S.W.	Ef
Chesham Road	Kingston		Am
Chesham Street	Chelsea and Westminster	S.W.	Ef
Cheshire Street	Bethnal Green	E.	Hd
Chesholm Road	Stoke Newington	N.	Hb
Cheshunt Road	West Ham	E.	Lc
Chesilton Road	Fulham	S.W.	Ch
Chesney Street	Battersea	S.W.	Eh
Chesnut Road	Woolwich		Ng
Chesson Road	Fulham	W.	Cg
Chester Buildings	Camberwell	S.E.	Gh
Chester House	Wimbledon	S.W.	Cl
Chester Mews	Lambeth	S.E.	Fg
Chester Mews	Paddington	W.	Ee
Chester Mews	St. Pancras	N.W.	Ed
Chester Mews	Westminster	S.W.	Ef
Chester Mews North	Westminster	S.W.	Ef
Chester Place	Paddington	W.	De
Chester Place	St. Pancras	N.W.	Ed
Chester Place	Woolwich		Ng
Chester Road	St. Pancras	N.	Eb

STREET OR PLACE.	BOROUGH.	P.D.	MAP.
Christie's Place	Lambeth	S.E.	Ff
Christopher Court	Stepney	E.	He
Christopher Place	Marylebone	W.	Ee
Christopher Street	Finsbury and Shoreditch	E.C.	Ge
Christopher Yard	Finsbury	E.C.	Ge
Chryssell Road	Lambeth	S.W.	Gh
Chubworthy St.	Deptford	S.E.	Jg
Chudleigh Road	Lewisham	S.E.	Jj
Chumleigh Street	Camberwell	S.E.	Gg
Church Alley	City of London	E.C.	Ge
Church Approach	Camberwell	S.E.	Gl
Church Avenue	Lewisham	S.E.	Hk
Church Avenue	Poplar	E.	Kd
Church Court	City of London	E.C.	Ge
Church Court	Westminster	S.W.	Ff
Church Crescent	Lewisham	S.E.	Kj
Church Crescent	Hackney	E.	Jc
Church Crescent	Kensington	W.	Df
Church End	Walthamstow U.D.	E.	Ka
Church End	Willesden U.D.	N.W.	Bc
Church End House	Willesden U.D.	N.W	Bc
Church Gardens	Ealing	W.	Sf
Church Grove	Hampton Wick U.D.		Sm
Church Grove	Islington	N.	Gc
Church Grove	Wimbledon	S.W.	Bm
Church Hill	Walthamstow U.D.	E.	Ka
Church Hill	Wimbledon	S.W.	Cl
Church House	Bromley		Lm
Church Lane	Battersea	S.W.	Dh
Church Lane	Chislehurst U.D.		Nm
Church Lane	Chislehurst U.D. & Foots Cray U.D.		Ol
Church Lane	Merton U.D.	S.W.	Cm
Church Lane	Ealing	W.	Sf
Church Lane	Greenwich		Lg
Church Lane	Hampstead	N.W.	Db
Church Lane	Islington	N.	Gc
Church Lane	Kingsbury U.D.		Aa
Church Lane	Leyton U.D.	E.	La
Church Lane	Stepney	E.	He
Church Lane	Walthamstow U.D.	E.	Ka
Church Lane	Wandsworth	S.W.	El
Church Lane	Willesden U.D.	N.W.	Bc
Church Parade	Acton U.D.	W.	Af
Church Passage	Bermondsey	S.E.	Jf
Church Passage	Islington	N.	Gc
Church Passage	Stepney	E.	He
Church Path	Barnes U.D.	S.W.	Ah
Church Path	Fulham	W.	Cg
Church Path	Poplar	E.	Je
Church Path	Stoke Newington	N.	Gb
Church Place	Chelsea	S.W.	Dg
Church Place	Greenwich	S.E.	Jg
Church Place	Paddington	W.	De
Church Place	Paddington	W.	Cd
Church Place	Westminster	S.W.	Fe
Church Road	Acton U.D.	W.	Af
Church Road	Barnes U.D.	S.W.	Bh
Church Road	Battersea	S.W.	Dh
Church Road	Beckenham U.D.		Jm
Church Road	Beckenham U.D.		Km
Church Road	Bromley		Lm
Church Road	Croydon R.D.	S.W.	Dm
Church Road	Croydon	S.E.	Gm
Church Road	Hackney	E.	Jc
Church Road	Hackney and Islington	N.	Gc
Church Road	Hackney	N.	Hb
Church Road	Ham U.D.		Sl
Church Road	Hammersmith	W.	Bf
Church Road	Hampstead	N.W.	Ec
Church Road	Hornsey	N.	Ea
Church Road	Kingston		Sm
Church Road	Lambeth	S.W.	Fj
Church Road	Lewisham	S.E.	Jk
Church Road	Leyton U.D.	E.	Kb
Church Road	Richmond		Sj
Church Road	Stoke Newington	N.	Gb
Church Road	Willesden U.D.	N.W.	Bc
Church Road	Wimbledon	S.W.	Cl
Church Row	Bethnal Green	E.	Hd
Church Row	Chislehurst U.D.		Nm
Church Row	City of London	E.	He
Church Row	City of London	E.C.	Ge
Church Row	Fulham	S.W.	Ch
Church Row	Hampstead	N.W.	Db
Church Row	Stepney	E.	Je
Church Row	Stoke Newington	N.	Gb
Church Row	Westminster	W.C.	Fe
Church Street	Bermondsey	S.E.	Hf
Church Street	Bethnal Green & Shoreditch	E.	Hd
Church Street	Camberwell	S.E.	Gh
Church Street	Chelsea	S.W.	Dg
Church Street	Chiswick U.D.	W.	Bg
Church Street	Westminster	W.	Fe
Church Street	Croydon R.D.		Dm
Church Street	Deptford and Greenwich	S.E.	Jg
Church Street	East Ham	E.	Mf
Church Street	Greenwich	S.E.	Kg
Church Street	Islington	N.	Gc
Church Street	Kensington	W.	Df
Church Street	Kingston		Sm
Church Street	Lambeth	S.W.	Fg
Church Street	Lewisham	S.E.	Kj
Church Street (part)	Marylebone	N.W.	Dd
Church Street (part)	Paddington	W.	Dd
Church Street	Stepney	E.	He
Church Street	Stoke Newington	N.	Gb
Church Street	West Ham	E.	Kd
Church Street	West Ham	E.	Ld
Church Street	West Ham	E.	Le
Church Street	Westminster	S.W.	Ff
Church Street	Westminster	S.W.	Eg
Church Street	Woolwich		Mg
Church Terrace	Hackney	E.	Jc
Church Terrace	Lambeth	S.E.	Ff
Church Terrace	Lewisham	S.E.	Kj
Church Terrace	Wandsworth	S.W.	Fh
Church Terrace	Woolwich		Ng
Church Walk	Chiswick U.D.	W.	Bg
Church Walk	Hendon U.D.	N.W.	Db
Church Walk	Kensington	W.	Df
Church Walk	Wandsworth	S.W.	Dj
Church Way	St. Pancras	N.W.	Fd
Church Wharf	Chiswick U.D.	W.	Bg
Churchfield Road	Ealing	W.	Se
Churchfield Road, East	Acton U.D.	W.	Ae
Churchfield Road, West	Acton U.D.	W.	Ae
Churchill Road	Hackney	E.	Jb
Churchill Road	St. Pancras	N.W.	Eb
Churchill Road	West Ham	E.	Le
Churchill Road	Willesden U.D.	N.W.	Bc
Churchley Road	Lewisham	S.E.	Hl
Church Manor Way	Woolwich		Of
Churchwell Path	Hackney	E.	Hb
Churchyard Bottom	Hornsey	N.	Ea
Churchyard Row	Southwark	S.E.	Gg
Churston Avenue	West Ham	E.	Ld
Churton Place	Westminster	S.W.	Fg
Churton Street	Westminster	S.W.	Fg
Chute Street	Wandsworth	S.W.	Dl
Cibber Road	Lewisham	S.E.	Jk
Cicada Road	Wandsworth	S.W.	Dj
Cicely Road	Camberwell	S.E.	Hh
Cillian Street	Lewisham	S.E.	Jj
Cinnamon Street	Stepney	E.	Hf
Cintra House	Penge U.D.	S.E.	Hm
Cintra Park	Penge U.D.	S.E.	Hm
Circular Road	Southwark	S.E.	Gf
Circular Road	Woolwich		Mg
Circus, The	City of London	E.	He
Circus, The	Greenwich	S.E.	Kh
Circus, The	St. Pancras	N.W.	Eb
Circus Place	City of London	E.C.	Ge
Circus Road	Marylebone	N.W.	Dd
Circus Street	Greenwich	S.E.	Kh
Circus Street	Marylebone	W.	Ee
Cirencester Mews	Paddington	W.	De

STREET OR PLACE.	BOROUGH.	P.D.	MAP.
Cobb's Street	Stepney	E.	He
Cobb's Yard	Stepney	E.	He
Cobden Road	Leyton U.D.	E.	Kb
Cobden Statue	St. Pancras	N.W.	Fd
Cobden Street	Camberwell	S.E.	Gg
Cobden Street	Poplar	E.	Ka
Cobham Road	Barking U.D.		Nd
Cobham Road	Kingston		Am
Cobham Street	Marylebone	N.W.	Ed
Coborn Road	Poplar & Stepney	E.	Jd
Coborn Road Station	**Stepney**	E.	Jd
Coborn Street	Poplar	E.	Jd
Cobourg Court	Stepney	E.	He
Cobourg Road	Camberwell	S.E.	Hg
Cobourg Row	Westminster	S.W.	Ff
Coburg Mews	Lambeth	S.E.	Fg
Coburg Place	Paddington	W.	De
Coburg Street	Finsbury	E.C.	Gd
Coburg Street	St. Pancras	N.W.	Fd
Cochrane Mews	Marylebone	N.W.	Dd
Cochrane Road	Wimbledon	S.W.	Cm
Cochrane Street	Marylebone	N.W.	Dd
Cock Alley	Shoreditch	E.	He
Cock Inn	East Ham	E.	Md
Cock Lane	City of London	E.C.	Ge
Cock Yard	Camberwell	S.E.	Gh
Cock Yard	Westminster	W.	Ee
Cockspur Street	Westminster	S.W.	Ff
Code Street	Bethnal Green	E.	He
Codrington Hill	Lewisham	S.E.	Jk
Codrington Mews	Kensington	W.	Ce
Coimbra Road	Wandsworth	S.W.	Dj
Coin Street	Lambeth	S.E.	Gf
Coke Street	Stepney	E.	He
Colbeck Mews	Kensington	W.	Dg
Colberg Place	Hackney	N.	Ha
Colby Road	Camberwell	S.E.	Gl
Colchester Road	Leyton U.D.	E.	Ka
Colchester Road	Walthamstow U.D.	E.	Ja
Colchester Street	City of London	E.C.	He
Colchester Street	Westminster	S.W.	Fg
Colchester Street	Stepney	E.	He
Cold Bath Buildings	Finsbury	E.C.	Fd
Cold Bath Square	Finsbury	E.C.	Fd
Cold Bath Street	Greenwich	S.E.	Kh
Cold Blow	Deptford	S.E.	Jg
Cold Blow Avenue	Croydon R.D.	S.E.	Em
Cold Blow Cottages	Deptford	S.E.	Jg
Coldbath Wood	Bexley U.D.		Nk
Goldborough Road	Lambeth	S.W.	Fh
Coldharbour	Poplar	E.	Kf
Coldharbour Lane	Camberwell and Lambeth	S.W.	Gj
Coldharbour Place	Lambeth	S.W.	Gh
Coldingwood Street	Stepney	E.	Je
Cole Place	Southwark	S.E.	Gf
Cole Street	Southwark	S.E.	Gf
Colebrook Avenue	Ealing	W.	Se
Colebrook Lodge	Fulham	S.W.	Cj
Colebrook Road	Walthamstow U.D.	E.	Ja
Colebrooke Row	Islington	N.	Gd
Coleford Road	Wandsworth	S.W.	Dj
Colegrave Road	West Ham	E.	Kc
Colegrove Road	Camberwell	S.E.	Hg
Coleherne Court	Kensington	S.W.	Dg
Coleherne Mansions	Kensington	S.W.	Dg
Coleherne Mews	Kensington	S.W.	Dg
Coleherne Road	Kensington	S.W.	Dg
Coleherne Terrace	Kensington	S.W.	Dg
Colehill Lane	Fulham	S.W.	Ch
Coleman Road	Camberwell	S.E.	Gg
Coleman Street	City of London	E.C.	Ge
Coleman Street	Islington	N.	Gc
Coleman Street	Stepney	E.	Hf
Coleman Street Bldgs.	City of London	E.C.	Ge
Coleman's Place	Lambeth	S.E.	Ff
Colenso Road	Croydon R.D.		Em
Colenso Road	Hackney	E.	Jb
Coleraine Road	Greenwich	S.E.	Lg
Coleridge Gardens	Hampstead	N.W.	De
Coleridge Road	Hornsey	N.	Fa
Coleridge Road	Islington	N.	Gb
Coles Wharf	Stepney	S.E.	Hf
Coleshill Mews	Westminster	S.W.	Ef
Colestown Street	Battersea	S.W.	Eh
Colet Gardens	Hammersmith	W.	Cg
Colfe Road	Lewisham	S.E.	Jk
Colford Road	Wandsworth	S.W.	Dj
Coligny Street	Wandsworth	S.W.	Dj
Colin Road	Willesden U.D.	N.W.	Bc
Colin Street	Poplar	E.	Ke
Colina Road	Tottenham U.D.	N.	Ga
Colinette Road	Wandsworth	S.W.	Cj
Coliseum	**Westminster**	W.C.	Fe
Coliston Road	Wandsworth	S.W.	Dj
Collas Mews	Hampstead	N.W.	Dc
College Avenue	Hackney	E.	Jb
College Crescent	Hampstead	N.W.	Db
College Gardens	Camberwell	S.E.	Gk
College Grove	St. Pancras	N.W.	Fc
College Hill	City of London	E.C.	Ge
College Lane	Hackney	E.	Jc
College Lane	St. Pancras	N.W.	Eb
College Mews	Westminster	S.W.	Ff
College Park Terrace	Hammersmith	W.	Cd
College Park Villas	Hammersmith	W.	Cd
College Place	Chelsea	S.W.	Eg
College Place	Hackney	E.	Jc
College Place	St. Pancras	N.W.	Fc
College Place East	Greenwich	S.E.	Kh
College Place West	Greenwich	S.E.	Kh
College Road	Camberwell	S.E.	Gk
College Road	Hampstead	N.W.	Ec
College Road	Leyton U.D. and Walthamstow U.D.	E.	Ka
College Road	Romford		Lm
College Road	Willesden U.D.	N.W.	Cd
College Slip	Bromley		Lm
College Street	Chelsea	S.W.	Eg
College Street	City of London	E.C.	Ge
College Street	Hackney	E.	Jc
College Street	Islington	N.	Gc
College Street	Lambeth	S.E.	Ff
College Street	Stepney	E.	Jd
College Street	Wandsworth	S.W.	Cj
College Terrace	Hackney	E.	Hb
College Terrace	Hampstead	N.W.	Dc
College View	Lewisham	S.E.	Jk
College Villas Road	Hampstead	N.W.	Dc
Collerston Cottages	Greenwich	S.E.	Lg
Collerston Road	Greenwich	S.E.	Lg
Collett Road	Bermondsey	S.E.	Hf
Collier Place	Barking U.D.		Nd
Collier Place	Finsbury	N.	Fd
Collier Street	Finsbury and Islington	N.	Fd
Collingbourne Street	Hammersmith	W.	Be
Collingham Place	Kensington	S.W.	Df
Collingham Road	Kensington	S.W.	Df
Collington Street	Greenwich	S.E.	Kg
Collingwood Road	Walthamstow U.D.	E.	Ja
Collingwood Street	Bethnal Green	E.	Hd
Collingwood Street	Chelsea	S.W.	Eg
Collingwood Street	Southwark	S.E.	Gf
Collingwood Street	Woolwich		Mf
Collins' Music Hall	**Islington**	N.	Gd
Collins Square	Lewisham	S.E.	Kh
Collins Street	Lewisham	S.E.	Kh
Collinson Street	Southwark	S.E.	Gf
Collins's Hill	Bromley R.D.		Om
Collison Road	Wandsworth	S.W.	Dj
Collitch Place	Southwark	S.E.	Gg
Colls Road	Camberwell	S.E.	Hh
Colman Street	Greenwich	S.E.	Kg
Colman's Buildings	Hammersmith	W.	Bf
Colmar Street	Stepney	E.	Jd
Colmer Road	Wandsworth	S.W.	Fm
Colnbrook Road	Stoke Newington	N.	Gb
Colnbrook Street	Southwark	S.E.	Gf
Colne Road	Hackney	E.	Jb
Cologne Road	Battersea	S.W.	Dj
Cologne Street	Stepney	E.	Je

STREET OR PLACE.	BOROUGH.	P.D.	MAP.
Colomb Street	Greenwich	S.E.	Kg
Colonial Avenue	City of London	E.	He
Colonial Wharf	Stepney	E.	Hf
Colonnade	St. Pancras	W.C.	Fd
Colosseum Terrace	St. Pancras	N.W.	Ed
Coltman Place	Greenwich	S.E.	Kg
Coltman Street	Greenwich	S.E.	Kg
Columbia Market	Bethnal Green	E.	Hd
Columbia Road	Bethnal Green	E.	Hd
Columbia Road	West Ham	E.	Le
Columbia Square	Bethnal Green	E.	Hd
Colva Mews	St. Pancras	N.	Eb
Colva Street	St. Pancras	N.	Eb
Colvestone Crescent	Hackney	E.	Hc
Colville Gardens	Kensington	W.	Ce
Colville House	Kensington	W.	Ce
Colville Mews	Kensington	W.	Ce
Colville Place	St. Pancras	W.	Fe
Colville Road	Acton U.D.	W.	Af
Colville Road	Kensington	W.	Ce
Colville Square	Kensington	W.	Ce
Colville Square Mews	Kensington	W.	Ce
Colville Sq. Terrace	Kensington	W.	Ce
Colville Terrace	Kensington	W.	Ce
Colvin Road	East Ham	E.	Md
Colvin Street	Hammersmith	W.	Cf
Colwell Road	Camberwell	S.E.	Hj
Colwick Street	Deptford	S.E.	Jg
Colworth Grove	Southwark	S.E.	Gg
Colworth Road	Leyton U.D.	E.	Ka
Colwyn Street	Lambeth	S.E.	Ff
Colyton Road	Camberwell	S.E.	Hj
Combedale Road	Greenwich	S.E.	Lg
Combermere Road	Lambeth	S.W.	Fh
Comberton Road	Hackney	E.	Hb
Comboss Road	Poplar	E.	Jc
Comely Bank Road	Walthamstow U.D.	E.	Ka
Comeragh Mews	Fulham	W.	Cg
Comeragh Road	Fulham	W.	Cg
Comet Street	Deptford	S.E.	Jh
Comerford Road	Lewisham	S.E.	Jj
Commerce Place	Lambeth	S.W.	Fj
Commercial Dock Passage	Bermondsey	S.E.	Jf
Commercial Dock Pier	Bermondsey	S.E.	Jf
Commercial Docks	Bermondsey	S.E.	Jf
Commercial Gasworks	Stepney	E.	Je
Commercial Gasworks	Poplar	E.	Ke
Commercial Place	Westminster	S.W.	Eg
Commercial Place	St. Pancras	N.W.	Ec
Commercial Road	Camberwell	S.E.	Hh
Commercial Road	Westminster	S.W.	Eg
Commercial Rd. (part)	Lambeth	S.E.	Ff
Commercial Rd. (part)	Southwark	S.E.	Gf
Commercial Rd. East	Stepney	E.	Je
Commercial Salerooms	City of London	E.C.	Ge
Commercial Salerooms Passage	City of London	E.C.	Ge
Commercial Street	Shoreditch and Stepney	E.	He
Commercial Yard	Bermondsey	S.E.	Jf
Commerell Street	Greenwich	S.E.	Kg
Commodore Street	Stepney	E.	Je
Common Street	Poplar	E.	Jf
Como Road	Lewisham	S.E.	Jk
Compass Hill	Richmond		Sj
Compayne Gardens	Hampstead	N.W.	Dc
Compton Avenue	East Ham	E.	Md
Compton Avenue	Islington	N.	Gc
Compton Buildings	City of London	E.C.	Ge
Compton Mews	St. Pancras	W.C.	Fd
Compton Place	St. Pancras	W.C.	Fd
Compton Place, east	St. Pancras	W.C.	Fd
Compton Road	Islington	N.	Gc
Compton Road	Willesden U.D.	N.W.	Cd
Compton Road	Wimbledon	S.W.	Cl
Compton Square	Islington	N.	Gc
Compton Street	Finsbury	E.C.	Gd
Compton Street	Islington	N.	Gc
Compton Street	St. Pancras	W.C.	Fd
Compton Street	West Ham	E.	Ld
Compton Terrace	Chiswick U.D.	W.	Ag
Compton Terrace	Islington	N.	Gc
Comrie Road	Lambeth	S.W.	Fj
Comus Place	Southwark	S.E.	Hg
Comyn Road	Battersea	S.W.	Ej
Conan Street	Fulham	S.W.	Cg
Condar Gardens	Hampstead	N.W.	Cb
Conder Street	Stepney	E.	Je
Conderton Road	Lambeth	S.E.	Gh
Conduit Avenue	Greenwich	S.E.	Kh
Conduit Court	Hackney	E.	Hb
Conduit Court	Westminster	W.C.	Fe
Conduit Mews	Paddington	W.	De
Conduit Place	Hackney	E.	Hb
Conduit Place	Paddington	W.	De
Conduit Road	Woolwich		Ng
Conduit Street	Westminster	W.	Ee
Conduit Street	Hackney	E.	Hb
Conduit Wood	Barnes and Richmond		Aj
Condy Road	Wandsworth	S.W.	Di
Conewood Street	Islington	N.	Gb
Conference Road	Woolwich		Og
Congo Road	Woolwich		Ng
Congress Hall	Hackney	E.	Hb
Congress Road	Woolwich		Og
Congreve Street	Southwark	S.E.	Gg
Conifers	Teddington U.D.		Sm
Conlger Road	Fulham	S.W.	Dh
Coningham Mews	Hammersmith	W.	Bf
Coningham Road	Hammersmith	W.	Bf
Coningsby Road	Ealing	W.	Sf
Coningsby Road	Hornsey	N.	Ga
Coniston Road	Lewisham	S.E.	Km
Coniston Road	Lewisham	S.E.	Hk
Conlan Street	Kensington	S.W.	Cd
Conley Road	Willesden U.D.	N.W.	Bc
Conley Street	Greenwich	S.E.	Kg
Connaught Gardens	Hornsey	N.	Ea
Connaught Mansions	Lambeth	S.W.	Fj
Connaught Mews	Paddington	W.	Ee
Connaught Place	Paddington	W.	Ee
Connaught Road	Hornsey	N.	Ga
Connaught Road	Leyton U.D.	E.	Ka
Connaught Road	West Ham	E.	Mf
Connaught Road	Willesden U.D.	N.W.	Bc
Connaught Rd. Station	**West Ham**	E.	Me
Connaught Square	Paddington	W.	Ee
Connaught Sq. Mews	Paddington	W.	Ee
Connaught Street	Paddington	W.	Ee
Connor Street	Hackney	E.	Jd
Conquering Hero, The	Croydon	S.E.	Gm
Conrad Street	Hackney	E.	Hc
Conroy Street	Lambeth	S.W.	Fg
Constance Cottages	Camberwell	S.E.	Gj
Constance Road	Camberwell	S.E.	Gj
Constance Street	West Ham	E.	Le
Constance Street	West Ham	E.	Mf
Constantine Road	Hampstead	N.W.	Eb
Constitution Hill	Westminster	S.W.	Ef
Constitution Hill	Woolwich		Mg
Constitution Hill	Woolwich		Mh
Content Street	Southwark	S.E.	Gg
Contentment Place	Shoreditch	N.	Hd
Convent of Good Shepherd	Hammersmith	W.	Cg
Convent of Sacred Heart	Wandsworth	S.W.	Dj
Conville Road	Fulham	S.W.	Ch
Conway Mews	St. Pancras	W.	Ed
Conway Road	Tottenham U.D.	N.	Ga
Conway Road	Wimbledon	S.W.	Cm
Conway Road	Woolwich		Ng
Conway Street	Hackney	E.	Jc
Conyer Street	Bethnal Green	E.	Jd
Conyers Road	Wandsworth	S.W.	Fl
Cook Street	Poplar	E.	Ke
Cooke Street	Barking U.D.		Nd
Cook's Place	Lewisham	S.E.	Hk
Cook's Place	Woolwich		Mf
Cook's Road	Southwark	S.E.	Gg
Cook's Road	West Ham	E.	Kd
Coolfyn Road	West Ham	E.	Le
Coolhurst Avenue	Hornsey	N.	Fa

STREET OR PLACE.	BOROUGH.	P.D.	MAP.
Cosbycote Avenue	Lambeth	S.E.	Gj
Cosmo Place	Holborn	W.C.	Fe
Cossall Road	Camberwell	S.E.	Hh
Costa Street	Camberwell	S.E.	Hh
Costerwood Street	Deptford	S.E.	Jg
Cosway Street	Marylebone	N.W.	Ee
Cotall Street	Poplar	E.	Je
Cotesbach Road	Hackney	E.	Hb
Cotford Place	Lambeth	S.E.	Fg
Cotham Street	Southwark	S.E.	Gg
Cotherstone Road	Wandsworth	S.W.	Fk
Cotleigh Road	Hampstead	N.W.	Dc
Cotman Street	Wandsworth	S.W.	Dj
Cottage Court	Stepney	E.	Je
Cottage Green	Camberwell	S.E.	Gg
Cottage Grove	Lambeth	S.W.	Fj
Cottage Grove	Southwark	S.E.	Gg
Cottage Grove	Stepney	E.	Jd
Cottage Lane	Finsbury	E.C.	Gd
Cottage Place	Battersea	S.W.	Dh
Cottage Place	Chelsea	S.W.	Ef
Cottage Place	Kensington	S.W.	Ef
Cottage Place	Stepney	E.	He
Cottage Row	Bermondsey	S.E.	Hf
Cottage Row	Poplar	E.	Ke
Cottage Street	Poplar	E.	Ke
Cotteford Road	Wandsworth	S.W.	El
Cottenham Park Road	Wimbledon	S.W.	Bm
Cottenham Road	Islington	N.	Fa
Cottenham Terrace	Islington	N.	Fa
Cottesbrook Street	Deptford	S.E.	Jg
Cotterell Buildings	Holborn	W.C.	Fe
Cottesmore Gardens	Kensington	W.	Df
Cottingham Road	Beckenham U.D.	S.E.	Jm
Cotton Street	City of London	E.C.	Ge
Cotton Street	Poplar	E.	Ke
Cotton Street	Stepney	E.	He
Cotton's Gardens	Shoreditch	E.	Hd
Cotton's Wharf	Bermondsey	S.E.	Gf
Cottrill Road	Hackney	E.	Hc
Coulgate Street	Deptford	S.E.	Jh
Coulson Street	Chelsea	S.W.	Eg
Coulter Road	Hammersmith	W.	Bf
Counter Street	Bermondsey	S.E.	Gf
Counter's Creek Road	Kensington	W.	Ce
Countess Road	St. Pancras	N.W.	Fb
County Club	Bromley		Lm
County Grove	Camberwell	S.E.	Gh
County Terrace Street	Southwark	S.E.	Gf
Coupland Terrace	Woolwich		Ng
Courage Place	Lambeth	S.E.	Ff
Courland Grove	Wandsworth	S.W.	Fh
Court Hill Road	Lewisham	S.E.	Kj
Court Lane	Bromley R.D.		Ml
Court Lane	Camberwell	S.E.	Hk
Court Road	Lambeth	S.E.	Gk
Court Road	Woolwich		Mk
Court Street	Stepney	E.	He
Court Downs Road	Beckenham U.D.		Jm
Courtenay Road	Walthamstow U.D.	E.	Ja
Courtenay Street	Lambeth	S.E.	Fg
Courtfield Gardens	Kensington	S.W.	Dg
Courtfield Mews	Kensington	S.W.	Dg
Courtfield Road	Kensington	S.W.	Dg
Courthope Road	St. Pancras	N.W.	Eb
Courthope Road	Wimbledon	S.W.	Cl
Courthope Villas	Wimbledon	S.W.	Cm
Courtland Terrace	Kensington	W.	Df
Courtnay Road	Beckenham U.D.	S.E.	Jm
Courtnay Road	Leyton U.D.	E.	Lb
Courtnell Street	Paddington	W.	De
Courtney Road	Croydon R.D.	S.W.	Dm
Courtney Road	Islington	N.	Gb
Courtrai Road	Lewisham	S.E.	Jj
Court Yard	Woolwich		Mk
Couthurst Road	Greenwich	S.E.	Lg
Coutts Road	Stepney	E.	Je
Covent Garden	Westminster	W.C.	Fe
Covent Garden Market	Westminster	W.C.	Fe
Covent Garden Piazza	Westminster	W.C.	Fe
Covent Garden Station	Westminster	W.C.	Fe
Covent Garden Theatre	Westminster	W.C.	Fe
Coventry Place	Bethnal Green	E.	He
Coventry Street	Bethnal Green	E.	He
Coventry Street	Westminster	W.	Fe
Coverdale Road	Barking U.D.		Nd
Coverdale Road	Hammersmith	W.	Cf
Coverdale Road	Willesden U.D.	N.W.	Cc
Coverton Road	Wandsworth	S.W.	Dl
Cow Bridge	Hackney	E.	Jb
Cow Lane	Bermondsey	S.E.	Jf
Cowan Street	Camberwell	S.E.	Gg
Cowcross Street	Finsbury	E.C.	Ge
Cowdrey Road	Wimbledon	S.W.	Dl
Cowdry Street	Hackney	E.	Jc
Cowick Road	Wandsworth	S.W.	El
Cowleaze House	Hendon U.D.	N.W.	Ba
Cowleaze Road	Kingston		Bm
Cowlett Road	Camberwell	S.E.	Hj
Cowley Place	St. Pancras	N.W.	Fd
Cowley Road	Lambeth	S.W.	Gh
Cowley Road	Leyton U.D.	E.	Lb
Cowley Road	Wanstead U.D.	E.	La
Cowley Street	Westminster	S.W.	Ff
Cowley Street	Stepney	E.	He
Cowper Mansions	Chelsea	S.W.	Eg
Cowper Road	Acton U.D.	W.	Ae
Cowper Road	Stoke Newington	N.	Hb
Cowper Road	Wimbledon	S.W.	Dl
Cowper Street	Finsbury	E.C.	Gd
Cowper's Court	City of London	E.C.	Ge
Cowper's Row	Wandsworth	S.W.	Fk
Cowthorpe Road	Lambeth	S.W.	Fh
Cox Court	City of London	E.C.	Ge
Cox's Quay	City of London	E.C.	Ge
Cox's Square	Stepney	E.	He
Cox's Walk	Camberwell	S.E.	Hk
Coxson Place	Bermondsey	S.E.	Hf
Coxwell Road	Woolwich		Ng
Crabtree Cottages	Fulham	S.W.	Cg
Crabtree Lane	Fulham	S.W.	Cg
Crab Hill	Beckenham U.D.		Km
Crab Hill	Lewisham	S.E.	Jk
Craddock Street	St. Pancras	N.W.	Ec
Craigerne Road	Greenwich	S.E.	Lh
Craignish Avenue	Croydon	S.W.	Fm
Craig's Court	Westminster	S.W.	Ff
Craigton Road	Woolwich		Mj
Crail Road	Southwark	S.E.	Gg
Crake Hall	Wandsworth	S.W.	Ck
Cramond Road	Woolwich		Mj
Crampton Road	Beckenham U.D. & Penge U.D.	S.E.	Hm
Crampton Street	Southwark	S.E.	Gg
Cranborne Court	Battersea	S.W.	Eh
Cranborne Road	Barking U.D.		Nd
Cranbourn Passage	Bermondsey	S.E.	Hf
Cranbourn Alley	Westminster	W.C.	Fe
Cranbourn Street	Westminster	W.C.	Fe
Cranbourne Road	Leyton U.D.	E.	Kb
Cranbrook Road	Bethnal Green	E.	Jd
Cranbrook Road	Chiswick U.D.	W.	Bg
Cranbrook Road	Croydon		Gm
Cranbrook Road	Deptford	S.E.	Jh
Cranbrook Road	East Ham	E.	Lf
Cranbrook Road	Walthamstow U.D.	E.	Ja
Cranbrook Road	Wimbledon	S.W.	Cm
Cranbrook Street	Bethnal Green	E.	Jd
Cranbrook Terrace	Southwark	S.E.	Hg
Cranbury Road	Fulham	S.W.	Dh
Crane Court	City of London	E.C.	Ge
Crane Grove	Islington	N.	Gc
Crane Street	Greenwich	S.E.	Kg
Cranfield Grove	Lambeth	S.E.	Gl
Cranfield Road	Deptford	S.E.	Jh
Cranfield Villas	Lambeth	S.E.	Gl
Cranford Street	Stepney	E.	Je
Cranham Road	Bermondsey	S.E.	Hg
Cranhurst Road	Willesden U.D.	N.W.	Cb
Cranleigh Road	Leyton U.D.	E.	Kb
Cranleigh Road	Tottenham U.D.	N.	Ga
Cranley Buildings	Holborn	E.C.	Ge
Cranley Gardens	Hornsey	N.	Fa
Cranley Gardens	Kensington	S.W.	Dg

STREET OR PLACE.	BOROUGH.	P.D.	MAP.
De Burgh Road	Wimbledon	S.W.	Dm
Decima Street	Bermondsey	S.E.	Gf
Decoy, The	Hendon U.D.	N.W.	Ca
Decoy Wood	Hendon U.D.	N.W.	Ca
De Crespigny Park	Camberwell	S.E.	Gh
Dee Road	West Ham	E.	Ld
Dee Street	Poplar	E.	He
Deerbrook Road	Lambeth	S.E.	Gk
Deerdale Road	Lambeth	S.E.	Gj
Deerhurst Road	Wandsworth	S.W.	Fl
Defoe Road	Stoke Newington	N.	Hb
Defoe Road	Wandsworth	S.W.	Dl
Defrene Road	Lewisham	S.E.	Jl
Dekker Road	Camberwell	S.E.	Gj
Delacourt Road	Greenwich	S.E.	Lh
Delafield Road	Greenwich		Lg
Delaford Road	Bermondsey and Camberwell	S.E.	Hg
Delaford Road	Fulham	S.W.	Cg
Delahaye Mews	Westminster	S.W.	Ff
Delahaye Street	Westminster	S.W.	Ff
Delamere Crescent	Paddington	W.	De
Delamere Mews	Paddington	W.	De
Delamere Road	Wimbledon	S.W.	Cm
Delamere Street	Paddington	W.	De
Delamere Terrace	Paddington	W.	De
Delancey Street	St. Pancras	N.W.	Ed
Delaney Court	Greenwich	S.E.	Kg
De Laune Street	Southwark	S.E.	Gg
Delaware Road	Paddington	W.	Dd
Delhi Street	Islington	N.	Fc
Delia Street	Wandsworth	S.W.	Dj
Dell, The	Croydon	S.E.	Gm
Dellow Street	Stepney	E.	He
Dells Mews	Westminster	S.W.	Fg
Deloraine Street	Deptford	S.E.	Jh
Delorme Street	Fulham	W.	Cg
Delph Street	Southwark	S.E.	Gf
Delta Street	Bethnal Green	E.	He
Delvan Street	Woolwich		Mg
Delverton Road	Southwark	S.E.	Gg
Delvino Road	Fulham	S.W.	Dh
De Morgan Road	Fulham	S.W.	Dh
Dempsey Street	Stepney	E.	He
Dempster Road	Wandsworth	S.W.	Dj
Denbigh Gardens	Richmond		Sj
Denbigh Mews	Westminster	S.W.	Fg
Denbigh Mews	Kensington	W.	Ce
Denbigh Place	Westminster	S.W.	Fg
Denbigh Road	Ealing	W.	Se
Denbigh Road	Kensington	W.	Ce
Denbigh Road	Poplar	E.	Jd
Denbigh Road	Willesden U.D.	N.W.	Bc
Denbigh Street	Westminster	S.W.	Fg
Denbigh Terrace	Kensington	W.	Ce
Denbigh Yard	Westminster	S.W.	Fg
Denby Street	Wandsworth	S.W.	Ek
Denewood Road	Hornsey	N.	Ea
Denford Street	Greenwich	S.E.	Lg
Denham Street	Greenwich	S.E.	Lg
Denham Yard	Westminster	W.C.	Fe
Denholme Road	Paddington	W.	Cd
Denehurst Gardens	Acton U.D.	W.	Af
Denison Road	Croydon R.D.	S.W.	Dl
Denison Road	Twickenham U.D.		Sj
Denman Road	Camberwell	S.E.	Hh
Denman Street	Bermondsey	S.E.	Gf
Denman Street	Westminster	W.	Fe
Denmark Avenue	Wimbledon	S.W.	Cm
Denmark Crescent	Islington	N.	Fd
Denmark Grove	Islington	N.	Fd
Denmark Hill	Camberwell and Lambeth	S.E.	Gh
Denmark Hill Station	**Camberwell**	S.E.	Gh
Denmark Place	Holborn	W.C.	Fe
Denmark Place	Lambeth	S.E.	Gl
Denmark Road	Ealing	W.	Se
Denmark Road	Camberwell and Lambeth	S.E.	Gh
Denmark Road	Islington	N.	Fd
Denmark Road	Kingston		Sm
Denmark Road	Willesden U.D.	N.W.	Cd
Denmark Road	Wimbledon	S.W.	Cm
Denmark Street	Holborn	W.C.	Fe
Denmark Street	Islington	N.	Gd
Denmark Street	Lambeth	S.E.	Gh
Denmark Street	Leyton U.D.	E.	Kb
Denmark Street	Stepney	E.	He
Denmark Street	West Ham	E.	Le
Denmark Terrace	Islington	N.	Fb
Denmark Terrace	Lewisham	S.E.	Jl
Denmark Villas	Bromley		Lm
Dennett's Grove	Deptford	S.E.	Jh
Dennett's Road	Deptford	S.E.	Hh
Denning Road	Hampstead	N.W.	Db
DenningtonPark Mans.	Hampstead	N.W.	Dc
Dennington Park Road	Hampstead	N.W.	Dc
Dennis Road	Wembley U.D.		Ab
Dennis Street	Islington	N.	Fd
Denny Street	Lambeth	S.E.	Gg
Dents Road	Battersea	S.W.	Ej
Denton Road	Hornsey	N.	Fa
Denton Road	Willesden U.D.	N.W.	Ac
Denton Street	Wandsworth	S.W.	Dj
Denver Road	Hackney	N.	Ha
Denyer Street	Chelsea	S.W.	Eg
Denzell Street	Westminster	W.C.	Ee
Denzil Road	Willesden U.D.	N.W.	Bb
Deodar Road	Wandsworth	S.W.	Cj
Depôt Street	Camberwell	S.E.	Jl
Deptford Bridge	Deptford	S.E.	Jh
Deptford Broadway	Deptford	S.E.	Jh
Deptford Cattle Mkt.	Greenwich	S.E.	Jg
Deptford Creek (part)	Greenwich and Deptford	S.E.	Kh
Deptford Cemetery	Lewisham	S.E.	Jj
Deptford Dry Dock	Greenwich	S.E.	Jh
Deptford Green	Greenwich	S.E.	Jg
Deptford Park	Deptford	S.E.	Jg
Deptford Road Station	**Bermondsey**	S.E.	Jg
Deptford Station	**Deptford**	S.E.	Jg
Deptford Ferry Road	Poplar	E.	Jg
Derby Buildings	St. Pancras	W.C.	Fd
Derby Mews	St. Pancras	N.W.	Fd
Derby Road	Barnes U.D.	S.W.	Aj
Derby Road	Hackney	E.	Jc
Derby Road	Hackney	N.	Hc
Derby Road	Tottenham U.D.	N.	Ga
Derby Road	Wimbledon	S.W.	Dm
Derby Street	Westminster	W.	Ef
Derby Street	St. Pancras	W.C.	Fd
Derby Villas	Camberwell	S.E.	Hj
Derby Villas	Lewisham	S.E.	Hk
Derbyshire Street	Bethnal Green	E.	Hd
Derinton Road	Wandsworth	S.W.	El
Dering Street	Westminster	W.	Ee
Dering Yard	Westminster	W.	Ee
Dermody Gardens	Lewisham	S.E.	Kj
Dermody Road	Lewisham	S.E.	Kj
Dermot Road	Battersea and Wandsworth	S.W.	Ek
Deronda Road	Lambeth	S.E.	Gk
Derrick Street	Bermondsey	S.E.	Jf
Derry Street	St. Pancras	W.C.	Fd
Dersingham Road	Hendon U.D.	N.W.	Cb
Derwent Buildings	Camberwell	S.E.	Hg
Derwent Grove	Camberwell	S.E.	Hj
Derwent Road	Penge U.D.	S.E.	Hm
Derwent Street	Greenwich	S.E.	Kg
Derwent Terrace	Greenwich	S.E.	Kg
Derwentwater House	Acton U.D.	W.	Ae
Derwentwater Road	Acton U.D.	W.	Ae
Desart Street	Poplar	E.	He
Desborough Street	Paddington	W.	De
Desenfans Road	Camberwell	S.E.	Gj
Desford Road	West Ham	E.	Ke
Desmond Street	Deptford	S.E.	Jg
Despard Road	Islington	N.	Fa
Detmold Road	Hackney	E.	Hb
Devas Street	Poplar	E.	Kd
De Vere Gardens	Kensington	W.	Df
De Vere Mansions	Kensington	W.	Df
De Vere Mans. West	Kensington	W.	Df
De Vere Mews	Kensington	W.	Df
Deverell Street	Southwark	S.E.	Gf

STREET OR PLACE.	BOROUGH.	P.D.	MAP.
Drysdale Place	Shoreditch	N.	Hd
Drysdale Road	Greenwich and Lewisham	S.E.	Kh
Drysdale Street	Shoreditch	N.	Hd
Ducal Street	Bethnal Green	E.	Hd
Du Cane Road	Hammersmith	W.	Be
Duchess Mews	Marylebone	W.	Ee
Duchess Street	Marylebone	W.	Ee
Duchess Theatre	**Wandsworth**	S.W.	Ek
Duchess of Bedford's Walk	Kensington	W.	Df
Duchess of Edinburgh	Bexley U.D.		Oh
Ducie Street	Lambeth	S.W.	Fj
Duck Lane	Westminster	W.	Fg
Duckett Road	Hornsey and Tottenham U.D.	N.	Ga
Duckett Street	Stepney	E.	Je
Duckett's Canal	Poplar	E.	Jd
Ducking Pond Mews	Westminster	W.	Ef
Ducksfoot Lane	City of London	E.C.	Ge
Dudding Hill	Willesden U.D.	N.W.	Bb
Dudding Hill Lane	Willesden U.D.	N.W.	Bb
Dudgeon's Wharf	Poplar	E.	Kg
Dudley Mews	Paddington	W.	De
Dudley Place	Paddington	W.	De
Dudley Gardens	Ealing	W.	Sf
Dudley Road	Wimbledon	S.W.	Dl
Dudley Street	Paddington	W.	De
Dudlington Road	Hackney	E.	Hb
Duff Street	Poplar	E.	Ke
Dufferin Avenue	Finsbury	E.C.	Gd
Dufferin Street	Finsbury	E.C.	Gd
Duffield Street	Battersea	S.W.	Eh
Duff's Fields	Poplar	E.	Ke
Dugald Street	Deptford	S.E.	Kg
Dugdale Street	Lambeth	S.E.	Gg
Duke Mews	Marylebone	W.	Ee
Duke of St. Albans	St. Pancras	N.	Eb
Duke of York's Column	Westminster	S.W.	Ff
Duke of York's Theatre	**Westminster**	W.C.	Fe
Duke Street	Bermondsey and Southwark	S.E.	Gf
Duke Street	City of London	E.C.	He
Duke Street	Westminster	S.W.	Ff
Duke Street	Westminster	W.C.	Fe
Duke Street	Lambeth	S.E.	Gf
Duke Street	Marylebone	N.W.	Ed
Duke Street	Marylebone and Westminster	W.	Ee
Duke Street	Poplar	E.	Ke
Duke Street	Richmond		Sj
Duke Street	Stepney	E.	He
Duke Street Mansions	Westminster	W.	Ee
Duke's Avenue	Chiswick U.D.	W.	Bg
Duke's Avenue	The Maldens and Coombe U.D.		Bm
Duke's Lane	Kensington	W.	Df
Duke's Lane Chambers	Kensington	W.	Df
Duke's Mews	Marylebone	N.W.	Ed
Duke Road	Chiswick U.D.	W.	Bg
Duke's Road	East Ham	E.	Md
Duke's Road	St. Pancras	W.C.	Fd
Duke's Yard	Westminster	W.	Ee
Duke's Head Passage	City of London	E.C.	Ge
Dukesthorpe Road	Lewisham	S.E.	Jl
Dulas Street	Islington	N.	Fa
Dulford Street	Kensington	W.	Ce
Dulka Road	Battersea	S.W.	Ej
Dulton Street	Battersea	S.W.	Eh
Dulwich Almshouses	Camberwell	S.E.	Gk
Dulwich College	Camberwell	S.E.	Gk
Dulwich Common	Camberwell	S.E.	Gk
Dulwich Common Lane	Camberwell	S.E.	Gk
Dulwich Park	Camberwell	S.E.	Hk
Dulwich Park Road	Camberwell	S.E.	Gl
Dulwich Road	Lambeth	S.E.	Gj
Dulwich Station	**Camberwell**	S.E.	Gk
Dulwich Wood	Camberwell	S.E.	Hl
Dulwich Wood House	Lewisham	S.E.	Hl
Dumbarton Road	Lambeth	S.W.	Fj
Dumbreck Road	Woolwich		Mj
Dumont Road	Stoke Newington	N.	Hb
Dumpton Place	St. Pancras	N.W.	Ec
Dunbar Avenue	Croydon	S.W.	Fm
Dunbar Road	West Ham	E.	Lc
Dunbar Street	Lambeth	S.E.	Gl
Dunboyne Street	Hampstead	N.W.	Eb
Duncan Avenue	Holborn	W.C.	Fd
Duncan Court	Poplar	E.	Ke
Duncan Mews	Islington	N.	Gd
Duncan Road	Hackney	E.	Hd
Duncan Road	Richmond		Sh
Duncan Square	Hackney	E.	Hd
Duncan Street	Hackney	E.	Hc
Duncan Street	Islington	N.	Gd
Duncan Street	Stepney	E.	He
Duncan Terrace	Shoreditch	N.	Gd
Duncannon Street	Westminster	W.C.	Fe
Duncombe Hill	Lewisham	S.E.	Jk
Duncombe Road	Islington	N.	Fa
Duncrievie Road	Lewisham	S.E.	Kj
Dundalk Street	Deptford	S.E.	Jj
Dundas Road	Camberwell	S.E.	Hh
Dundee Road	West Ham	E.	Ld
Dundee Street	Stepney	E.	Hf
Dundonald Road	Willesden U.D.	N.W.	Cc
Dundonald Road	Wimbledon	S.W.	Cm
Dundonald Street	Westminster	S.W.	Fg
Dunedin Road	Leyton U.D.	E.	Kb
Dunford Road	Islington	N.	Fb
Dungarvan Avenue	Wandsworth	S.W.	Bj
Dunk Street	Stepney	E.	He
Dunkeld Street	Poplar	E.	He
Dunlace Road	Hackney	E.	Jb
Dunloe Street	Shoreditch	E.	Hd
Dunlop Place	Bermondsey	S.E.	Hf
Dunmore Road	Willesden U.D.	N.W.	Cc
Dunmow Place	Lambeth	S.E.	Fg
Dunmow Road	Leyton U.D. and West Ham	E.	Kb
Dunn Street	Hackney	E.	Hb
Dunollie Place	St. Pancras	N.W.	Fb
Dunollie Road	St. Pancras	N.W.	Fb
Dunoon Road	Lewisham	S.E.	Hk
Dunraven Road	Hammersmith	W.	Be
Duncrievie Road	Lewisham	S.E.	Kk
Dunsany Road	Hammersmith	W.	Cf
Dunsmure Road	Hackney & Stoke Newington	N.	Ha
Dunstable Court	Holborn	W.C.	Fe
Dunstable Road	Richmond		Sj
Dunstan Place	Stepney	E.	Je
Dunstan Street	Shoreditch	E.	Hc
Dunstan's Road	Camberwell	S.E.	Hk
Dunster Court	City of London	E.C.	Ge
Dunster Gardens	Willesden U.D.	N.W.	Cc
Dunston Road	Shoreditch	E.	Hd
Dunton Road	Leyton U.D.	E.	Ka
Duntshill Road	Wandsworth	S.W.	Dk
Dunvegan Road	Woolwich		Mj
Dupont Road	Merton U.D.	S.W.	Cm
Dupont Street	Stepney	E.	Je
Dupree Road	Greenwich	S.E.	Lg
Durand Gardens	Lambeth	S.W.	Fh
Durant Street	Bethnal Green	E.	Hd
Durban Road	Beckenham U.D.		Jm
Durban Road	Lambeth	S.E.	Gl
Durban Road	West Ham	E.	Kd
Durell Road	Fulham	S.W.	Ch
Durham Grove	Hackney	E.	Jc
Durham Hill Lane	Lewisham	S.E.	Kl
Durham Road	Ealing	W.	Sf
Durham Road	Islington	N.	Fb
Durham Road	Lambeth	S.E.	Gl
Durham Road	Stepney	E.	Je
Durham Road	Wandsworth	S.W.	Bk
Durham Road	West Ham	E.	Le
Durham Road	Wimbledon	S.W.	Bm
Durham Road	Wimbledon	S.W.	Ck
Durham Road	Woolwich		Ng
Durham Street	Westminster	W.C.	Fe
Durham Street	Lambeth	S.E.	Fg
Durham Terrace	Paddington	W.	De
Durham Villas	Kensington	W.	Df

STREET OR PLACE.	BOROUGH.	P.D.	MAP.
East Dulwich Station	**Camberwell**	S.E.	Gj
Eastern Dock	Stepney	E.	Hf
Eastern Empire	**Poplar**	E.	Kd
Eastern Road	Lewisham	S.E.	Jj
Eastern Road	Walthamstow U.D.	E.	Ka
Eastern Road	West Ham	E.	Ld
East Ferry Road	Poplar	E.	Kg
Eastfield Street	Stepney	E.	Je
East Ham Manor Way	East Ham	E.	Me
East Ham Level	East Ham	E.	Me
East Ham Sewage Wks.	Barking U.D.	E.	Nd
East Harding Street	City of London	E.C.	Ge
East Heath Road	Hampstead	N.W.	Db
East India Avenue	City of London	E.C.	Ge
East India Dock Road	Poplar and Stepney	E.	Je
East India Docks	Poplar	E.	Ke
East India Dockwall Road	Poplar	E.	Ke
Eastlake Road	Lambeth	S.E.	Gh
Eastlands	Camberwell	S.E.	Hk
Eastley Mews	Marylebone	W.	Ee
East London Cemetery	West Ham	E.	Ld
East London District Reservoir	Walthamstow U.D.	E.	Ja
East London District Water Works	Leyton U.D.	E.	Jb
Eastman's Street	Bethnal Green	E.	He
Eastmearn Road	Lambeth	S.E.	Gk
East Mount Street	Stepney	E.	He
Eastnor Place	St. Pancras	N.W.	Fd
Easton Place	Finsbury	W.C.	Fd
Easton Street	Finsbury	W.C.	Fd
East Putney Station	**Fulham**	S.W.	Cj
East Sheen Avenue	Barnes U.D.	S.W.	Aj
East Sheen Lodge	Barnes U.D.	S.W.	Aj
East Surrey Grove	Camberwell	S.E.	Hg
Eastward Street	Poplar	E.	Je
East Wickham	Bexley U.D.		On
East Wickham House	Bexley U.D.		Oh
Eastwood Street	Wandsworth	S.W.	Em
Eatington Road	Leyton U.D.	E.	Ka
Eaton Court	Westminster	S.W.	Ef
Eaton Grove	Islington	N.	Fb
Eaton Lane	Westminster	S.W.	Fg
Eaton Livery Stables	Westminster	S.W.	Ef
Eaton Mews	Westminster	S.W.	Ef
Eaton Mews North	Westminster	S.W.	Ef
Eaton Mews South	Westminster	S.W.	Ef
Eaton Mews West	Westminster	S.W.	Ef
Eaton Place	Chelsea and Westminster	S.W.	Ef
Eaton Place	Hackney	E.	Hc
Eaton Place	Stepney	E.	He
Eaton Road	Hendon U.D.	N.W.	Ca
Eaton Road	Lambeth	S.W.	Gj
Eaton Rise	Ealing	W.	Se
Eaton Row	Westminster	S.W.	Ef
Eaton Square	Westminster	S.W.	Ef
Eaton Street	Lambeth	S.E.	Gf
Eaton Terrace	Chelsea and Westminster	S.W.	Ef
Eaton Terrace	Marylebone	N.W.	Dd
Eaton Terrace Mews	Westminster	S.W.	Ef
Eaton Villas	Marylebone	N.W.	Dd
Eatonville Road	Wandsworth	S.W.	Ek
Ebbsfleet Road	Hampstead	N.W.	Cb
Ebenezer Place	Poplar	E.	Ke
Ebenezer Place	Stepney	E.	Je
Ebenezer Road	Hendon U.D.	N.W.	Db
Ebenezer Street	Shoreditch	N.	Gd
Ebner Street	Wandsworth	S.W.	Dj
Ebor Street	Bethnal Green & Shoreditch	E.	Hd
Eborall Place	Chelsea	S.W.	Eg
Ebsworth Road	Lewisham	S.E.	Jk
Ebury Buildings	Westminster	S.W.	Eg
Ebury Gardens	Westminster	S.W.	Eg
Ebury Mews	Westminster	S.W.	Eg
Ebury Mews East	Westminster	S.W.	Eg
Ebury Square	Westminster	S.W.	Eg
Ebury Street	Westminster	S.W.	Ef
Ebury Street	Islington	N.	Fb
Eccles Road	Battersea	S.W.	Ej
Ecclesbourne Road	Islington	N.	Gc
Eccleston Mews	Westminster	S.W.	Eg
Eccleston Place	Westminster	S.W.	Eg
Eccleston Square	Westminster	S.W.	Eg
Eccleston Square Mews	Westminster	S.W.	Eg
Eccleston Street	Westminster	S.W.	Ef
Eccleston Street East	Westminster	S.W.	Ef
Eckersley Street	Bethnal Green	E.	He
Eckett Street	Bermondsey	S.E.	Hf
Eckington Gardens	Deptford	S.E.	Jh
Eckington Road	Tottenham U.D.	N.	Ga
Eckstein Road	Battersea	S.W.	Ej
Eda Road	Lewisham	S.E.	Jj
Edale Road	Deptford	S.E.	Jg
Edbrooke Road	Paddington	W.	Dd
Eddington Street	Islington	N.	Fa
Eddiscombe Road	Fulham	S.W.	Ch
Eddystone Road	Lewisham	S.E.	Jj
Eden Grove	Islington	N.	Fc
Eden Place	Islington	N.	Fc
Eden Road	Lambeth	S.E.	Gl
Eden Road	Walthamstow U.D.	E.	Ka
Eden Street	Kingston		Sm
Eden Street	St. Pancras	N.W.	Fd
Eden Terrace	Hackney	E.	Ha
Edenbridge Road	Hackney	E.	Jc
Edenham Mews	Kensington	W.	Cd
Edenham Street	Kensington	W.	Cd
Edenhurst Avenue	Fulham	S.W.	Ch
Edenvale Street	Fulham	S.W.	Dh
Ederline Avenue	Croydon	S.W.	Fm
Edgar Place	Bethnal Green	E.	Hd
Edgar Road	Poplar	E.	Kd
Edgarly Terrace	Fulham	S.W.	Ch
Edge Hill	Wimbledon	S.W.	Cm
Edge Hill	Woolwich		Mg
Edge Hill Road	Ealing	W.	Se
Edge Street	Kensington	W.	Df
Edge Street	Wandsworth	S.W.	Dj
Edge Terrace	Kensington	W.	Df
Edgecombe Hall	Wandsworth	S.W.	Ck
Edgel Street	Wandsworth	S.W.	Dj
Edgeley Road	Wandsworth	S.W.	Fh
Edgeley Street	Wandsworth	S.W.	Fh
Edgington Road	Wandsworth	S.W.	Fm
Edgware Place	Paddington	W.	Dd
Edgware Road (part)	Hendon U.D.	N.W.	Cb
Edgware Road (part)	Willesden U.D.	N.W.	Cb
Edgware Road (part)	Marylebone	W.	Ee
Edgware Road (part)	Paddington	W.	Dd
Edgware Road Station	**Marylebone**	N.W.	Ee
Edinburgh Mansions	Westminster	S.W.	Ff
Edinburgh Road	Kensington	W.	Cd
Edinburgh Road	Stepney	E.	Je
Edinburgh Road	West Ham	E.	Ld
Edinburgh Road	Walthamstow U.D.	E.	Ja
Edison Road	Hornsey	N.	Fa
Edith Grove	Chelsea	S.W.	Dg
Edith Road	Camberwell	S.E.	Hh
Edith Road	Fulham and Hammersmith	W.	Cf
Edith Road	Leyton U.D. and West Ham	E.	Kb
Edith Road	West Ham	E.	Kc
Edith Road	Wimbledon	S.W.	Dl
Edith Row	Fulham	S.W.	Dh
Edith Street	Shoreditch	E.	Hd
Edith Terrace	Chelsea	S.W.	Dg
Edith Villas	Fulham	W.	Cg
Edithna Road	Lambeth	S.W.	Fg
Edmond's Place	Battersea	S.W.	Dj
Edmond's Place	Shoreditch	N.	Gd
Edmund Passage	City of London	E.C.	Ge
Edmund Street	Camberwell	S.E.	Gg
Edna Road	Merton U.D.	S.W.	Cm
Edna Street	Battersea	S.W.	Eh
Edney Street	Lewisham	S.E.	Hl
Edric Road	Deptford	S.E.	Jg
Edward Court	Bethnal Green	E.	Hd

STREET OR PLACE.	BOROUGH.	P.D.	MAP.
Edward Road	Acton U.D.	W.	Ad
Edward Road	Beckenham U.D.	S.E.	Jm
Edward Road	Walthamstow U.D.	E.	Ja
Edward Square	Islington	N.	Fd
Edward Street	Westminster	S.W.	Fg
Edward Street	Westminster	W.	Fe
Edward Street	Deptford	S.E.	Jg
Edward Street	St. Pancras	N.W.	Fd
Edward Street	Shoreditch	N.	Gd
Edward Street	Southwark	S.E.	Gf
Edward Street	Stepney	E.	Jd
Edward Street	Wandsworth	S.W.	El
Edward Street	West Ham	E.	Le
Edward Terrace	Chislehurst U.D.		Mm
Edward Terrace	Hampstead	N.W.	Db
Edwardes Place	Kensington	W.	Df
Edwards Lane	Stoke Newington	N.	Gb
Edward Road	Bromley		Lm
Edwards Road	Stepney	E.	Jd
Edwards Square	Kensington	W.	Cf
Edwards Sq. Cottages	Kensington	W.	Cf
Edwin Road	Camberwell	S.E.	Hg
Edwin's Row	Camberwell	S.E.	Hg
Edwin Street	West Ham	E.	Le
Eel Brook Common	Fulham	S.W.	Dh
Eele Place	Stepney	E.	He
Effie Mews	Fulham	S.W.	Dg
Effie Road	Fulham	S.W.	Dg
Effingham Road	Hornsey and Tottenham U.D.	N.	Ga
Effingham Road	Lewisham	S.E.	Kj
Effingham Street	Westminster	S.W.	Eg
Effort Street	Wandsworth	S.W.	Dl
Effra Parade	Lambeth	S.W.	Gj
Effra Road	Lambeth	S.W.	Fj
Effra Road	Wimbledon	S.W.	Dl
Effra Terrace	Lambeth	S.W.	Gj
Egan Street	Camberwell	S.E.	Hg
Egbert Place	St. Pancras	N.W.	Ec
Egbert Street	St. Pancras	N.W.	Ec
Egerton Crescent	Kensington	S.W.	Ef
Egerton Gardens	Ealing	W.	Se
Egerton Gardens	Kensington	S.W.	Ef
Egerton Mansions	Kensington	S.W.	Ef
Egerton Mews	Kensington	S.W.	Ef
Egerton Place	Kensington	S.W.	Ef
Egerton Place Stables	Kensington	S.W.	Ef
Egerton Road	Greenwich	S.E.	Kh
Egerton Road	Hackney	N.	Ha
Egerton Road	Wandsworth	S.W.	Fm
Egerton Terrace	Kensington	S.W.	Ef
Eglantine Road	Wandsworth	S.W.	Dj
Egleton Road	Poplar	E.	Kd
Eglinton Road	Poplar	E.	Jd
Eglinton Road	Woolwich		Mh
Egliston Road	Wandsworth	S.W.	Ch
Egmont Street	Deptford	S.E.	Jh
Eider Street	Southwark	S.E.	Gg
Elaine Grove	St. Pancras	N.W.	Eb
Elam Street	Lambeth	S.E.	Gh
Eland Road	Battersea	S.W.	Eh
Elbe Street	Fulham	S.W.	Dh
Elborough Street	Wandsworth	S.W.	Ck
Elbow Lane	Stepney	E.	He
Elbow Street	Stepney	E.	He
Elcho Street	Battersea	S.W.	Eg
Elcom Street	Paddington	W.	Cd
Elcot Avenue	Camberwell	S.E.	Hh
Elder Avenue	Hornsey	N.	Fa
Elder Place	Lambeth	S.E.	Gl
Elder Road	Lambeth	S.E.	Gl
Elder Street	Stepney	E.	He
Elderfield Road	Hackney	E.	Jb
Elderslie Road	Woolwich		Mj
Elderton Road	Lewisham	S.E.	Jl
Eldon Road	Hampstead	N.W.	Db
Eldon Road	Kensington	W.	Df
Eldon Road	Walthamstow U.D.	E.	Ja
Eldon Road	West Ham	E.	Le
Eldon Street	City of London and Shoreditch	E.C.	Ge
Eldon Terrace	Deptford	S.E.	Jg
Eldrick Road	Barking U.D.		Nd
Eldridge Road	Bermondsey	S.E.	Hf
Eleanor Grove	Barnes U.D.	S.W.	Bh
Eleanor Road	Hackney	E.	Hc
Eleanor Road	West Ham	E.	Lc
Eleanor Road	Woolwich		Mg
Eleanor Street	Poplar	E.	Jd
Electric Avenue	Lambeth	S.W.	Fj
Electric Lane	Lambeth	S.W.	Fj
Electric Mansions	Lambeth	S.W.	Fj
Elephant Road	Southwark	S.E.	Gf
Elephant Lane	Bermondsey	S.E.	Jf
Elephant Yard	City of London	E.C.	Ge
Elephant and Castle	Southwark	S.E.	Gf
Elephant & Castle Sta.	**Southwark**	S.E.	Gf
Elephant and Castle Theatre	**Southwark**	S.E.	Gf
Elers Road	Ealing	W.	Sf
Elfin Road	Camberwell	S.E.	Gh
Elfindale Road	Camberwell	S.E.	Gj
Elfort Road	Islington	N.	Gb
Elgar Street	Bermondsey	S.E.	Jf
Elgin Avenue	Paddington	W.	Dd
Elgin Crescent	Kensington	W.	Ce
Elgin Mews North	Paddington	W.	Dd
Elgin Mews South	Kensington	W.	Ce
Elgin Mews South	Paddington	W.	Dd
Elgin Road	Merton U.D.	S.W.	Cm
Elgin Terrace	Paddington	W.	Dd
Elgin Terrace	Woolwich		Mg
Eli Street	Fulham	W.	Cg
Elginor Road	Fulham	S.W.	Cg
Elibank Road	Woolwich		Mj
Elim Place	Bermondsey	S.E.	Gf
Elim Street	Bermondsey	S.E.	Gf
Eliot Bank	Lewisham	S.E.	Hk
Eliot Hill	Lewisham	S.E.	Kh
Eliot Park	Lewisham	S.E.	Kh
Eliot Place	Lewisham	S.E.	Kh
Eliot Vale	Lewisham	S.E.	Kh
Eliza Place	Camberwell	S.E.	Hg
Elizabeth Court	Stepney	E.	Je
Elizabeth Mews	Hampstead	N.W.	Ec
Elizabeth Place	Bermondsey	S.E.	Hf
Elizabeth Place	Poplar	E.	Ke
Elizabeth Place	Stepney	E.	He
Elizabeth Place	Wandsworth	S.W.	Dj
Elizabeth Place	Woolwich		Nh
Elizabeth Place	Finsbury	E.C.	Gd
Elizabeth Place	Southwark	S.E.	Gf
Elizabeth Road	Richmond		Ah
Elizabeth Road	Tottenham U.D.	N.	Ha
Elizabeth Street	Westminster	S.W.	Ef
Elizabeth Street	East Ham and Woolwich	E.	Mf
Elizabeth Street	Southwark	S.E.	Gg
Elizabeth Street	Stepney	E.	He
Elizabeth Terrace	Hackney	E.	Ha
Elizabeth Terrace	Woolwich		Mk
Elizabethan Charity School	Fulham	S.W.	Dh
Elkington Buildings	Westminster	W.	Fe
Ella Road	Islington	N.	Fa
Elland Road	Camberwell	S.E.	Hj
Ellen Court	Stepney	E.	He
Ellen Place	Stepney	E.	He
Ellen Street	Stepney	E.	He
Ellenborough House	Wandsworth	S.W.	Bj
Ellenborough Road	Islington	N.	Fb
Ellerby Street	Fulham	S.W.	Ch
Ellerdale Road	Hampstead	N.W.	Db
Ellerdale Street	Lewisham	S.E.	Kj
Ellerker Gardens	Richmond		Sj
Ellerslie Road	Hammersmith	W.	Be
Ellerslie Road	Lambeth	S.W.	Fj
Ellerthorp Street	Poplar	E.	Ke
Ellerton Road	Wandsworth	S.W.	Dk
Ellery Street	Camberwell	S.E.	Hh
Ellesmere Road	Bethnal Green	E.	Jd
Ellesmere Road	Chiswick U.D.	W.	Ag
Ellesmere Road	West Ham	E.	Lf
Ellesmere Road	Willesden U.D.	N.W.	Bb

STREET OR PLACE.	BOROUGH.	P.D	MAP.
Ellesmere Street	Poplar	E.	Ke
Ellingfort Road	Hackney	E.	Hc
Ellingham Road	Hammersmith	W.	Bf
Ellingham Road	Leyton U. D.	E.	Kb
Ellington Street	Islington	N.	Fc
Eliot Bank	Lewisham	S.E.	Hk
Elliots Road	Chiswick U.D.	W.	Bf
Elliott Road	Lambeth	S.W.	Gh
Elliott's Court	City of London	E.C.	Ge
Elliott's Place	Islington	N.	Gc
Elliott's Row	Southwark	S.E.	Gf
Ellis Street	Chelsea	S.W.	Ef
Elliscombe Road	Greenwich		Lg
Ellison Road	Barnes U.D.	S.W.	Bh
Ellison Road	Wandsworth	S.W.	Fm
Ellison Street	City of London	E.	He
Ellora Road	Wandsworth	S.W.	Fl
Elm Court	City of London	E.C.	Fe
Elm Court	Croydon R.D.		Em
Elm Crescent	Kingston		Sm
Elm Gardens	Hammersmith	W.	Cf
Elm Grove	Acton U.D.	W.	Be
Elm Grove	Camberwell	S.E.	Hh
Elm Grove	Hendon U.D.	N.W.	Cb
Elm Grove	Hornsey	N.	Fa
Elm Grove	Hammersmith	W.	Cf
Elm Grove	Kingston		Sm
Elm Grove	Lambeth	S.E.	Gk
Elm Grove	Wimbledon	S.W.	Cm
Elm Grove	Woolwich		Ng
Elm Lane	Lewisham	S.E.	Jk
Elm Lodge	Richmond		Sk
Elm Lodge	St. Pancras	N.	Ea
Elm Park	Chelsea	S.W.	Dg
Elm Park	Lambeth	S.W.	Fj
Elm Park Gardens	Barnes U.D.	S.W.	Bh
Elm Park Gardens	Chelsea	S.W.	Dg
Elm Park Gdns. Mews	Chelsea	S.W.	Dg
Elm Park Road	Chelsea	S.W.	Dg
Elm Park Road	Leyton U.D.	E.	Ja
Elm Park Terrace	Chelsea	S.W.	Dg
Elm Place	Kensington	S.W.	Dg
Elm Place	St. Pancras	W.C.	Fd
Elm Place East	Kensington	S.W.	Dg
Elm Road	Barnes U.D.	S.W.	Ah
Elm Road	Beckenham U.D.		Jm
Elm Road	Chislehurst U.D. & Foots Cray U.D.		Ol
Elm Road	Hampstead	N.W.	Db
Elm Road	Leyton U.D.	E.	Kb
Elm Road	The Maldens and Coombe U.D.		Am
Elm Road	St. Pancras	N.W.	Fc
Elm Road	Walthamstow U.D.	E.	Ka
Elm Road	West Ham	E.	Lc
Elm Road	Wembley U.D.		Sb
Elm Road	Kingston		Sm
Elm Row	Stepney	E.	Je
Elm Row South	Stepney	E.	Je
Elm Street	Holborn and St. Pancras	W.C.	Fd
Elm Street	Woolwich		Ng
Elm Terrace	Deptford	S.E.	Jg
Elm Terrace	Hendon U.D.	N.W.	Db
Elmar Road	Tottenham U.D.	N.	Ga
Elmbourne Road	Wandsworth	S.W.	El
Elmcroft Street	Hackney	E.	Hb
Elmdale Road	Wandsworth	S.W.	El
Elmdale Street	Fulham	W.	Cg
Elmer Lodge	Hendon U.D.	N.W.	Ca
Elmer Road	Lewisham	S.E.	Kk
Elmers End Road	Beckenham U.D. & Penge U.D.	S.E.	Hm
Elmfield	Bromley		Mm
Elmfield	Wandsworth	S.W.	Fl
Elmfield Mansions	Wandsworth	S.W.	Ek
Elmfield Road	Bromley		Lm
Elmfield Road	Wandsworth	S.W.	Ek
Elmgrove Road	Barnes U.D.	S.W.	Bh
Elmgrove Road	Ealing	W.	Sf
Elmhurst	Ealing	W.	Se
Elmhurst	Wimbledon	S.W.	Cl
Elmhurst Road	Croydon R.D.		Em
Elmhurst Road	West Ham	E.	Lc
Elmington Road	Camberwell	S.E.	Gh
Elmington Terrace	Camberwell	S.E.	Gh
Elmira Street	Lewisham	S.E.	Kj
Elmley House	Wandsworth	S.W.	Ck
Elmore Road	Leyton U.D.	E.	Kb
Elmore Street	Islington	N.	Gc
Elms, The	Acton U.D.	W.	Ae
Elms, The	Barnes U.D.	S.W.	Ah
Elms, The	Camberwell	S.E.	Hk
Elms, The	Chislehurst U.D.		Om
Elms, The	Croydon	S.E.	Gm
Elms, The	Ealing	W.	Sd
Elms, The	Hampstead	N.W.	Da
Elms, The	Hampton Wick U.D.		Sm
Elms, The	Hendon U.D.	N.W.	Ca
Elms, The	Wandsworth	S.W.	Bk
Elms Mews	Paddington	W.	De
Elms Mews	Wandsworth	S.W.	Fj
Elms Road	Camberwell	S.E.	Gj
Elms Road	Wandsworth	S.W.	Fj
Elmsfield Road	Walthamstow U.D.	E.	Ja
Elmsleigh Road	Wandsworth	S.W.	Dj
Elmsleigh Terrace	Wandsworth	S.W.	Dj
Elmsley Place	Hackney	E.	Hc
Elmstead Court	Bromley		Ml
Elmstead Glade	Bromley		Mm
Elmstead Grange	Bromley		Mm
Elmstead Knoll	Bromley R.D.		Ml
Elmstead Lane (part)	Bromley R.D.		Ml
Elmstead Lane (part)	Bromley		Mm
Elmstead Lane	Woolwich		Lk
Elmstead Lodge	Bromley		Mm
Elmstead Woods Sta.	**Bromley**		Mm
Elmstead Wood	Bromley		Ml
Elmstead Wood	Bromley		Mm
Elmstone Road	Fulham	S.W.	Ch
Elm Tree Road	Marylebone	N.W.	Dd
Elmwood Road	Croydon R.D.		Em
Elm Wood Road	Camberwell	S.E.	Gj
Elnathan Mews	Paddington	W.	Dd
Elphick Street	West Ham	E.	Le
Elphinstone Street	Islington	N.	Gb
Elric Street	Hammersmith	W.	Cg
Elsa Street	Stepney	E.	Je
Elsburg Street	Hampstead	N.W.	Ec
Elsdale Street	Hackney	E.	Jc
Elsdon Street	Wandsworth	S.W.	El
Elsenham Street	Wandsworth	S.W.	Ck
Elsham Road	Kensington	W.	Cf
Elsham Road	Leyton U.D.	E.	Kb
Elsie Road	Camberwell	S.E.	Hj
Elsiemaud Street	Lewisham	S.E.	Jj
Elsinore Road	Lewisham	S.E.	Jk
Elsley Road	Battersea	S.W.	Eh
Elspeth Road	Battersea	S.W.	Ej
Elsted Street	Southwark	S.E.	Gg
Elstree Hill	Lewisham	S.E.	Km
Elstree Road	Lewisham		Km
Elswick Road	Lewisham	S.E.	Kh
Elswick Street	Fulham	S.W.	Dh
Elsworthy Road	Hampstead	N.W.	Ec
Elsworthy Terrace	Hampstead	N.W.	Ec
Eltham Bottom	Woolwich		Mh
Eltham Bridge	Woolwich		Lj
Eltham Common	Woolwich		Mh
Eltham Court	Woolwich		Mk
Eltham Green	Woolwich		Lj
Eltham High Street	Woolwich		Mj
Eltham Lodge	Woolwich		Mk
Eltham Park	Woolwich		Mj
Eltham Park Road	Woolwich		Mj
Eltham Pottery	Woolwich		Nj
Eltham Road	Lewisham and Woolwich	S.E.	Lj
Eltham Station	**Woolwich**		Mk
Eltham Street	Southwark	S.E.	Gg
Elthiron Road	Fulham	S.W.	Dh
Elthorne Road	Islington	N.	Fa

STREET OR PLACE.	BOROUGH.	P.D.	MAP.
Elthruda Road	Lewisham	S.E.	Kj
Elton Road	Kingston		Sm
Elton Street	Islington	N.	Hb
Eltringham Street	Wandsworth	S.W.	Dj
Elvaston Mews	Kensington	S.W.	Df
Elvaston Place	Kensington	S.W.	Df
Elverson Road	Greenwich and Lewisham	S.E.	Kh
Elverson Road South	Greenwich	S.E.	Kh
Elvino Road	Lewisham	S.E.	Jl
Elwell Road	Lambeth and Wandsworth	S.W.	Fh
Elwin Street	Bethnal Green	E.	Hd
Elwood Street	Islington	N.	Gb
Ely Court	Holborn	E.C.	Ge
Ely Mews	Holborn	E.C.	Ge
Ely Place	Holborn	E.C.	Ge
Ely Place	Lambeth	S.W.	Fg
Ely Place	Shoreditch	N.	Hd
Ely Place	Southwark	S.E.	Gf
Ely Place	Stepney	E.	He
Ely Road	Leyton U.D.	E.	Ka
Ely Terrace	Stepney	E.	Je
Elyne Road	Hornsey	N.	Ga
Elysium Mews	Fulham	S.W.	Ch
Emanuel Avenue	Acton U.D.	W.	Ae
Emanuel School	Battersea	S.W.	Dj
Emba Street	Bermondsey	S.E.	Hf
Embankment Gardens	Chelsea	S.W.	Eg
Embankment Mans.	Chelsea	S.W.	Eg
Embankment Resids.	Chelsea	S.W.	Eg
Embankment Station	**Westminster**	W.C.	Ff
Embden Road	Fulham	S.W.	Dh
Embleton Road	Lewisham	S.E.	Kj
Emerald Street	Holborn	W.C.	Fe
Emerson Street	Southwark	S.E.	Gf
Emery Hill	Westminster	S.W.	Ff
Emery Street	Southwark	S.E.	Gf
Emery's Place	Stepney	E.	He
Emily Place	Islington	N.	Gb
Emily Road	Camberwell	S.E.	Hg
Emily Street	Paddington	W.	De
Emily Street	West Ham	E.	Le
Emlyn Road	Chiswick U.D. & Hammersmith	W.	Bf
Emma Place	Kensington	W.	Df
Emma Road	West Ham	E.	Ld
Emma Street	Battersea	S.W.	Dj
Emma Street	Bethnal Green	E.	Hd
Emma Street	West Ham	E.	Mf
Emma Terrace	Camberwell	S.E.	Gg
Emmanuel Road	Wandsworth	S.W.	Ek
Emmett Street	Poplar & Stepney	E.	Jf
Emmott Street	Stepney	E.	Je
Emperor's Gate	Kensington	S.W.	Df
Empire Music Hall	**Battersea**	S.W.	Dh
Empire Music Hall	**Westminster**	W.C.	Fe
Empress Music Hall	**Lambeth**	S.W.	Fj
Empress Street	Southwark	S.E.	Gg
Empson Street	Poplar	E.	Ke
Emscote Road	Lewisham	S.E.	Kj
Emsworth Street	Wandsworth	S.W.	Fk
Emu Road	Battersea	S.W.	Eh
Ena Road	Wandsworth	S.W.	Fl
Enbrook Street	Paddington	W.	Cd
Endell Street	Holborn & Westminster	W.C.	Fe
Enderby's Wharf	Greenwich	S.E.	Kg
Endive Street	Stepney	E.	Je
Endlesham Road	Wandsworth	S.W.	Ek
Endsleigh Gardens	St. Pancras	N.W.	Fd
Endsleigh Street	St. Pancras	W.C.	Fd
Endsleigh Terrace	St. Pancras	N.W.	Fd
Endwell Road	Deptford	S.E.	Jh
Endymion Road	Hornsey & Tottenham U.D.	N.	Ga
Endymion Road	Lambeth	S.W.	Fj
Enfield Road	Acton U.D.	W.	Af
Enfield Road	Brentford U.D.		Sg
Enfield Road	Hackney	N.	Hc
Enfield Road	Hornsey	N.	Fa
Engadine Street	Wandsworth	S.W.	Ck
Engate Street	Lewisham	S.E.	Kj
Engine Court	Westminster	S.W.	Ff
Engineer Road	Woolwich		Mg
England's Lane	Hampstead	N.W.	Ec
Englefield Road	Hackney and Islington	N.	Gc
Engleheart Road	Lewisham	S.E.	Kk
Englewood Road	Wandsworth	S.W.	Ej
English Ground	Southwark	S.E.	Gf
Enid Street	Bermondsey	S.E.	Hf
Enkel Street	Islington	N.	Fb
Enmore Road	Wandsworth	S.W.	Cj
Ennerdale Road	Richmond		Ah
Ennersdale Road	Lewisham	S.E.	Kj
Ennis Road	Hornsey	N.	Ga
Ennis Road	Woolwich		Ng
Enniskillen Road	West Ham	E.	Ld
Ennismore Gardens	Westminster	S.W.	Ef
Ennismore Gardens Mews	Westminster	S.W.	Ef
Ennismore Mews	Westminster	S.W.	Ef
Enoch Court	City of London & Stepney	E.	He
Enslin Road	Woolwich		Mk
Ensoll Street	Lewisham	S.E.	Hl
Ensor Mews	Kensington	S.W.	Dg
Entick Street	Stepney	E.	He
Epirus Mansions	Fulham	S.W.	Cg
Epirus Road	Fulham	S.W.	Cg
Epping Forest	Leyton U.D.	E.	La
Epple Road	Fulham	S.W.	Ch
Epsom Road	Leyton U.D.	E.	Ka
Equity Buildings	St. Pancras	N.W.	Fd
Erasmus Street	Westminster	S.W.	Fg
Erebus Terrace	Hammersmith	W.	Ce
Eresby Mews	Hampstead	N.W.	Dc
Eresby Road	Hampstead	N.W.	Dc
Eric Road	Willesden U.D.	N.W.	Bc
Eric Street	Stepney	E.	Jd
Eridge Road	Acton U.D.	W.	Af
Erivane Street	Fulham	S.W.	Dh
Erkenwald Road	Barking U.D.		Nd
Erlam Road	Deptford	S.E.	Hg
Erlanger Road	Deptford	S.E.	Jh
Ermine Road	Lewisham	S.E.	Jj
Ernald Avenue	East Ham	E.	Md
Ernest Place	Bethnal Green	E.	Hd
Ernest Place	Lambeth	S.E.	Gl
Ernest Road	West Ham	E.	Lb
Ernest Road	West Ham	E.	Le
Ernest Street	Bermondsey	S.E.	Hf
Ernest Street	Kensington	W.	Df
Ernest Street	Lambeth	S.E.	Gl
Ernest Street	Stepney	E.	Je
Erpingham Road	Wandsworth	S.W.	Ch
Errington Road	Paddington	W.	Cd
Erridge Road	Merton U.D.	S.W.	Dm
Errol Street	Finsbury	E.C.	Gd
Erskine Hill	Hendon U.D.	N.W.	Da
Erskine Mews	St. Pancras	N.W.	Ec
Erskine Road	St. Pancras	N.W.	Ec
Erskine Road	Walthamstow U.D.	E.	Ja
Escott Cottages	Poplar	E.	Jf
Esher Street	Lambeth	S.E.	Fg
Esher Street	Westminster	S.W.	Fg
Eskdale Villas	Lambeth	S.W.	Fh
Esmeralda Road	Bermondsey	S.E.	Hg
Esmond Road	Acton U.D.	W.	Bf
Esmond Road	Willesden U.D.	N.W.	Cc
Esmond Street	Wandsworth	S.W.	Cj
Esparto Street	Wandsworth	S.W.	Dk
Essendine Road	Paddington	W.	Dd
Essex County Cricket Ground	Leyton U.D.	E.	Ka
Essex Court	City of London	E.C.	Fe
Essex Gardens	Tottenham U.D.	N.	Ga
Essex Grove	Croydon	S.E.	Gm
Essex Park Mews	Acton U.D.	W.	Bf
Essex Place	Lewisham	S.E.	Kh
Essex Place	Shoreditch	E.	Hc
Essex Road	Acton U.D.	W.	Ae
Essex Road	Islington	N.	Gc
Essex Road	Leyton U.D.	E.	Ka

STREET OR PLACE.	BOROUGH.	P.D.	MAP.
Essex Road	Walthamstow U.D.	E.	Ja
Essex Road	Willesden U.D.	N.W.	Bc
Essex Street	City of London	E.C.	Ge
Essex Street	Hackney	E.	Hc
Essex Street	Shoreditch	N.	Hd
Essex Street	Stepney	E.	He
Essex Street	West Ham	E.	Lc
Essex Street	Westminster	W.C.	Fe
Essex Villas	Kensington	W.	Df
Essex Wharf	Hackney	E.	Jb
Esslan Street	Stepney	E.	Je
Estcourt Road	Battersea	S.W.	Ej
Estcourt Road	Fulham	S.W.	Cg
Este Road	Battersea	S.W.	Eh
Estolle Road	St. Pancras	N.W.	Eb
Esther Road	Leyton U.D.	E.	Ka
Eswyn Road	Wandsworth	S.W.	El
Etchingham Road	Leyton U.D.	E.	Kb
Etham Place	Southwark	S.E.	Gf
Etham Street	Southwark	S.E.	Gf
Ethard Road	Camberwell	S.E.	Hg
Ethel Road	West Ham	E.	Le
Ethel Street	Lambeth	S.E.	Gf
Ethel Street	Southwark	S.E.	Gg
Ethel Street	West Ham	E.	Le
Ethelbert Road	Bromley		Lm
Ethelbert Road	Wimbledon	S.W.	Cm
Ethelbert Street	Wandsworth	S.W.	Ek
Ethelburga Street	Battersea	S.W.	Eh
Etheiden Road	Hammersmith	W.	Bf
Ethelgar Road	Woolwich		Nk
Ethelm Road	Lambeth	S.E.	Ff
Ethelred Street	Lambeth	S.E.	Fg
Etherley Road	Tottenham U.D.	N.	Ga
Etherow Street	Camberwell	S.E.	Hk
Etherston	Wandsworth	S.W.	Fl
Etherton Street	Southwark	S.E.	Gg
Etloe Road	Leyton U.D.	E.	Jb
Eton Avenue	Hampstead	N.W.	Ec
Eton Grove	Lewisham	S.E.	Kj
Eton Mews	Hampstead	N.W.	Ec
Eton Mews West	Hampstead	N.W.	Ec
Eton Place	Hackney	E.	Jc
Eton Place	Hampstead	N.W.	Ec
Eton Road	Hampstead	N.W.	Ec
Eton Road	Woolwich		Ng
Eton Street	Richmond		Sj
Eton Street	St. Pancras	N.W.	Ec
Eton Street	Woolwich		Ng
Eton Villas	Hampstead	N.W.	Ec
Eton Villas	Willesden U.D.	N.W.	Cd
Etropol Road	Hackney	E.	Jb
Etruria Street	Battersea	S.W.	Fh
Etta Street	Deptford	S.E.	Jg
Ettrick Street	Poplar	E.	Ke
Eugenia Road	Bermondsey and Deptford	S.E.	Jg
Eunice Road	Chiswick U.D.	W.	Bf
Eureka Road	Kingston		Am
Europa Place	Battersea	S.W.	Dh
Europa Place	Finsbury	E.C.	Gd
Eustace Place	Woolwich		Mg
Eustace Road	Fulham	S.W.	Dg
Euston Buildings	St. Pancras	N.W.	Fd
Euston Empire	**St. Pancras**	W.C.	Fd
Euston Grove	St. Pancras	N.W.	Fd
Euston Road	Marylebone & St. Pancras	N.W.	Fd
Euston Road Station	**St. Pancras**	N.W.	Fd
Euston Square	St. Pancras	N.W.	Fd
Euston Station	**St. Pancras**	N.W.	Fd
Euston Street	St. Pancras	N.W.	Fd
Evandale Road	Lambeth	S.W.	Gh
Evangelist Court	City of London	E.C.	Ge
Eve Road	Leyton U.D.	E.	Kb
Eve Road	Wembley U.D.		Sc
Eve Road	West Ham	E.	Kd
Evelina Road	Camberwell and Deptford	S.E.	Hh
Evelina Road	Penge U.D.	S.E.	Hm
Evelyn Buildings	Deptford	S.E.	Jh
Evelyn Gardens	Kensington	S.W.	Dg
Evelyn Mews	Deptford	S.E.	Jg
Evelyn Road	Acton U.D.	W.	Af
Evelyn Road	Richmond		Sh
Evelyn Road	Walthamstow U.D.	E.	Ka
Evelyn Road	West Ham	E.	Lf
Evelyn Road	Wimbledon	S.W.	Dl
Evelyn Street	Bermondsey, Deptford, Greenwich	S.E.	Jg
Evelyn Street	Shoreditch	N.	Ga
Everard Street	Stepney	E.	He
Everest Road	Woolwich		Mj
Everett Street	Battersea	S.W.	Fg
Everilda Street	Islington	N.	Fc
Evering Road	Hackney	E.&N.	Hb
Everington Street	Fulham	W.	Cg
Everleigh Street	Islington	N.	Fa
Eversfield Road	Richmond		Sh
Eversholt Mews	St. Pancras	N.W.	Ec
Eversholt Street	St. Pancras	N.W.	Ec
Evershot Road	Islington	N.	Fa
Eversleigh Road	Battersea	S.W.	Eh
Eversleigh Road	East Ham	E.	Md
Eversley Buildings	Westminster	W.C.	Fe
Eversley Road	Croydon	S.E.	Gm
Eversley Road	Greenwich		Lg
Eve's Place	Southwark	S.E.	Gf
Evesham Mews	Hammersmith	W.	Ce
Evesham Road	West Ham	E.	Lc
Evesham Street	Hammersmith	W.	Ce
Ewald Road	Fulham	S.W.	Dg
Ewart Road	Lewisham	S.E.	Jk
Ewelme Road	Lewisham	S.E.	Hk
Ewer Street	Southwark	S.E.	Gf
Ewhurst Court	Southwark	S.E.	Ge
Ewhurst Road	Lewisham	S.E.	Jj
Ewing Road	Stepney	E.	Jd
Exbury Road	Lewisham	S.E.	Jk
Excelsior Road	Kingston		Sm
Exchange Buildings	City of London	E.	He
Exchange Court	Westminster	W.C.	Fe
Exeter House	Fulham	S.W.	Cj
Exeter Mews	Hampstead	N.W.	Db
Exeter Road	Walthamstow U.D.	E.	Ja
Exeter Road	West Ham	E.	Le
Exeter Road	Willesden U.D.	N.W.	Cc
Exeter Street	Marylebone	N.W.	Ed
Exeter Street	Westminster	W.C.	Fe
Exhibition Road	Kensington and Westminster	S.W.	Df
Exing Road	West Ham	E.	Le
Exmoor Street	Kensington	W.	Cd
Exmouth Place	Hackney	E.	Hc
Exmouth Road	Walthamstow U.D.	E.	Ja
Exmouth Street	Finsbury	W.C.	Gd
Exmouth Street	St. Pancras	N.W.	Fd
Exmouth Street	Stepney	E.	He
Exon Street	Southwark	S.E.	Gg
Exton Street	Lambeth	S.E.	Ff
Eynella Road	Camberwell	S.E.	Hk
Eynham Road	Hammersmith	W.	Ce
Eyot Gardens	Hammersmith	W.	Bg
Eyot Terrace	Hammersmith	W.	Bg
Eyre Court	Holborn	E.C.	Gd
Eyre Place	Holborn	E.C.	Gd
Eyre Street Hill	Holborn	E.C.	Gd
Eyre Terrace	Holborn	E.C.	Gd
Eythorn Road	Lambeth	S.W.	Gh
Ezra Street	Bethnal Green	E.	Hd
Fabian Road	Fulham	S.W.	Cg
Fabian Street	East Ham	E.	Me
Façade, The	Islington	N.	Fa
Façade, The	Lewisham	S.E.	Hk
Factory Place	Poplar	E.	Kg
Factory Road	West Ham	E.	Mf
Factory Square	Wandsworth	S.W.	Fm
Fair Street	Bermondsey	S.E.	Hf
Fair Street	Stepney	E.	Je
Fairbank Street	Shoreditch	N.	Gd
Fairbridge Road	Islington	N.	Fb

STREET OR PLACE.	BOROUGH	P.D.	MAP.
Fairclough Stree	Stepney	E.	Ha
Fairfax Mews	Hampstead	N.W.	Dc
Fairfax Road	Chiswick U.D.	W.	Bf
Fairfax Road	Hampstead	N.W.	Dc
Fairfax Road	Hornsey&Tottenham U.D.	N.	Ga
Fairfax Road	Teddington U.D.		Sm
Fair Field	Kingston		Sm
Fairfield Gardens	Hornsey	N.	Fa
Fairfield Place	Kingston		Sm
Fairfield Road	Beckenham U.D.		Jm
Fairfield Road	Bromley		Lm
Fairfield Road	Greenwich		Lg
Fairfield Road	Hornsey	N.	Fa
Fairfield Road	Kingston		Sm
Fairfield Road	Lewisham	S.E.	Ll
Fairfield Road	Poplar	E.	Jd
Fairfield South	Kingston		Sm
Fairfield Street	Wandsworth	S.W.	Dj
Fairfield West	Kingston		Sm
Fairfoot Road	Poplar	E.	Jd
Fairford Grove	Lambeth	S.E.	Gg
Fairhazel Gardens	Hampstead	N.W.	Dc
Fairholme Road	Fulham	W.	Cg
Fairholt Road	Stoke Newington	N.	Ga
Fairland Road	West Ham	E.	Lc
Fairland Road South	West Ham	E.	Lc
Fairlawn	The Maldens and Coombe U.D.		Al
Fairlawn	Wandsworth	S.W.	Ck
Fairlawn	Wandsworth	S.W.	Fj
Fair Lawn	Lewisham	S.E.	Hk
Fair Lawn	Lewisham	S.E.	Hl
Fairlawn Avenue	Acton U.D.	W.	Af
Fairlawn Grove	Acton U.D.	W.	Af
Fairlawn Park	Lewisham	S.E.	Jl
Fairlawn Road	Wimbledon	S.W.	Cm
Fairlight Avenue	Willesden U.D.	N.W.	Bd
Fairlight Road	Wandsworth	S.W.	Dl
Fairlop Place	Marylebone	N.W.	Dd
Fairlop Road	Leyton U.D.	E.	Ka
Fairmead Road	Islington	N.	Fb
Fairmile Avenue	Wandsworth	S.W.	Fl
Fairmount Road	Lambeth	S.W.	Fj
Fairseat	St. Pancras	N.	Ea
Fairthorn Road	Greenwich		Lg
Fairview Place	Wandsworth	S.W.	Fk
Fairview Road	Croydon	S.W.	Fm
Fairview Road	Tottenham U.D.	N.	Ha
Fairy Street	Hackney	N.	Hb
Faith Street	Stepney	E.	He
FaithfulVirginConv'nt	Croydon	S.E.	Gl
Fakenham Street	Islington	N.	Fc
Falcon, The	Willesden U.D.	N.W.	Cd
Falcon Avenue	City of London	E.C.	Ge
Falcon Court	City of London	E.C.	Ge
Falcon Court	Stoke Newington	N.	Gb
Falcon Grove	Battersea	S.W.	Eh
Falcon Place	Stoke Newington	N.	Gb
Falcon Road	Battersea	S.W.	Dh
Falcon Square	City of London	E.C.	Ge
Falcon Street	City of London	E.C.	Ge
Falcon Terrace	Battersea	S.W.	Eh
Falcon Wharf	Southwark	S.E.	Ge
Falconburg Mews	Westminster	W.	Fe
Falconburgh Court	Westminster	W.C.	Fe
Falconwood	Woolwich		Nh
Falka Place	Southwark	S.E.	Gf
Falkland Park	Croydon	S.E.	Gm
Falkland Place	St. Pancras	N.W.	Ec
Falkland Road	St. Pancras	N.W.	Fb
Fallsbrook Road	Wandsworth	S.W.	Em
Falmouth Chambers	Southwark	S.E.	Gf
Falmouth Road	Southwark	S.E.	Gf
Falmouth Street	West Ham	E.	Kc
Fallsbrook Road	Wandsworth	S.W.	Em
False Point	Barking U.D.		Oe
Falstaff Yard	Southwark	S.E.	Gf
Fane Street	Fulham	W.	Cg
Fann Court	Finsbury	E.C.	Ge
Fann Street	City of London and Finsbury	E.C.	Ge
Fanny Road	Barnes U.D.	S.W.	Bg
Fanshaw Street	Shoreditch	E.C.	Gd
Fantassie Street	Greenwich	S.E.	Kg
Fanthorpe Road	Wandsworth	S.W.	Ch
Faraday Mansions	Fulham	W.	Cg
Faraday Road	Acton U.D.	W.	Ae
Faraday Road	Kensington	W.	Ce
Faraday Road	Wimbledon	S.W.	Dl
Faraday Street	Southwark	S.E.	Gg
Farleigh Road	Hackney	N.	Hb
Farley Road	Lewisham	S.E.	Kk
Farlow Road	Wandsworth	S.W.	Ch
Farlton Road	Wandsworth	S.W.	Dk
Farm Avenue	Wandsworth	S.W.	Fl
Farm Lane	Fulham	S.W.	Dg
Farm Street	Westminster	W.	Ee
Farmcote Place	Bermondsey	S.E.	Hf
Farmdale Road	Greenwich	S.E.	Lg
Farmer Road	Leyton U.D.	E.	Ka
Farmer Street	Kensington	W.	Df
Farmer's Road	Camberwell and Lambeth and Southwark	S.E.	Gg
Farmer's Row	Stepney	E.	Je
Farmiloe Road	Leyton U.D.	E.	Ja
Farnaby Road	Beckenham U.D. and Bromley		Km
Farnan Road	Wandsworth	S.W.	Fl
Farnborough Road	Lewisham	S.E.	Hk
Farncombe Street	Bermondsey	S.E.	Hf
Farndon Row	Camberwell	S.E.	Gh
Farnham Place	Southwark	S.E.	Gf
Farnham Royal	Lambeth	S.E.	Fg
Faroe Road	Hammersmith	W.	Cf
Farquhar Road	Camberwell and Lambeth	S.E.	Hl
Farrance Street	Stepney	E.	Je
Farrant Street	Paddington	W.	Cd
Farrar Street	Lambeth	S.W.	Gh
Farrar's Buildings	City of London	E.C.	Fe
Farringdon Avenue	City of London	E.C.	Ge
Farringdon Market	City of London	E.C.	Ge
Farringdon Road (part)	Finsbury	E.C.	Fd
Farringdon Road (part)	Holborn	E.C.	Ge
Farringdon Road (part)	Holborn	W.C.	Ge
Farringdon Road bldgs.	Finsbury	E.C.	Ge
Farringdon Street	City of London	E.C.	Ge
Farringdon St. Station	**Finsbury**	E.C.	Ge
Farringford Road	West Ham	E.	Kc
Farwig Lane	Bromley		Lm
Fashion Court	Stepney	E.	He
Fashion Street	Stepney	E.	He
Fashoda Place	Fulham	S.W.	Dh
Fashoda Street	Greenwich	S.E.	Kg
Fassett Road	Hackney	E.	Hc
Fassett Road	Kingston		Sm
Fassett Square	Hackney	E.	Hc
Fauconberg Road	Chiswick U.D.	W.	Ag
Faulkner Street	Deptford	S.E.	Hh
Faunce Street	Southwark	S.E.	Gg
Faustin Place	Bermondsey	S.E.	Jf
Favart Road	Fulham	S.W.	Dh
Faversham Road	Lewisham	S.E.	Jk
Favonia Street	Poplar	E.	Kd
Fawcett Mews	Kensington	S.W.	Dg
Fawcett Road	Deptford	S.E.	Jg
Fawcett Street	Kensington	S.W.	Dg
Fawcett's Court	Bermondsey	S.E.	Hf
Fawe Park Road	Wandsworth	S.W.	Cj
Fawe Street	Poplar	E.	Ke
Fawley Road	Hampstead	N.W.	Dc
Fawnbrake Avenue	Lambeth	S.E.	Gj
Faygate Road	Wandsworth	S.W.	Fk
Fayland Avenue	Wandsworth	S.W.	El
Fayland Road	Wandsworth	S.W.	El
Fearnley Road	Camberwell	S.E.	Gh
Fearon Street	Greenwich	S.E.	Lg
Feathers Court	City of London	E.C.	Ge
Feathers Hotel	Ealing	W.	Se
FeatherstoneBuildings	Holborn	W.C.	Fe
Featherstone Court	Finsbury	E.C.	Gd
Featherstone Street	Finsbury	E.C.	Gd
Federation Road	Woolwich		Og
Felday Road	Lewisham	S.E.	Kj

STREET OR PLACE.	BOROUGH.	P.D.	MAP.
Felden Street	Fulham	S.W.	Ch
Feldheim	Wandsworth	S.W.	Ck
Felix Avenue	Hornsey	N.	Fa
Felix Street	Bethnal Green	E.	Hd
Felix Street	Lambeth	S.E.	Ff
Felixstowe Road	Willesden U.D.	N.W.	Cd
Fell Street	City of London	E.C.	Ge
Fellbrigg Road	Camberwell	S.E.	Hj
Fellbrigg Street	Bethnal Green	E.	He
Fellows Road	Hampstead	N.W.	Ec
Fellows Street	Shoreditch	E.	Hd
Fells Square	Southwark	S.E.	Gf
Felma Court	Greenwich	S.E.	Jg
Felmingham Road	Beckenham U.D.		Hm
Felsham Road	Wandsworth	S.W.	Ch
Felstead Street	Hackney	E.	Jc
Feltham Mews	Chelsea	S.W.	Ef
Feltham Road	Croydon R.D.		Em
Felton Street	Shoreditch	N.	Gd
Fen Court	City of London	E.C.	Ge
Fen Street	Hackney	E.	Jc
Fenchurch Avenue	City of London	E.C.	Ge
Fenchurch Buildings	City of London	E.C.	Ge
Fenchurch Street	City of London	E.C.	Ge
Fenchurch St. Station	**City of London**	E.C.	He
Fendall Street	Bermondsey	S.E.	Hf
Fendick Road	Camberwell	S.E.	Hg
Fenelon Road	Kensington	W.	Cg
Fenham Road	Camberwell	S.E.	Hh
Fenner Road	Bermondsey	S.E.	Hf
Fenning Street	Southwark	S.E.	Gf
Fenning's Wharf	Bermondsey	S.E.	Gf
Fenstanton	Lambeth	S.W.	Fk
Fentiman Road	Lambeth	S.W.	Fg
Fenton Street	Greenwich	S.E.	Kg
Fenton's Avenue	West Ham	E.	Ld
Fenwick Grove	Camberwell	S.E.	Hj
Fenwick Place	Lambeth	S.W.	Fj
Fenwick Road	Camberwell	S.E.	Hj
Fenwick Street	Woolwich		Mh
Ferdinand Place	St. Pancras	N.W.	Ec
Ferdinand Street	St. Pancras	N.W.	Ec
Fergus Road	Islington	N.	Gc
Ferguson Cottages	Islington	N.	Gc
Ferguson's Wharf	Poplar	E.	Jg
Ferme Park Sidings	Hornsey	N.	Fa
Ferme Park Road	Hornsey	N.	Fa
Fermor Road	Lewisham	S.E.	Jk
Fermoy Road	Paddington	W.	Cd
Fern Bank	Barnes U.D.	S.W.	Aj
Fernbank	Wandsworth	S.W.	Bk
Fern Road	West Ham	E.	Lc
Fern Street	Poplar	E.	Je
Fern Villas	Greenwich	S.E.	Lg
Fernbrook Road	Lewisham	S.E.	Kj
Ferncliffe Road	Hackney	E.	Hc
Ferncroft Avenue	Hampstead	N.W.	Db
Ferndale Road	Lambeth	S.W.	Fj
Ferndale Road	Leyton U.D.	E.	Lb
Ferndale Road	Tottenham U.D.	N.	Ha
Ferndale Road	Walthamstow U.D.	E.	Ka
Ferndale Road	West Ham	E.	Lc
Ferndale Street	East Ham	E.	Me
Ferndean Road	Lambeth	S.E.	Gj
Fern Dene	Kingsbury U.D.		Aa
Fernhead Road	Paddington	W.	Cd
Fernhead Yard	Paddington	W.	Cd
Fernhill	Ham U.D.		Sl
Fernhill Street	Woolwich	E.	Mf
Fernhill Terrace	Woolwich	E.	Mf
Fernholme Road	Camberwell	S.E.	Jj
Fernhurst Road	Fulham	S.W.	Ch
Fernlea Road	Wandsworth	S.W.	Ek
Fernleigh Villas	Camberwell	S.E.	Hk
Fernshaw Road	Chelsea	S.W.	Dg
Fernshaw Mansions	Chelsea	S.W.	Dg
Fernside	Hendon U.D.	N.W.	Da
Fernside Road	Wandsworth	S.W.	Ek
Fernthorpe Road	Wandsworth	S.W.	El
Ferntower Road	Islington	N.	Gc
Fernwood	Wandsworth	S.W.	Ck
Ferrand Street	Bermondsey	S.E.	Gf

STREET OR PLACE.	BOROUGH.	P.D.	MAP.
Ferrers Road	Wandsworth	S.W.	Fl
Ferrier Court	Stepney	E.	Jd
Ferrier Street	Wandsworth	S.W.	Dj
Ferris Court	Stepney	E.	Jd
Ferris Road	Camberwell	S.E.	Hj
Ferron Road	Hackney	E.	Hb
Ferry Hill	Hackney	E.	Ha
Ferry Lane	Brentford U.D.		Jg
Ferry Lane	Richmond		Ag
Ferry Passage	Hackney	E.	Ha
Ferry Place	Greenwich	S.E.	Kg
Ferry Road	Barnes U.D.	S.W.	Bh
Ferry Road	Poplar	E.	Jg
Ferry Street	Lambeth	S.E.	Ff
Ferry Street	Poplar	E.	Kg
Festing Road	Wandsworth	S.W.	Ch
Fetter Lane	City of London	E.C.	Ge
Ffinch Street	Deptford	S.E.	Jg
Field Court	Holborn	W.C.	Fe
Field Mews	Wandsworth	S.W.	Dj
Field Place	Battersea	S.W.	Dh
Field Place	Hampstead	N.W.	Db
Field Road	Fulham	W.	Cg
Field Road	Walthamstow U.D.	E.	Ja
Field Road	West Ham	E.	Lb
Field Street	St. Pancras	W.C.	Fd
Fielders Meadow	Fulham	S.W.	Ch
Fieldgate Street	Stepney	E.	He
Fieldhead	Wimbledon	S.W.	Cl
Fieldhouse Road	Wandsworth	S.W.	Fk
Fielding Road	Acton U.D.	W.	Bf
Fielding Road	Hammersmith	W.	Cf
Fieldway Crescent	Islington	N.	Gc
Fife Road	Barnes U.D.	S.W.	Aj
Fife Road	Kingston		Sm
Fife Road	West Ham	E.	Le
Fife Terrace	Islington	N.	Fd
Fife's Court	Bethnal Green	E.	Hd
Fifth Avenue	Paddington	W.	Cd
Figg's Marsh	Croydon R.D.		Em
Figtree Court	City of London	E.C.	Fe
Filey Avenue	Hackney	N.	Ha
Fillebrook Road	Leyton U.D.	E.	Ka
Filmer Road	Fulham	S.W.	Ch
Finborough Road	Croydon R.D.	S.W.	Em
Finborough Road	Kensington	S.W.	Dg
Finborough Road	Wandsworth	S.W.	El
Finch Lane	City of London	E.C.	Ge
Finch Street	Stepney	E.	He
Fincham Street	Wandsworth	S.W.	Ej
Finchley Place	Hampstead and Marylebone	N.W.	Dd
Finchley Road (part)	Hampstead	N.W.	Db
Finchley Road (part)	Hendon U.D.	N.W.	Da
Finchley Road (part)	Marylebone	N.W.	Dc
Finchley Road	Southwark	S.E.	Gg
Finchley Road Station	**Hampstead**	N.W.	Dc
Finch's Court	Poplar	E.	Jd
Finck Street	Lambeth	S.E.	Ff
Finden Road	East Ham	E.	Lc
Findhorn Street	Poplar	E.	Ke
Findock Mews	Paddington	W.	Dd
Findon Road	Hammersmith	W.	Bf
Fingal Street	Greenwich	S.E.	Lg
Finland Street	Deptford	S.E.	Jj
Finlay Street	Fulham	S.W.	Ch
Finnemore's Court	Bermondsey	S.E.	Hf
Finnemore's Place	Bermondsey	S.E.	Hf
Finnis Street	Bethnal Green	E.	Hd
Finsbury Avenue	Shoreditch	E.C.	Ge
Finsbury Chambers	City of London	E.C.	Ge
Finsbury Circus	City of London	E.C.	Ge
Finsbury Market	Shoreditch	E.C.	Ge
Finsbury Park	Hornsey	N.	Ga
Finsbury Park Avenue	Tottenham U.D.	N.	Ga
Finsbury Park Road	Stoke Newington	N.	Gb
Finsbury Park Station	**Islington**	N.	Gb
Finsbury Pavement	City of London and Finsbury	E.C.	Ge
Finsbury Square	Deptford	S.E.	Jg
Finsbury Square	Finsbury	E.C.	Ge
Finsbury Station	**City of London**	E.C.	Ge

STREET OR PLACE.	BOROUGH.	P.D.	MAP.
Finsbury Street	Finsbury	E.C.	Ge
Finsden Road	Lambeth	S.E.	Gj
Fir Hill	Lewisham	S.E.	Jl
Fir Hill House	Lewisham	S.E.	Jl
Fir Street	Lewisham	S.E.	Hl
Firbank Road	Camberwell	S.E.	Hh
Fircroft Road	Wandsworth	S.W.	Ek
Firmin Place	Stepney	E.	Je
Firs, The	Barnes U.D.	S.W.	Aj
Firs, The	Hampstead	N.W.	Da
Firs, The	Wimbledon	S.W.	Bm
Firsby Road	Hackney	N.	Ha
First Archway	Wandsworth	S.W.	Ej
First Avenue	Acton U.D.	W.	Bf
First Avenue	Barnes U.D.	S.W.	Bh
First Avenue	Paddington	W.	Cd
First Avenue	Walthamstow U.D.	E.	Ka
First Avenue	West Ham	E.	Ld
First Road	Woolwich		Mg
First Street	Chelsea	S.W.	Ef
Firth Gardens	Fulham	S.W.	Ch
Fish Street Hill	City of London	E.C.	Ge
Fisher Street	Barking U.D.		Nd
Fisher Street	Holborn	W.C.	Fe
Fisher Street	West Ham	E.	Le
Fisher's Alley	Stepney	E.	He
Fisher's Buildings	Southwark	S.E.	Gf
Fisher's Court	Bermondsey	S.E.	Jf
Fisher's Lane	Chiswick U.D.	W.	Bf
Fisher's Rents	Deptford	S.E.	Jh
Fisher's Wharf	Poplar	E.	Jf
Fishmonger's Almshs.	Wandsworth	S.W.	Dj
Fishmonger's Hall St.	City of London	E.C.	Ge
Fishpond Wood	Wimbledon	S.W.	Bl
Fitchett's Court	City of London	E.C.	Ge
Fitzalan Street	Lambeth	S.E.	Ff
Fitzgeorge Avenue	Fulham	W.	Cg
Fitzgerald Avenue	Barnes U.D.	S.W.	Bh
Fitzgerald Road	Barnes U.D.	S.W.	Ah
Fitzgerald Road	Wanstead U.D.	E.	La
Fitzjames Avenue	Fulham	W.	Cg
Fitzjohn's Avenue	Hampstead	N.W.	Db
Fitzroy Court	St. Pancras	W.	Fd
Fitzroy Mews	St. Pancras	W.	Fd
Fitzroy Park	St. Pancras	N.	Ea
Fitzroy Place	St. Pancras	N.W.	Ed
Fitzroy Road	St. Pancras	N.W.	Ec
Fitzroy Square	St. Pancras	W.	Fd
Fitzroy Street	St. Pancras	W.	Fd
Fitzroy Yard	St. Pancras	N.W.	Ec
Fitzwilliam Road	Wandsworth	S.W.	Fh
Five Bell Alley	Stepney	E.	Je
Fladbury Road	Tottenham U.D.	N.	Ga
Fladgate Road	Leyton U.D.	E.	La
Flamborough Street	Stepney	E.	Je
Flamsteed Hill	Greenwich	S.E.	Kh
Flanchford Road	Hammersmith	W.	Bf
Flanders Road	Chiswick U.D.	W.	Bf
Flask Lane	Westminster	S.W.	Eg
Flask Walk	Hampstead	N.W.	Db
Flats, The	Wandsworth	S.W.	Ek
Flavell Road	Wandsworth	S.W.	Dj
Flaxman Road	Lambeth	S.E.	Gh
Flaxton Road	Woolwich		Nh
Fleet Lane	City of London	E.C.	Ge
Fleet Mews	Hampstead	N.W.	Eb
Fleet Road	Hampstead	N.W.	Eb
Fleet Row	Holborn	E.C.	Gd
Fleet Street	City of London	E.C.	Fe
Fleet Street Hill	Bethnal Green	E.	Hd
Fleetwood Street	Stoke Newington	N.	Hb
Fleming Road	Southwark	S.E.	Gg
Flemming Street	Shoreditch	N.	Hd
Fletcher Road	Acton U.D.	W.	Af
Fletcher Row	Finsbury	E.C.	Gd
Fletching Road	Hackney	E.	Hb
Fleur-de-Lis Court	City of London	E.C.	Ge
Fleur-de-Lis Court	City of London	E.	He
Fleur-de-Lis Street	Stepney	E.	He
Flight's Yard	Islington	N.	Gd
Flint Avenue	Poplar	E.	Ke
Flint Street	Poplar	E.	Ke
Flint Street	Southwark	S.E.	Gg
Flinton Street	Southwark	S.E.	Gg
Flint's Court	Bethnal Green	E.	Jd
Flockton Street	Bermondsey	S.E.	Hf
Flockton Road	Lewisham	S.E.	Kj
Flodden Road	Camberwell and Lambeth	S.E.	Gh
Flood Street	Chelsea	S.W.	Eg
Flora Gardens	Hammersmith	W.	Bf
Flora Terrace	Southwark	S.E.	Gg
Floral Street	Westminster	W.C.	Fe
Florefield Road	Hackney	E.	Hc
Florence Cottages	Deptford	S.E.	Jh
Florence Grove	Deptford	S.E.	Jh
Florence Road	Bromley		Lm
Florence Road	Deptford	S.E.	Jh
Florence Road	Ealing	W.	Se
Florence Road	East Ham	E.	Ld
Florence Road	Hornsey	N.	Fa
Florence Road	Kingston		Sm
Florence Road	Leyton U.D.	E.	Kb
Florence Road	West Ham	E.	Ld
Florence Road	West Ham	E.	Le
Florence Road	Wimbledon	S.W.	Dm
Florence Street	Deptford	S.E.	Jh
Florence Street	Islington	N.	Gc
Florence Street	West Ham	E.	Le
Florence Street East	Deptford	S.E.	Jh
Florence Terrace	Hackney	N.	Hb
Florence Terrace	Ham U.D.	S.W.	Bk
Florence Villas	Lewisham	S.E.	Jk
Florian Road	Wandsworth	S.W.	Cj
Florida Street	Bethnal Green	E.	Hd
Floriston Street	Stepney	E.	Je
Florys, The	Wandsworth	S.W.	Ck
Floss Street	Wandsworth	S.W.	Ch
Flower & Dean Street	Stepney	E.	He
Flower House	Lewisham	S.E.	Kl
Flower Walk, The	Westminster	S.W.	Df
Flowermead	Wandsworth	S.W.	Ck
Flower's Mews	Islington	N.	Fa
Flying Horse Yard	Southwark	S.E.	Gf
Foley Avenue	Hampstead	N.W.	Db
Foley Street	Marylebone	W.	Fe
Foley Street	Wandsworth	S.W.	Eh
Folkestone Gardens	Deptford	S.E.	Jg
Folkestone Road	East Ham	E.	Md
Folkestone Road	Walthamstow U.D.	E.	Ka
Folkestone Road	West Ham	E.	Kd
Foilett Street	Poplar	E.	Ke
Folly Pond	Greenwich	S.E.	Kh
Folly Street	Poplar	E.	Kf
Folly Wall	Poplar	E.	Kf
Fontaine Road	Wandsworth	S.W.	Fm
Fontarabia Road	Battersea	S.W.	Ej
Fontenoy Road	Wandsworth	S.W.	Ek
Fonthill Mews	Islington	N.	Fa
Fonthill Road	Islington	N.	Fa
Foote's Row	Battersea	S.W.	Fg
Foots Cray Bridge	Foots Cray U.D.		Om
Foots Cray Lane	Bexley U.D. and Foots Cray U.D.		Ol
Foots Cray Place	Foots Cray U.D.		Ol
Foots Cray Road	Foots Cray U.D. and Woolwich		Nk
Forburg Road	Hackney	N.	Ha
Ford Road	Poplar	E.	Jd
Ford Square	Stepney	E.	He
Ford Street	Poplar	E.	Jd
Ford Street	West Ham	E.	Le
Fordel Road	Lewisham	S.E.	Kk
Fordham Villas	Wandsworth	S.W.	Bh
Fordham Street	Stepney	E.	He
Fordham's Grove	Islington	N.	Gc
Fordhook	Ealing	W.	Ae
Fordingley Road	Paddington	W.	Cd
Ford's Market	West Ham	E.	Le
Ford's Park Road	West Ham	E.	Le
Ford's Place	Battersea	S.W.	Dh
Fordwych Road	Hampstead	N.W.	Cb
Fordwych Road North	Hampstead	N.W.	Ce'
Fordyce Hill	Lewisham	S.E.	Kj

STREET OR PLACE.	BOROUGH.	P.D.	MAP.
Fordyce Road	Lewisham	S.E.	Kj
Fore Street	City of London	E.C.	Ge
Fore Street Avenue	City of London	E.C.	Ge
Foreign Cattle Market	Greenwich	S.E.	Jg
Foreign Office	**Westminster**	S.W.	Ff
Foreign Street	Lambeth	S.E.	Gh
Forest Drive	Leyton U.D.	E.	Ka
Forest Gate	West Ham	E.	Lc
Forest Gate Station	**West Ham**	E.	Lc
Forest Glade	Leyton U.D.	E.	Ka
ForestGrammarSchool	Walthamstow U.D.	E.	La
Forest Hill House	Lewisham	S.E.	Hk
Forest Hill Road (part)	Camberwell	S.E.	Hj
Forest Hill Road (part)	Lewisham	S.E.	Hk
Forest Hill Station	**Lewisham**	S.E.	Hk
Forest House	Leyton U.D.	E.	Ka
Forest Lane	West Ham	E.	Lc
Forest Lodge	Bexley U.D.		Nk
Forest Lodge	Woolwich		Mh
Forest New Road	Walthamstow U.D.	E.	Ka
Forest Rise	Leyton U.D.	E.	Ka
Forest Road	Hackney	E.	Hc
Forest Road	Leyton U.D.	E.	Ka
Forest Road	Richmond		Ag
Forest Road	West Ham	E.	Lb
Forest Street	West Ham	E.	Lc
Forest View Avenue	Leyton U.D.	E.	Ka
Forester's Arms, The	Bexley U.D.		Oh
Forester Road	Camberwell	S.E.	Hj
Forester Street	Stepney	E.	Jd
Forester's Terrace	Wandsworth	S.W.	El
Forester's Hall Place	Finsbury	E.C.	Gd
Forfar Road	Battersea	S.W.	Eh
Formosa Street	Paddington	W.	Dd
Forsbrook Street	Hammersmith	W.	Ce
Forster Road	Walthamstow U.D.	E.	Ja
Forston Street	Shoreditch	N.	Gd
Fort Buildings	Bermondsey	S.E.	Hf
Fort Cottages	Islington	N.	Gb
Fort Passage	Bermondsey	S.E.	Hf
Fort Road	Bermondsey	S.E.	Hg
Fort Street	Stepney	E.	He
Fortescue Avenue	Hackney	E.	Hc
Fortescue Road	Croydon R.D.	S.W.	Dm
Fortess Grove	St. Pancras	N.W.	Fb
Fortess Mews	St. Pancras	N.W.	Fb
Fortess Road	St. Pancras	N.W.	Fb
Forthbridge Road	Battersea	S.W.	Ej
Fortnam Road	Islington	N.	Fb
Fortune Gate Road	Willesden U.D.	N.W.	Bc
Fortune Green	Hampstead	N.W.	Db
Fortune Green Road	Hampstead	N.W.	Db
Forty Lane	Barnes U.D.	S.W.	Ah
Forty Lane	Wembley U.D.		Ab
Forty Acre Lane	West Ham	E.	Le
Fosbury Mews	Paddington	W.	De
Foscote Mews	Paddington	W.	Dd
Foskett Road	Fulham	S.W.	Ch
Foss Road	Wandsworth	S.W.	Dl
Fossdene Road	Woolwich		Lg
Fosskett's Buildings	Stepney	E.	Jd
Foster Court	Southwark	S.E.	Gf
Foster House	Hendon U.D.	N.W.	Ca
Foster Lane	City of London	E.C.	Ge
Foster Road	West Ham	E.	Ld
Foster Street	Bethnal Green	E.	He
Foston Place	Southwark	S.E.	Gf
Foubert's Place	Westminster	W.	Fe
Foulden Road	Hackney	N.	Hb
Foulis Terrace	Kensington	S.W.	Dg
Foulser Road	Wandsworth	S.W.	El
Foulsham Road	Croydon		Sm
Founder's Court	City of London	E.C.	Ge
Foundling Hospital	St. Pancras	W.C.	Fd
Foundry Row	Shoreditch	N.	Gd
Foundry Walk	Shoreditch	N.	Gd
Fountain Court	City of London	E.C.	Ge
Fountain Dock	Bermondsey	S.E.	Hf
Fountain Gardens	Lambeth	S.E.	Fg
Fountain Place	Croydon R.D.		Dm
Fountain Road	Camberwell	S.E.	Hl
Fountain Road	Croydon R.D.		Em
Fountain Road	Wandsworth	S.W.	Dl
Fountain Street	Bethnal Green	E.	Hd
Fountain Street	Lambeth	S.W.	Fh
Fountain Villas	Wandsworth	S.W.	Dk
Fountain Yard	Hackney	E.	Hb
Fountayne Road	Hackney	N.	Hb
Fournier Street	Stepney	E.	He
Fourth Avenue	Paddington	W.	Cd
Fowell Street	Kensington	W.	Ce
Fowke's Buildings	City of London	E.C.	Ge
Fowler Road	West Ham	E.	Lb
Fowler Street	Camberwell	S.E.	Gh
Fowler Street	Islington	N.	Gc
Fox Court	Holborn	E.C.	Fd
Fox Grove	Beckenham U.D.		Km
Fox Hill	Croydon		Hm
Fox Lane	Hackney	E.	Hc
Fox Reservoir	Ealing	W.	Bd
Fox Street	Bethnal Green	E.	Hd
Fox Street	West Ham	E.	Le
Fox Terrace	Finsbury	E.C.	Ge
Foxberry	Chislehurst U.D.		Nl
Foxberry Road	Deptford and Lewisham	S.E.	Jj
Foxberry Wood	Chislehurst U.D.		Nl
Foxbourne Road	Wandsworth	S.W.	Ek
Foxbury Road	Bromley		Lm
Foxcroft Road	Woolwich		Nh
Fox and Goose	Ealing	W.	Bd
Fox Grove Road	Beckenham U.D.		Km
Foxham Road	Islington	N.	Fb
Foxhill Gardens	Croydon	S.E.	Hm
Foxley Road	Lambeth	S.W.	Gg
Foxlow Street	Bermondsey	S.E.	Hf
Foxmore Street	Battersea	S.W.	Eh
Fox's Buildings	Southwark	S.E.	Gf
Fox's Court	Southwark	S.E.	Gf
Fox's Lane	Stepney	E.	Hf
Foxwell Street	Deptford	S.E.	Jh
Foyle Road	Greenwich	S.E.	Lg
Fradley Road	Wandsworth	S.W.	El
Fram Place	Lambeth	S.E.	Gf
Framfield Road	Islington	N.	Gb
Frampton Park Road	Hackney	E.	Hc
Frances Street	Battersea	S.W.	Dh
Frances Street	Lambeth	S.E.	Ff
Frances Street	Woolwich		Mg
Franche Court Road	Wandsworth	S.W.	Dl
Francis Court	Finsbury	E.C.	Ge
Francis Court	Lambeth	S.E.	Fg
Francis Grove	Wimbledon	S.W.	Cm
Francis Road	Leyton U.D.	E.	Kb
Francis Street	Chelsea	S.W.	Eg
Francis Street	East Ham and Woolwich	E.	Mf
Francis Street	Fulham	S.W.	Cg
Francis Street	Hackney	E.	Hb
Francis Street	Holborn and St. Pancras	W.C.	Fe
Francis Street	Islington	N.	Fc
Francis Street	Paddington	W.	De
Francis Street	West Ham	E.	Kc
Francis Street	West Ham	E.	Le
Francis Street	Westminster	S.W.	Ff
Francis Street	Woolwich		Ng
Francis Terrace	Islington	N.	Fb
Francis Terrace	Shoreditch	E.	Hd
Franciscan Road	Wandsworth	S.W.	El
Franconia Road	Wandsworth	S.W.	Fj
Frank Street	Lambeth	S.E.	Fg
Frank Street	West Ham	E.	Le
Frankfort Mews	Paddington	W.	Cd
Frankfurt Road	Camberwell	S.E.	Gj
Frankham Street	Deptford	S.E.	Jh
Franklin Grove	Deptford	S.E.	Hh
Franklin Road	Kingston		Am
Franklin Road	Penge U.D.	S.E.	Hm
Franklin Street	Poplar	E.	Kd
Franklin Street	St. Pancras	N.W.	Ec
Franklin Street	Tottenham U.D.	N.	Ha
Franklin's Row	Chelsea	S.W.	Eg

STREET OR PLACE.	BOROUGH.	P.D.	MAP.
Gale Street	Barking U.D.		Od
Gale Street	Poplar	E.	Ke
Galen Place	Holborn	W.C.	Fe
Galena Road	Hammersmith	W.	Bf
Gale's Gardens	Bethnal Green	E.	Hd
Galesbury Road	Wandsworth	S.W.	Dj
Gallery Road	Camberwell	S.E.	Gk
Gallery Wall Road	Bermondsey	S.E.	Hg
Gallia Road	Islington	N.	Gc
Gallions Hotel	Woolwich	E.	Nf
Gallions Reach	Woolwich		Nf
Gallions Station	**Woolwich**	E.	Nf
Galiosson Road	Woolwich		Ng
Gallows Hill	The Maldens and Coombe U.D.		Am
Galt Street	Stepney	E.	Je
Galton Street	Paddington	W.	Cd
Galveston Road	Wandsworth	S.W.	Cj
Galway Street	Finsbury	E.C.	Gd
Gambia Street	Southwark	S.E.	Gf
Gambole Road	Wandsworth	S.W.	Dl
Gamuel Road	Walthamstow U.D.	E.	Ja
Gandy Court	Poplar	E.	Jd
Gange Street	Camberwell	S.E.	Gd
Ganton Street	Westminster	W.	Fe
Gap Road	Wimbledon	S.W.	Dl
Garden Cottages	Hackney	E.	Ha
Garden Cottages	Wandsworth	S.W.	Dj
Garden Court	City of London	E.C.	Fe
Garden Court	Finsbury	E.C.	Ge
Garden Lane	Wandsworth	S.W.	Fk
Garden Place	Bethnal Green	E.	Hd
Garden Road	Bromley		Lm
Garden Road	Marylebone	N.W.	Dd
Garden Road	Richmond		Ah
Garden Row	Camberwell	S.E.	Gk
Garden Row	Fulham	S.W.	Dg
Garden Row	Greenwich	S.E.	Kh
Garden Row	Lambeth	S.W.	Fh
Garden Row	Southwark	S.E.	Gf
Garden Street	Camberwell	S.E.	Hg
Garden Street	Stepney	E.	Je
Garden Street	Westminster	S.W.	Fg
Garden Terrace	Kensington	W.	Cd
Garden Villas	Islington	N.	Gc
Garden Wood	Bromley		Lm
Gardener's Road	Bethnal Green	E.	Jd
Gardens, The	Camberwell	S.E.	Hj
Gardens, The	Lambeth	S.W.	Fh
Garden Wharf Lane	Battersea	S.W.	Dh
Gardner's Lane	City of London	E.C.	Ge
Gardnor Road	Hampstead	N.W.	Db
Gardom Street	Stepney	E.	He
Gareth Place	Bermondsey	S.E.	Gf
Garfield Buildings	Holborn	W.C.	Fd
Garfield Road	Battersea	S.W.	Eh
Garfield Road	West Ham	E.	Le
Garfield Road	Wimbledon	S.W.	Dl
Garfield Villas	Marylebone	N.W.	Dc
Garford Street	Poplar	E.	Je
Garibaldi Street	Woolwich		Ng
Garkskel Road	Woolwich		Nk
Garland Hill	Bromley R.D.		Om
Garland Street	Woolwich		Nh
Garlick Hill	City of London	E.C.	Ge
Garlies Road	Lewisham	S.E.	Jl
Garlinge Road	Hampstead	N.W.	Cc
Garnault Mews	Finsbury	E.C.	Gd
Garner Street	Bethnal Green	E.	Hd
Garnet Road	Willesden U.D.	N.W.	Bc
Garnies Street	Camberwell	S.E.	Hg
Garrard's Road	Wandsworth	S.W.	Fl
Garratt Green	Wandsworth	S.W.	Dk
Garratt Lane	Wandsworth	S.W.	Dj-k
Garratt Mill	Wandsworth	S.W.	Dk
Garratt Park	Wandsworth	S.W.	Dk
Garratt Street	Finsbury	E.C.	Gd
Garrick Grove	Chiswick U.D.	W.	Bg
Garrick Road	Hendon U.D.	N.W.	Ba
Garrick Street	Westminster	W.C.	Fe
Garrick Theatre	**Westminster**	W.C.	Fe
Garsdale Road	Camberwell	S.E.	Hh

STREET OR PLACE.	BOROUGH.	P.D.	MAP.
Garth Street	Stepney	B.	Je
Garthorne Road	Lewisham	S.E.	Jk
Garthwaite Road	Fulham	S.W.	Dh
Gartmoor Gardens	Wandsworth	S.W.	Ck
Garton Street	Wandsworth	S.W.	Dj
Garvan Road	Fulham	W.	Cg
Garway Road	Paddington	W.	De
Gascoigne Place	Bethnal Green	E.	Hd
Gascoigne Road	Barking U.D.		Nd
Gascony Avenue	Hampstead	N.W.	Dc
Gascoyne Road	Hackney	E.	Jc
Gaselee Street	Poplar	E.	Kf
Gasfields	Rotherhithe	S.E.	Jf
Gaskarth Road	Wandsworth	S.W.	Ej
Gasholder Place	Lambeth	S.E.	Fg
Gaskell Street	Lambeth and Wandsworth	S.W.	Fh
Gas Light & Coke Co.	Westminster	S.W.	Ff
Gas Light & Coke Co's. Works, Beckton	Barking U.D. and East Ham	E.	Ne
Gas Light & Coke Co's. Works	Battersea	S.W.	Fg
Gas Light & Coke Co's. Works	Fulham	S.W.	Dh
Gas Light & Coke Co's. Works	Kensington	W.	Cd
Gas Light & Coke Co's. Works	Poplar	E.	Je
Gas Light & Coke Co's. Works	West Ham	E.	Kd
Gas Light & Coke Co's. Works	Westminster	S.W.	Fg
Gaspar Mews	Kensington	S.W.	Dg
Gassiot Road	Wandsworth	S.W.	El
Gastein Road	Fulham	W.	Cg
Gastein Road	Willesden U.D.	N.W.	Bc
Gastigny Place	Finsbury	E.C.	Gd
Gataker Street	Bermondsey	S.E.	Hf
Gatcomb Road	Islington	N.	Fb
Gate House	Hornsey	N.	Ea
Gate Street	Holborn	W.C.	Fe
Gateley Road	Lambeth	S.W.	Fh
Gate House Street	Westminster	S.W.	Ff
Gatesborough Street	Shoreditch	E.C.	Hd
Gateshead Place	Stepney	E.	He
Gatestone Road	Croydon	S.E.	Gm
Gateway, The	Camberwell	S.E.	Hh
Gatliff Road	Westminster	S.W.	Eg
Gatling Road	Woolwich		Og
Gatton Road	Wandsworth	S.W.	El
Gatward's Buildings	Finsbury	E.C.	Gd
Gauden Road	Wandsworth	S.W.	Fh
Gaverick Street	Poplar	E.	Jg
Gavin Street	Woolwich		Ng
Gawber Street	Bethnal Green	E.	Hd
Gawthorne Street	Poplar	E.	Jd
Gay Street	Wandsworth	S.W.	Ch
Gayford Road	Hammersmith	W.	Bf
Gayhurst Road	Hackney	E.	Hc
Gayton Crescent	Hampstead	N.W.	Db
Gayton Lodge	Wandsworth	S.W.	Ck
Gayton Road	Hampstead	N.W.	Db
Gayville Road	Battersea	S.W.	Ej
Gaywood Street	Southwark	S.E.	Gf
Gaza Street	Southwark	S.E.	Gg
Geary Road	Willesden U.D.	N.W.	Bb
Gedling Street	Bermondsey	S.E.	Hf
Gee Street	Finsbury	E.C.	Gd
Gee Street	St. Pancras	N.W.	Fd
Geere Road	West Ham	E.	Ld
Gees Court	Marylebone	W.	Ee
Geldart Road	Camberwell	S.E.	Hh
Geldeston Road	Hackney	E.	Hb
Gellatly Road	Deptford	S.E.	Hh
G.P.O. Submarine Cable Depot	Woolwich		Mf
General Steam Navigation Co's. Works	Greenwich	S.E.	Kg
Genesta Road	Woolwich		Nh
Geneva Road	Kingston		Sm
Geneva Road	Lambeth	S.W.	Gj
Geneva Terrace	Lambeth	S.W.	Gj
Genoa Avenue	Wandsworth	S.W.	Cj

STREET OR PLACE.	BOROUGH.	P.D.	MAP.
Glaskin Road	Hackney	E.	Hc
Glaskin Street	Hackney	E.	Hc
Glaskin Villas	Hackney	E.	Jb
Glass Street	Bethnal Green	E.	Hd
Glasshouse Alley	City of London	E.C.	Ge
Glasshouse Buildings	Stepney	E.	He
Glasshouse Street	Lambeth	S.E.	Fg
Glasshouse Street	Stepney	E.	He
Glasshouse Street	Stepney	E.	Je
Glasshouse Street	Westminster	W.	Fe
Glasshouse Yard	City of London	E.C.	Ge
Glasshouse Yard	Stepney	E.	Je
Glasslyn Road	Hornsey	N.	Fa
Glass Mill Lane	Bromley		Lm
Glastonbury Street	Hampstead	N.W.	Dc
Glaucus Street	Poplar	E.	Ke
Glazbury Road	Fulham	W.	Cg
Glebe, The	Camberwell	S.E.	Gh
Glebe, The	Lewisham	S.E.	Kh
Glebe Mansions	Chelsea	S.W.	Eg
Glebe Place	Chelsea	S.W.	Eg
Glebe Place	Stoke Newington	N.	Gd
Glebe Road	Barnes U.D.	S.W.	Bh
Glebe Road	Bermondsey	S.E.	Hf
Glebe Road	Poplar	E.	Jd
Glebe Road	Romford		Ln
Glebe Street	Chiswick U.D.	W.	Bg
Glebe Street	Hornsey	N.	Fa
Gledhow Gardens	Kensington	S.W.	Da
Gledhow Terrace	Kensington	S.W.	Da
Gledstanes Road	Fulham	W.	Cg
Glen, The	Kingsbury U.D.		Aa
Glen Road	Walthamstow U.D.	E.	Ja
Glen Street	St. Pancras	N.W.	Ed
Glenalvon Street	Woolwich		Mg
Glenarm Road	Hackney	E.	Jb
Glenavon Road	West Ham	E.	Kc
Glenbrook Road	Hampstead	N.W.	Dc
Glenburnie Road	Wandsworth	S.W.	Ek
Glenburnie Terrace	Wandsworth	S.W.	Ek
Glencairn Road	Wandsworth	S.W.	Fm
Glencoe Road	Croydon R.D.		Em
Glencoe Street	Poplar	E.	Ke
Glendal Street	Lambeth	S.W.	Fj
Glendarvon Street	Wandsworth	S.W.	Ch
Glendower Place	Kensington	S.W.	Dg
Gleneagle Road	Wandsworth	S.W.	Fl
Gleneldon Road	Wandsworth	S.W.	Fl
Glenelg Road	Lambeth	S.W.	Fj
Glenesk Road	Woolwich		Mj
Glenfarg Road	Lewisham	S.E.	Kk
Glenfield Avenue	Ealing	W.	Sf
Glenfield Road	Wandsworth	S.W.	Fk
Glenforth Street	Greenwich	S.E.	Lg
Glengall Cottages	Camberwell	S.E.	Hg
Glengall Dry Dock	Poplar	E.	Jf
Glengall Mews	Camberwell	S.E.	Hg
Glengall Road	Camberwell	S.E.	Hg
Glengall Road	Poplar	E.	Jf
Glengall Road	Willesden U.D.	N.W.	Cc
Glengall Terrace	Camberwell	S.E.	Hg
Glengarry Road	Camberwell	S.E.	Hj
Glenhouse Road	Woolwich		Mj
Glenhurst Road	Brentford U.D.		Sg
Glenilla Street	Hampstead	N.W.	Ec
Glenister Road	Greenwich	S.E.	Kg
Glenister Street	East Ham	E.	Mf
Glenlea Road	Woolwich		Mj
Glenlock Street	Hampstead	N.W.	Ec
Glenluce Road	Greenwich	S.E.	Lg
Glenlyon Road	Woolwich		Mj
Glenmore Road	Hampstead	N.W.	Ec
Glennie Road	Lambeth	S.E.	Fl
Glenparke Road	West Ham	E.	Lc
Glenrosa Street	Fulham	S.W.	Dh
Glenrose	Chislehurst U.D.		Mm
Glenroy Street	Hammersmith	W.	Ce
Glensdale Road	Deptford	S.E.	Jj
Glenshaw Mansions	Lambeth	S.W.	Fh
Glenshero	Wandsworth	S.W.	Ck
Glenshiel Road	Woolwich		Mj
Glenside Road	Woolwich		Ng
Glentham Road	Barnes U.D.	S.W.	Bg
Glenthorn Road	Kingston		Sm
Glenthorne Road	Hammersmith	W.	Bf
Glenthorne Road	Walthamstow U.D.	E.	Ja
Glenton Road	Lewisham	S.E.	Kj
Glentworth Street	Marylebone	N.W.	Ed
Glenure Road	Woolwich		Mj
Glenview Road	Lewisham	S.E.	Kj
Glenville Grove	Deptford	S.E.	Jh
Glenville Road	Kingston		Am
Glenwood Road	Lewisham	S.E.	Jk
Glenwood Road	Tottenham U.D.	N.	Ga
Gliddon Road	Fulham and Hammersmith	W.	Cf
Globe Alley	Stepney	E.	Je
Globe Buildings	Stepney	E.	Jd
Globe Chambers	Southwark	S.E.	Gf
Globe Court	Lambeth	S.E.	Fg
Globe Court	Shoreditch	E.	Hd
Globe Court	Stepney	E.	Hf
Globe Crescent	West Ham	E.	Lc
Globe Dock	Bermondsey	S.E.	Jf
Globe Lane	Woolwich		Mf
Globe Place	West Ham	E.	Lc
Globe Road (part)	Bethnal Green	E.	Hd
Globe Road (part)	Stepney	E.	Jd
Globe Road Station	**Stepney**	E.	Jd
Globe Square	Bermondsey	S.E.	Jf
Globe Street	Southwark	S.E.	Gf
Globe Street	Stepney	E.	Hf
Globe Wharf	Bermondsey	S.E.	Je
Globe Wharf	Stepney	E.	Hf
Globe Stairs Pier	Bermondsey	S.E.	Jf
Gloucester Court	Stepney	E.	Je
Gloucester Crescent	Islington	N.	Gd
Gloucester Crescent	St. Pancras	N.W.	Ec
Gloucester Gardens	Paddington	W.	De
Gloucester Gate	St. Pancras	N.W.	Ed
Gloucester Grove East	Kensington	S.W.	Dg
Gloucester Lodge	Hendon U.D.	N.W.	Ca
Gloucester Mews	Paddington	W.	De
Gloucester Mews	St. Pancras	N.W.	Ed
Gloucester Mews East	Marylebone	W.	Ee
Gloucester Mews East	Paddington	W.	De
Gloucester Mews West	Marylebone	W.	Ee
Gloucester Mews West	Paddington	W.	De
Gloucester Passage	Bethnal Green	E.	Hd
Gloucester Place	Bethnal Green	E.	Hd
Gloucester Place	Camberwell	S.E.	Hg
Gloucester Place	Deptford	S.E.	Jh
Gloucester Place	Greenwich	S.E.	Kh
Gloucester Place	Marylebone	W.	Ee
Gloucester Road	Acton U.D.	W.	Af
Gloucester Road	Camberwell	S.E.	Gg
Gloucester Road	Hackney	E.	Jc
Gloucester Road	Islington	N.	Fb
Gloucester Road	Kensington	S.W.	Df
Gloucester Road	Kingston		Am
Gloucester Road	Leyton U.D.	E.	Ja
Gloucester Road	Richmond		Ag
Gloucester Road	St. Pancras	N.W.	Ec
Gloucester Road	Stoke Newington	N.	Ga
Gloucester Road Sta.	**Kensington**	S.W.	Df
Gloucester Row	Shoreditch	E.C.	Gd
Gloucester Square	Paddington	W.	De
Gloucester Street	Finsbury	E.C.	Gd
Gloucester Street	Holborn	W.C.	Fe
Gloucester Street	Lambeth	S.E.	Gf
Gloucester Street	Lambeth	S.E.	Gl
Gloucester Street	Marylebone	W.	Ee
Gloucester Street	Westminster	S.W.	Eg
Gloucester Terrace	Greenwich	S.E.	Kh
Gloucester Terrace	Kensington	S.W.	Dg
Gloucester Terrace	Paddington	W.	De
Gloucester Terrace	St. Pancras	N.W.	Ed
Gloucester Terrace	Westminster	S.W.	Ef
Gloucester Villas	St. Pancras	N.W.	Ed
Gloucester Walk	Kensington	W.	Df
Glover Street	Shoreditch	E.C.	Gd
Glover's Hall Court	City of London	E.C.	Ge
Glover's Island	Richmond		Sk
Glycena Road	Battersea	S.W.	Eh

STREET OR PLACE.	BOROUGH.	P.D.	MAP.
Goswell Terrace	Finsbury	E.C.	Gd
Gotha Street	Hackney	E.	Hd
Gothic Villas	Camberwell	S.E.	Hk
Gough Cottages	Woolwich		Mg
Gough Gallery	St. Pancras	N.W.	Fd
Gough Road	West Ham	E.	Lb
Gough Square	City of London	E.C.	Ge
Gough Street	Poplar	E.	Je
Gough Street	St. Pancras	W.C.	Fd
Gould Hill	Stepney	E.	Je
Goulden Street	Battersea	S.W.	Eh
Goulston Lane	Stepney	E.	He
Goulston Street	Stepney	E.	He
Goulston's Buildings	Bermondsey	S.E.	Hg
Goulton Road	Hackney	E.	Hb
Gourley Road	Tottenham U.D.	N.	Ha
Gourock Road	Woolwich		Mj
Govan Street	Shoreditch	E.	Hd
Governesses Institute	Chislehurst U.D.		Nm
Govey Place	Stepney	E.	Jd
Gowan Avenue	Fulham	S.W.	Ch
Gowan Road	Willesden U.D.	N.W.	Bc
Gower Mews	Holborn	W.C.	Fd
Gower Place	St. Pancras	W.C.	Fd
Gower Road	West Ham	E.	Lc
Gower Street	Holborn and St. Pancras	W.C.	Fd
Gower Street Station	**St. Pancras**	W.C.	Fd
Gowers Walk	Stepney	E.	He
Gowland Place	Beckenham U.D.		Jm
Gowlett Road	Camberwell	S.E.	Hj
Gowlett Terrace	Camberwell	S.E.	Hj
Gowrie Road	Battersea	S.W.	Ej
Grace Road	West Ham	E.	Kc
Grace Street	Islington	N.	Fc
Grace Street	Poplar	E.	Kd
Gracechurch Street	City of London	E.C.	Ge
Gracedale Road	Wandsworth	S.W.	El
Grace's Alley	Stepney	E.	He
Graces Road	Camberwell	S.E.	Gh
Graydon Street	Woolwich		Mg
Grafton Crescent	St. Pancras	N.W.	Ec
Grafton Gardens	Tottenham U.D.	N.	Ga
Grafton Green	St. Pancras	N.W.	Ec
Grafton Mansions	St. Pancras	N.W.	Fd
Grafton Mews	St. Pancras	W.	Fd
Grafton Place	St. Pancras	N.W.	Fd
Grafton Road	Acton U.D.	W.	Ae
Grafton Road	Islington	N.	Fb
Grafton Road	St. Pancras	N.W.	Ec
Grafton Road	The Maldens & Coombe U.D.		Bm
Grafton Road North	West Ham	E.	Ld
Grafton Road South	West Ham	E.	Ld
Grafton Square	Wandsworth	S.W.	Fh
Grafton Street	St. Pancras	W. & W.C.	Fd
Grafton Street	Stepney	E.	Jd
Grafton Street	Westminster	W.	Ee
Grafton Terrace	Greenwich	S.E.	Kg
Grafton Terrace	St. Pancras	N.W.	Ec
Grafton Yard	Hampstead and St. Pancras	N.W.	Ec
Graham Avenue	Croydon R.D.		Em
Graham Avenue	Ealing	W.	Sf
Graham Mansions	Hackney	E.	Hc
Graham Road	Acton U.D.	W.	Af
Graham Road	Croydon R.D.		Em
Graham Road	Hackney	E.	Hc
Graham Road	West Ham	E.	Ld
Graham Road	Wimbledon	S.W.	Cm
Graham Street	Finsbury and Islington	N.	Gd
Graham Street	Southwark	S.E.	Gg
Graham Street	Westminster	S.W.	Eg
Grail Street	Hackney	E.	Jc
Grainger Street	Camberwell	S.E.	Hg
Granard Lodge	Wandsworth	S.W.	Bj
Granard Road	Battersea	S.W.	Ek
Granary Street	Lewisham	S.E.	Jl
Granby Gardens	Lambeth	S.E.	Ff
Granby Mews	St. Pancras	N.W.	Fd
Granby Row	Bethnal Green	E.	Hd
Granby Street	Bethnal Green	E.	Hd
Granby Street	St. Pancras	N.W.	Fd
Grand Drive	Merton U.D.	S.W.	Cm
Grand Hotel Buildings	Westminster	W.C.	Ff
Grand Junction Dist. Reservoirs	Brentford U.D.		Sg
Grand Junction Mews	Paddington	W.	Ee
Grand Junction Road	Paddington	W.	De
Grand Junction Terrace	Paddington	W.	Ee
Grand Parade	Wandsworth	S.W.	Ek
Grand Surrey Wharf	Bermondsey	S.E.	Jf
Grand Theatre	**Battersea**	S.W.	Ej
Grand Theatre	**Fulham**	S.W.	Ch
Grand Theatre	**Islington**	N.	Gd
Grandison Road	Battersea	S.W.	Ej
Granfield Street	Battersea	S.W.	Dh
Grange, The	Acton U.D.	W.	Be
Grange, The	Bermondsey	S.E.	Hf
Grange, The	Camberwell	S.E.	Hk
Grange, The	Chislehurst U.D.		Nm
Grange, The	Ealing	W.	Se
Grange, The	Hampstead	N.W.	Db
Grange, The	Hampstead	N.W.	Dc
Grange, The	Hornsey	N.	Ea
Grange, The	Kingston		Am
Grange, The	Wandsworth	S.W.	Ek
Grange, The	Wimbledon	S.W.	Cl
Grange, The	Woolwich		Nl
Grange Brick and Tile Works	Willesden U.D.	N.W.	Cc
Grange Court	Westminster	W.C.	Fe
Grange Hill	Croydon	S.E.	Gm
Grange Hotel	Ealing	W.	Se
Grange Park	Ealing	W.	Se
Grange Park Road	Leyton U.D.	E.	Ka
Grange Road	Bermondsey	S.E.	Hf
Grange Road	Chiswick U.D.	W.	Ag
Grange Road	Croydon	S.E.	Gm
Grange Road	Ealing	W.	Se
Grange Road	Hornsey	N.	Ea
Grange Road	Islington	N.	Gc
Grange Road	Kingston		Sm
Grange Road	Leyton U.D.	E.	Ka
Grange Road	St. Pancras	N.W.	Ec
Grange Road	Wandsworth	S.W.	Eh
Grange Road	West Ham	E.	Ld
Grange Road	Willesden U.D.	N.W.	Cc
Grange Street	St. Pancras	N.W.	Ec
Grange Street	Shoreditch	N.	Gd
Grange Walk	Bermondsey	S.E.	Hf
Grange Wood	Croydon	S.E.	Gm
Grange Yard	Bermondsey	S.E.	Hf
Grangecourt Road	Stoke Newington	N.	Ga
Grangehill Road	Woolwich		Mj
Granite Wharf	Greenwich	S.E.	Kg
Granite Wharf	Stepney	E.	Hf
Gransden Avenue	Hackney	E.	Hc
Gransden Road	Hammersmith	W.	Bf
Grant House	Southwark	S.E.	Gg
Grant Road	Battersea	S.W.	Dh
Grantham House	Fulham	S.W.	Cj
Grantham Place	Southwark	S.E.	Gg
Grantham Place	Westminster	W.	Ee
Grantham Road	Lambeth	S.W.	Fg
Grantham Terrace	Shoreditch	N.	Gd
Grantully Road	Paddington	W.	Dd
Granville Buildings	Shoreditch	E.C.	Ga
Granville Chambers	Marylebone	W.	Ee
Granville Park	Lewisham	S.E.	Kh
Granville Place	Finsbury	W.C.	Fd
Granville Place	Marylebone	W.	Ee
Granville Road	Foots Cray U.D.		Ol
Granville Road	Hendon U.D.	N.W.	Ca
Granville Road	Hornsey	N.	Fa
Granville Road	Lewisham	S.E.	Kj
Granville Road	Walthamstow U.D.	E.	Ka
Granville Road	Wandsworth	S.W.	Ck
Granville Road	West Ham	E.	Le
Granville Road	Willesden U.D.	N.W.	Dd
Granville Road	Wimbledon	S.W.	Dm
Granville Square	Finsbury	W.C.	Fd
Granville Street	Finsbury	W.C.	Fd

STREET OR PLACE.	BOROUGH.	P.D.	MAP.
Grosvenor Road .	Chiswick U.D. .	W.	Ag
Grosvenor Road . .	East Ham . .	E.	Md
Grosvenor Road . .	Islington . .	N.	Gc
Grosvenor Road . .	Leyton U.D. .	E.	Ka
Grosvenor Road . .	Richmond . .		Sj
Grosvenor Road . .	Southwark . .	S.E.	Gg
Grosvenor Road . .	West Ham . .	E.	Lc
Grosvenor Road . .	Westminster .	S.W.	Fg
Grosvenor Road Bldgs.	Westminster .	S.W.	Fg
Grosvenor RoadStation	**Westminster** .	S.W.	Eg
Grosvenor Square .	Westminster	W.	Ee
Grosvenor Street . .	Islington . .	N.	Gd
Grosvenor Street . .	Stepney . .	E.	Je
Grosvenor Street . .	Westminster .	S.W.	Ef
Grosvenor Street . .	Westminster .	S.W.	Ff
Grosvenor Street . .	Westminster .	W.	Ee
Grosvenor Terrace .	Southwark . .	S.E.	Gg
Grosvenor Terrace .	Wandsworth .	S.W.	Dk
Grosvenor Wharf .	Poplar . . .	E.	Kg
[illegible] . . .	Wandsworth .	S.W.	Dk
Grote's Bldgs. . .	Lewisham . .	S.E.	Kh
Grote's Place . .	Lewisham . .	S.E.	Kh
Grotto Passage . .	Marylebone .	W.	Ee
Grotto Place . . .	Mary ebone .	W.	Ee
Ground Street . .	Southwark . .	S.E.	Gf
Grove, The . . .	Camberwell .	S.E.	Gk
Grove, The . . .	Ealing . . .	W.	Se
Grove, The . . .	Greenwich . .	S.E.	Kh
Grove, The . . .	Hackney . .	E.	Jc
Grove, The . . .	Hammersmith .	W.	Cf
Grove, The . . .	Hampstead . .	N.W.	Db
Grove, The . . .	Hornsey and St. Pancras . .	N.	Ea
Grove, The . . .	Hornsey . .	N.	Fa
Grove, The . . .	Kensington . .	S.W.	Dg
Grove, The . . .	Kingston . .		Sm
Grove, The . . .	Lambeth . .	S.W.	Fg
Grove, The . . .	Lewisham . .	S.E.	Hk
Grove, The . . .	Lewisham . .	S.E.	Kj
Grove, The . . .	Merton U.D. .	S.W.	Bm
Grove, The . . .	Wandsworth .	S.W.	Dj
Grove, The . . .	Wandsworth .	S.W.	Ck
Grove, The . . .	Wandsworth .	S.W.	Fl
Grove, The . . .	West Ham . .	E.	Kc
Grove, The . . .	Woolwich . .		Mj
Grove, The . . .	Woolwich . .		Mk
Grove Buildings . .	Hackney . .	E.	Hc
Grove Buildings . .	Stepney . .	E.	Jd
Grove Cottages . .	Camberwell .	S.E.	Hg
Grove Cottages . .	Chelsea . .	S.W.	Eg
Grove Court . . .	Stepney . .	E.	Je
Grove Crescent . .	Camberwell .	S.E.	Gh
Grove Crescent . .	Kingston . .		Sm
Grove Crescent Road .	Kingston . .		Sm
Grove Crescent Road .	West Ham . .	E.	Kc
Grove End . . .	Chiswick U.D. .		Ag
Grove End . . .	St. Pancras . .	N.W.	Eb
Grove End Road . .	Marylebone . .	N.W.	Dd
Grove End Rd. Mans. .	Marylebone . .	N.W.	Dd
Grove Green Road .	Leyton U.D. .	E.	Ka
Grove Hall Lunatic Asylum . . .	Poplar . . .	E.	Kd
Grove Hill Road . .	Camberwell .	S.E.	Gh
Grove Hotel . . .	Merton U.D. .	S.W.	Dm
Grove House . . .	Acton U.D. . .	W.	Af
Grove House . . .	Brentford U.D. .		Ag
Grove House . . .	Camberwell .	S.E.	Hk
Grove House . . .	Ham U.D. . .		Sk
Grove House . . .	Hendon U.D. .	N.W.	Ca
Grove House . . .	Wandsworth .	S.W.	Bj
Grove Lane . . .	Camberwell .	S.E.	Gh
Grove Lane . . .	Hackney . .	E.	Hc
Grove Lane . . .	Hackney . .	N.	Ha
Grove Lane . . .	Kingston . .		Sm
Grove Lane Mews .	Camberwell .	S.E.	Gh
Grove Lodge . . .	Hampstead . .	N.W.	Db
Grove Lodge . . .	Richmond . .		Sj
Grove Mews . . .	Hammersmith .	W.	Cf
Grove Mews . . .	Kensington . .	W.	Ce
Grove Mews . . .	Stepney . .	E.	Jd
Grove Park . . .	Camberwell .	S.E.	Gh
Grove Park . . .	Lewisham . .	S.E.	Kk
Grove Park . . .	Wanstead U.D. .	E.	La

STREET OR PLACE.	BOROUGH.	P.D.	MAP.
Grove Park Gardens .	Chiswick U.D .	W.	Ag
Grove Park Hotel. .	Chiswick U.D. .		Ag
Grove Park Road. .	Bromley R.D. .		Ll
Grove Park . .	Camberwell .	S.E.	Gh
Grove Park Road. .	Chiswick U.D. .	W.	Ag
Grove Park Road. .	Lewisham . .	S.E.	Kk
Grove Park Road. .	Leyton U.D. .	E.	Kb
Grove Park Road. .	Tottenham U.D.	N.	Ha
Grove Park Station .	**Lewisham** . .	S.E.	Ll
Grove Park Terrace .	Chiswick U.D. .	W.	Ag
Grove Park Terrace .	Lewisham . .	S.E.	Kk
Grove Passage . .	Hackney . .	E.	Hc
Grove Place . . .	Acton U.D. . .	W.	Ae
Grove Place . . .	Bethnal Green .	E.	Hd
Grove Place . . .	Camberwell . .	S.E.	Gh
Grove Place . . .	Hampstead . .	N.W.	Db
Grove Place . . .	Kensington . .	W.	Ce
Grove Place . . .	Stepney . .	E.	He
Grove Road . . .	Acton U.D. . .	W.	Af
Grove Road . . .	Barnes U.D. .	S.W.	Bh
Grove Road . . .	Bethnal Green & Stepney . .	E.	Jd
Grove Road . . .	Brentford U.D. .		Sg
Grove Road . . .	Croydon R.D. .		Em
Grove Road . . .	Ealing . . .	W.	Se
Grove Road . . .	Islington . .	N.	Fb
Grove Road . . .	Lambeth . .	S.W.	Fh
Grove Road . . .	Leyton U.D. .	E.	La
Grove Road . . .	Marylebone. .	N.W.	Dd
Grove Road . . .	Richmond . .		Sj
Grove Road . . .	Tottenham U.D.	N.	Ga
Grove Road . . .	Walthamstow U.D. . .	E.	Ka
Grove Road . . .	Wandsworth .	S.W.	Ek
Grove Road . . .	Wanstead U.D. .	E.	La
Grove Road . . .	Willesden U.D. .	N.W.	Cc
Grove Road . . .	Wimbledon. .	S.W.	Dm
Grove Road . . .	Woolwich . .		Og
Grove Street . . .	Bethnal Green .	E.	Hd
Grove Street . . .	Deptford and Greenwich .	S.E.	Jg
Grove Street . . .	Islington . .	N.	Fc
Grove Street . . .	Stepney . .	E.	He
Grove Terrace . .	Camberwell .	S.E.	Hh
Grove Terrace . .	Fulham . .	W.	Cg
Grove Terrace . .	Hampstead . .	N.W.	Db
Grove Terrace . .	St. Pancras . .	N.W.	Eb
Grove Vale . . .	Camberwell .	S.E.	Hj
Grove Villas . . .	Poplar . . .	E.	Ke
Grove Walk . . .	Shoreditch . .	N.	Gd
Grove Wood . . .	Hendon U.D. .	N.W.	Ca
Grovedale Road . .	Islington . .	N.	Fa
Grovedale Street . .	Hammersmith .	W.	Ce
Grub Street . . .	Westminster .	S.W.	Ff
Grummant Road . .	Camberwell .	S.E.	Hh
Grundy Street . .	Poplar . . .	E.	Ke
Guards Memorial .	Westminster .	W.C.	Ff
Gubyon Avenue . .	Lambeth . .	S.E.	Gj
Guelph Street . .	Wandsworth .	S.W.	Dk
Guerin Street . .	Poplar . . .	E.	Jd
Guernsey Grove . .	Lambeth . .	S.E.	Gk
Guernsey Road . .	Leyton U.D. .	E.	Kb
Guest Street . . .	Finsbury . .	E.C.	Gd
Guibal Road . . .	Lewisham and Woolwich .	S.E.	Lk
Guildersfield Road .	Wandsworth .	S.W.	Fm
Guildford Cottages .	Camberwell .	S.E.	Gh
Guildford Place . .	Camberwell .	S.E.	Gh
Guildford Road . .	Greenwich . .	S.E.	Kh
Guildford Road . .	Lambeth . .	S.W.	Fg
Guildford Road . .	Poplar . . .	E.	Ke
Guildford Street . .	Lambeth . .	S.E.	Ff
Guilford Mews . .	St. Pancras . .	W.C.	Fd
Guilford Street . .	Holborn and St. Pancras . .	W.C.	Fd
Guildhall . . .	**City of London** .	E.C.	Ge
Guildhall . . .	**Westminster** .	S.W.	Ff
Guildhall Buildings .	City of London .	E.C.	Ge
Guildhall Yard . .	City of London .	E.C.	Ge
Guilsboro Road . .	Willesden U.D. .	N.W.	Bc
Guinness Buildings .	Bethnal Green .	E.	Hd
Guinness Buildings .	Lambeth . .	S.E.	Fg
Guion Road . . .	Fulham . .	S.W.	Ch

STREET OR PLACE.	BOROUGH.	P.D.	MAP.
Gun Square	City of London	E.	He
Gun Street	Southwark	S.E.	Gf
Gun Street	Stepney	E.	He
Gun Wharf	Stepney	E.	Hf
Gun Yard	City of London	E.	He
Gundolf Street	Lambeth	S.E.	Fg
Gunmaker's Lane	Poplar	E.	Jd
Gunnersbury Avenue	Ealing	W.	Af
Gunnersbury House	Brentford U.D.	W.	Af
Gunnersbury Lane	Acton U.D., Brentford U.D., and Ealing	W.	Af
Gunnersbury Lodge	Ealing	W.	Af
Gunnersbury Park	Brentford U.D.	W.	Af
Gunnersbury Station	**Chiswick U.D.**		Ag
Gunning Street	Woolwich		Ng
Gunpowder Alley	City of London	E.C.	Ge
Gunter Grove	Chelsea	S.W.	Dg
Gunterstone Road	Fulham	W.	Cg
Gunton Road	Hackney	E.	Hb
Gurdon Road	Greenwich		Lg
Gurney Road	West Ham	E.	Kc
Gurney Street	Southwark	S.E.	Gg
Gurney Terrace	Camberwell	S.E.	Gg
Gurrey Lodge	Hendon U.D.	N.W.	Db
Guthrie Street	Chelsea	S.W.	Dg
Gutter Lane	City of London	E.C.	Ge
Gutteridge Court	Bermondsey	S.E.	Hf
Guy Court	Bermondsey	S.E.	Gf
Guy Street	Bermondsey	S.E.	Gf
Guy's Buildings	Shoreditch	N.E.	Hd
Gaysoliff Road	Lewisham	S.E.	Kj
Gwalior Road	Wandsworth	S.W.	Ch
Gwendolen Avenue	Wandsworth	S.W.	Cj
Gwendoline Avenue	West Ham	E.	Ld
Gwendwr Road	Fulham	W.	Cg
Gwyder Road	Bromley		Lm
Gwynant	Wimbledon	S.W.	Ck
Gwynne Road	Battersea	S.W.	Dh
Gye Street	Lambeth	S.E.	Fg
Gye Street Buildings	Lambeth	S.E.	Fg
Haarlem Mansions	Hammersmith	W.	Cf
Haarlem Road	Hammersmith	W.	Cf
Haberdasher Square	City of London	E.C.	Ge
Haberdasher Street	Shoreditch	N.	Gd
Haberdashers' School	Acton U.D.	W.	Ae
Haberson Road	West Ham	E.	Ld
Hack Road	West Ham	E.	Le
Hack Street	Poplar	E.	Ke
Hackford Road	Lambeth	S.W.	Fh
Hackney College	Hampstead	N.W.	Db
Hackney Common	Hackney	E.	Jc
Hackney Cut	Hackney	E.	Jb
Hackney Downs	Hackney	E.	Hb
Hackney Downs Junction Station	**Hackney**	E.	Hc
Hackney Empire	**Hackney**	E.	Hc
Hackney Grove	Hackney	E.	Hc
Hackney Marsh	Hackney	E.	Jb
Hackney Road	Bethnal Green & Shoreditch	E.	Hd
Hackney Workhouse	Hackney	E.	Jc
Hackney Wick	Hackney&Poplar	E.	Jc
Haddo Place	Greenwich	S.E.	Kg
Haddo Street	Greenwich	S.E.	Kg
Haddon Lodge	Woolwich		Mh
Haddon's Gardens	Shoreditch	E.C.	Gd
Hadfield Terrace	Camberwell	S.E.	Hh
Hadland Street	Bermondsey	S.E.	Hf
Hadley Street	St. Pancras	N.W.	Ec
Hadlow Place	Penge U.D.	S.E.	Hm
Hadlow Road	Foots Cray U.D.		Ol
Hadyn Park Flats	Hammersmith	W.	Bf
Hadyn Park Road	Hammersmith	W.	Bf
Hafer Road	Battersea	S.W.	Ej
Hafton Road	Lewisham	S.E.	Kk
Haggerston Road	Hackney & Shoreditch	E.	Hc
Haggerston Station	**Shoreditch**	E.	Hc
Hague Buildings	Bethnal Green	E.	Hd
Hague Place	Bethnal Green	E.	Hd
Hague Street	Bethnal Green	E.	Hd
Ha Ha Road	Woolwich		Mg

STREET OR PLACE.	BOROUGH.	P.D.	MAP.
Haig Road	West Ham	E.	Ld
Hailsham Avenue	Wandsworth	S.W.	Fk
Hainault Road	Leyton U.D.	E.	Ka
Hainault Road	Woolwich		Nk
Haines Street	Battersea	S.W.	Fg
Hainthorpe Road	Lambeth	S.E.	Gl
Halberstadt Mansions	Holborn	W.C.	Fe
Halcomb Place	Shoreditch	N.	Gd
Halcomb Street	Shoreditch	N.	Gd
Haldane Flats	Fulham	S.W.	Cg
Haldane Road	Fulham	S.W.	Cg
Haldon Road	Wandsworth	S.W.	Cj
Hale Gardens	Acton U.D.	W.	Ae
Hale Street	Deptford	S.E.	Jh
Hale Street	Islington	N.	Gd
Hale Street	Poplar	E.	Ke
Halesworth Road	Lewisham	S.E.	Kh
Half Acre	Brentford U.D.		Sg
Half Moon, The	Camberwell	S.E.	Gj
Half Moon Court	Holborn	E.C.	Fe
Half Moon Crescent	Islington	N.	Fd
Half Moon Lane	Camberwell	S.E.	Gj
Half Moon Passage	City of London	E.C.	Ge
Half Moon Passage	Stepney	E.	He
Half Moon Street	Westminster	W.	Ef
Half Moon Yard	Southwark	S.E.	Gf
Half Nichol Street	Bethnal Green	E.	Hd
Halford Road	Fulham	S.W.	Dg
Halford Road	Leyton U.D.	E.	Ka
Halford Road	Richmond		Sj
Halfway Street	Bexley U.D.		Nk
Halfway Street	Woolwich		Nk
Halfway Street	Bexley U.D.		Nk
Halfwilk Court	Stepney	E.	He
Halidon Street	Hackney	E.	Jb
Halkin Mews North	Westminster	S.W.	Ef
Halkin Place	Westminster	S.W.	Ef
Halkin Street	Westminster	S.W.	Ef
Halkin Street West	Westminster	S.W.	Ef
Hall Park	Paddington	W.	Dd
Hall Place	Croydon R.D.		Dm
Hall Place	Paddington	W.	Dd
Hall Place West	Paddington	W.	De
Hall Road	Camberwell	S.E.	Hj
Hall Road	Marylebone	N.W.	Dd
Hall Road	Wanstead U.D.	E.	La
Hall Street	Finsbury	E.C.	Gd
Hall Street	Stepney	E.	Je
Hallam Street	Marylebone	W.	Ee
Hallett's Place	Finsbury	E.C.	Gd
Halley Road	East Ham	E.	Lc
Halley Street	Stepney	E.	Je
Hallfield Road	Hampstead	N.W.	Db
Halliford Street	Islington	N.	Gc
Hall's Yard	Holborn	W.C.	Fe
Hallsville Road	West Ham	E.	Le
Halons Road	Woolwich		Mk
Halpin Place	Southwark	S.E.	Gg
Halsbury Road	Hammersmith	W.	Be
Halse Street	Islington	N.	Fc
Halsey Street	Chelsea	S.W.	Ef
Halsmere Road	Lambeth	S.E.	Gh
Halstead Street	Lambeth	S.W.	Fh
Halston Road	Willesden U.D.	N.W.	Cd
Halstow Road	Greenwich	S.E.	Lg
Halton Road	Islington	N.	Gc
Halton Street	Marylebone	N.W.	Dd
Halton Cross Street	Islington	N.	Gc
Ham, The	Brentford		Sg
Ham Common	Ham U.D.		Sl
Ham Fields	Ham U.D.		Sl
Ham Gate	Ham U.D.		Sl
Ham House	Richmond		Sk
Ham Lodge	Ham U.D.		Sl
Ham Park Road	West Ham	E.	Lc
Ham Road	Kingston		Sm
Ham Square	Wandsworth	S.W.	Dj
Ham Street	Ham U.D.		Sk
Ham Yard	Westminster	W.	Fe
Hambalt Road	Wandsworth	S.W.	Fj
Hamble Street	Fulham	S.W.	Dh
Hambro Road	Wandsworth	S.W.	Fl
Hamburg Street	Hackney	E.	Hc

STREET OR PLACE.	BOROUGH.	P.D.	MAP.
Hart's Court	City of London	E.C.	Ge
Hart's Lane	Deptford	S.F.	Jh
Hart's Place	Bethnal Green	E.	Hd
Hartshorn Alley	City of London	E.C.	Ge
Hartswood Road	Hammersmith	W.	Bf
Hartville Road	Woolwich		Ng
Hartwell Cottages	Hackney	E.	Hc
Hartwell Street	Hackney	E.	Hc
Harty Street	Finsbury	E.C.	Gd
Harvard Avenue Hill	Lewisham	S.E.	Jl
Harvard Road	Chiswick U.D.	W.	Ag
Harvard Road	Lewisham	S.E.	Kj
Harvest Road	Willesden U.D.	N.W.	Cd
Harvey Road	Camberwell	S.E.	Gh
Harvey Road	Hornsey	N.	Fa
Harvey Road	Leyton U.D.	E.	La
Harvey Street	Shoreditch	N.	Gd
Harvey's Bldgs.	Westminster	W.C.	Fe
Harvist Mews	Islington	N.	Fb
Harvist Road	Islington	N.	Fb
Harwar Street	Shoreditch	E.	Hd
Harwood Mansions	Fulham	S.W.	Dh
Harwood Road	Fulham	S.W.	Dh
Harwood Terrace	Fulham	S.W.	Dh
Hasborough Street	Paddington	W.	De
Haselrigge Road	Wandsworth	S.W.	Fj
Haseltine Road	Lewisham	S.E.	Jl
Hasker Street	Chelsea	S.W.	Ef
Haslam Place	Camberwell	S.E.	Hh
Haslemere Road	West Ham	E.	Lc
Haslett's buildings	Woolwich		Mg
Hassard Place	Bethnal Green	E.	Hd
Hassard Street	Bethnal Green	E.	Hd
Hassendean Road	Greenwich	S.E.	Lh
Hassett Road	Hackney	E.	Jc
Hassop Road	Willesden U.D.	N.W.	Cb
Hassop Road	Woolwich		Ng
Hastings Buildings	Bermondsey	S.E.	Hg
Hastings Road	Ealing	W.	Se
Hastings Road	West Ham	E.	Ld
Hastings Road West	Ealing	W.	Se
Hastings Street	St. Pancras	W.C.	Fd
Haston Street	St. Pancras	N.W.	Ec
Hat and Mitre Court	Finsbury	E.C.	Ge
Hatch Road	Croydon	S.W.	Fm
Hatch Row	Lambeth	S.E.	Gf
Hatcham Park Road	Deptford	S.E.	Jh
Hatcham Road	Camberwell	S.E.	Hg
Hatchard Road	Islington	N.	Fb
Hatcliffe Street	Greenwich	S.E.	Kg
Hatfield	Barking U.D.		Od
Hatfield Avenue	Croydon		Fm
Hatfield Place	Southwark	S.E.	Gf
Hatfield Road	Acton U.D.	W.	Bf
Hatfield Street	Finsbury	E.C.	Gd
Hatfield Street	Fulham	W.	Cg
Hatfield Street	Southwark	S.E.	Gf
Hatfield Terrace	Battersea	S.W.	Dh
Hatherington Road	Lambeth	S.W.	Fj
Hatherley Gardens	Hornsey	N.	Fa
Hatherley Grove	Bexley U.D.		Ol
Hatherley Grove	Paddington	W.	De
Hatherley Road	Foots Cray U.D.		Ol
Hatherley Road	Richmond		Sh
Hatherley Street	Hackney	E.	Hb
Hathway Road	Deptford	S.E.	Hh
Hat in Tun Yard	Holborn	E.C.	Ge
Hatley Road	Islington	N.	Fb
Hatterick Street	Bermondsey	S.E.	Hf
Hatton Court	City of London	E.C.	Ge
Hatton Garden	Holborn	E.C.	Ge
Hatton Street	Marylebone	N.W.	Ee
Hatton Wall	Holborn	E.C.	Ge
Hatton Yard	Holborn	E.C.	Ge
Haubeck Road	Battersea	S.W.	Ej
Haunch of Venison Yd.	Westminster	W.	Ee
Havannah Street	Poplar	E.	Jf
Havelock Place	Bethnal Green	E.	Jd
Havelock Place	Woolwich		Mg
Havelock Road	Hackney	E.	Hc
Havelock Road	Hammersmith	W.	Cf
Havelock Road	Wimbledon	S.W.	Dl
Havelock Street	Islington	N.	Fc
Havelock Street	Lewisham	S.E.	Hk
Havelock Terrace	Battersea	S.W.	Eh
Haven Green	Ealing	W.	Se
Haven Lane	Ealing	W.	Se
Haverhill Road	Wandsworth	S.W.	Ek
Havering Street	Stepney	E.	Je
Haversham Grange	Richmond		Sj
Haverstock Hill	Hampstead and St. Pancras	N.W.	Ec
Haverstock Hill Sta.	**St. Pancras**	N.W.	Ec
Haverstock Place	St Pancras	N.W.	Ec
Haverstock Road	Hampstead	N.W.	Eb
Haverstock Road	St. Pancras	N.W.	Ec
Haverstock Street	Islington	N.	Gd
Haverstock Street	St. Pancras	N.W.	Ec
Haveswood	Wandsworth	S.W.	El
Havill Street	Camberwell	S.E.	Gh
Haward Street	Battersea	S.W.	Fg
Hawarden Grove	Lambeth	S.E.	Gk
Hawarden Road	Walthamstow U.D.	E.	Ja
Hawarden Villas	Lambeth	S.E.	Gk
Hawbridge Road	Leyton U.D.	E.	Ka
Hawes Road	Bromley		Lm
Hawgood Street	Poplar	E.	Ke
Hawk Road	Lambeth	S.E.	Gl
Hawker Street	Camberwell	S.E.	Hh
Hawksley Road	Stoke Newington	N.	Gb
Hawkesley Road	Walthamstow U.D.	E.	Ja
Hawkins Street	Stepney	E.	He
Hawksmoor Street	Fulham	S.W.	Cg
Hawley Crescent	St. Pancras	N.W.	Ec
Hawley Mews	St. Pancras	N.W.	Ec
Hawley Road	St. Pancras	N.W.	Ec
Hawley Street	St. Pancras	N.W.	Ec
Hawks Road	Kingston		Sm
Hawkshead Road	Willesden U.D.	N.W.	Bc
Hawkslade Road	Camberwell	S.E.	Jj
Hawkstone Road	Bermondsey	S.E.	Jg
Hawkwood	Chislehurst U.D.		Nm
Hawstead Road	Lewisham	S.E.	Kk
Hawthorn Grove	Penge U.D.	S.E.	Hm
Hawthorn Place	Stepney	E.	Jd
Hawthorn Road	Willesden U.D.	N.W.	Bc
Hawthorn Street	Islington	N.	Hc
Hawthorns	Ealing	W.	Se
Hawthorns	Hendon U.D.	N.W.	Da
Hay Hill	Westminster	W.	Ee
Hay Street	Shoreditch	E.	Hd
Haycroft Road	Lambeth	S.W.	Fj
Hayday Road	West Ham	E.	Le
Hayden's Corner	Camberwell	S.E.	Gk
Haydock Road	Deptford	S.E.	Hg
Haydon Passage	Stepney	E.	He
Haydon Square	Stepney	E.	He
Haydon Street	City of London and Stepney	E.	He
Haydon's Road	Wimbledon	S.W.	Dl
Haydon's Road Station	**Wimbledon**	S.W.	Dl
Haydon's Park Road	Wimbledon	S.W.	Dl
Hayes Bldgs.	Lewisham	S.E.	Hk
Hayes Place	Marylebone	N.W.	Ed
Hayfield Passage	Stepney	E.	Je
Hayfield Place	Stepney	E.	Je
Hayfield Yard	Stepney	E.	Je
Hayles Buildings	Southwark	S.E.	Gf
Hayles Street	Southwark	S.E.	Gf
Hayley Street	Lewisham	S.E.	Kj
Hayling Road	Stoke Newington	N.	Hc
Hayman Street	Islington	N.	Gc
Haymarket	Westminster	S.W.	Fe
Haymarket Theatre	**Westminster**	S.W.	Fe
Haymerle Road	Camberwell	S.E.	Hg
Hayne Road	Beckenham U.D.		Jm
Hayne Street	City of London and Finsbury	E.C.	Ge
Hays Lane	Bermondsey	S.E.	Gf
Hay's Mews	Westminster	W.	Ee
Hay's Place	Camberwell	S.E.	Hh
Hay's Street	Westminster	W.	Ee
Hay's Wharf	Bermondsey	S.E.	Gf
Haysleigh Gardens	Penge U.D.	S.E.	Hm

STREET OR PLACE.	BOROUGH.	P.D.	MAP.
Hayter Road	Lambeth	S.W.	Fj
Hayward's Place	Finsbury	E.C.	Ge
Hazel Road	Willesden U.D.	N.W.	Cd
Hazelbank Road	Lewisham	S.E.	Kk
Hazeldean Road	Willesden U.D.	N.W.	Bc
Hazeldene Road	Chiswick U.D.	W.	Ag
Hazeldon Road	Lewisham	S.E.	Jj
Hazelhurst Road	Wandsworth	S.W.	Dl
Hazelmere Road	Hornsey	N.	Fa
Hazelmere Road	Willesden U.D.	N.W.	Dc
Hazelrigge Road	Wandsworth	S.W.	Fj
Hazelville Road	Islington	N.	Fa
Hazelwood House	Woolwich		Mh
Hazelwood Road	Walthamstow U.D.	E.	Ja
Hazlebourne Road	Wandsworth	S.W.	Ej
Hazlebury Road	Fulham	S.W.	Dh
Haslemere Road	Camberwell	S.E.	Hg
Hazlewell Road	Fulham	S.W.	Cj
Hazlewood	Croydon	S.E.	Gm
Hazlewood Crescent	Kensington	W.	Cd
Hazlitt Mews	Hammersmith	W.	Cf
Hazlitt Road	Hammersmith	W.	Cf
Headley Street	Camberwell	S.E.	Hh
Headworth Road	Wandsworth	S.W.	Dk
Heald Street	Deptford	S.E.	Jh
Healey Street	St. Pancras	N.W.	Ec
Heaman Street	Lambeth	S.W.	Fg
Hearn Road	Chiswick U.D.	W.	Ag
Hearn Street	Shoreditch	E.C.	Hd
Hearn Street	Shoreditch	N.	He
Hearn Street	West Ham	E.	Le
Hearn's Buildings	Southwark	S.E.	Gg
Hearnville Road	Wandsworth	S.W.	Ek
Heath Brow	Hampstead	N.W.	Db
Heath Drive	Hampstead	N.W.	Db
Heath Grove	Penge U.D.	S.E.	Hm
Heath House	Hampstead	N.W.	Db
Heath Lodge	Hampstead	N.W.	Db
Heath Road	Croydon		Sm
Heath Road	Wandsworth	S.W.	Eh
Heath Side	Hampstead	N.W.	Db
Heath Side Road	Hampstead	N.W.	Db
Heath Street	Barking U.D.		Nd
Heath Street	Hampstead	N.W.	Db
Heath Street	Stepney	E.	Je
Heath Terrace	Wandsworth	S.W.	Fh
Heath Villas Road	Woolwich		Ng
Heathcock Court	Westminster	W.C.	Fe
Heathcote Road	Twickenham U.D.		Bj
Heathcote Street	St. Pancras	W.C.	Fd
Heathcote Yard	St. Pancras	W.C.	Fd
Heather Road	Lewisham	S.E.	Lk
Heatherley Street	Hackney	E.	Hb
Heathfield	Wandsworth	S.W.	Dj
Heathfield	Wimbledon	S.W.	Ck
Heathfield Avenue	Wandsworth	S.W.	Dj
Heathfield Cottages	Wandsworth	S.W.	Dj
Heathfield Gardens	Chiswick U.D.	W.	Ag
Heathfield House	Wimbledon	S.W.	Ck
Heathfield Lodge	Acton U.D.	W.	Af
Heathfield Lodge	Chislehurst U.D.		Nm
Heathfield Park	Willesden U.D.	N.W.	Cc
Heathfield Road	Acton U.D.	W.	Af
Heathfield Road	Bromley		Km
Heathfield Road	Wandsworth	S.W.	Dj
Heathfield Street	Kensington	W.	Ce
Heathfield Terrace	Chiswick U.D.	W.	Ag
Heathfield Terrace	Woolwich		Ng
Heathhurst Road	Hampstead	N.W.	Eb
Heathland Road	Stoke Newington	N.	Ga
Heathlands	Hampstead	N.W.	Db
Heath's Buildings	Greenwich	S.E.	Kg
Heathside	Wandsworth	S.W.	Ck
Heathview Gardens	Wandsworth	S.W.	Bk
Heathville Road	Islington	N.	Fa
Heathwall Street	Battersea	S.W.	Eh
Heathwood Gardens	Greenwich		Mg
Heaton Place	West Ham	E.	Kc
Heaton Road	Camberwell	S.E.	Hh
Heaton Road	Croydon R.D.		Em
Heaver Road	Battersea	S.W.	Dh
Heavitree Road	Woolwich		Ng

STREET OR PLACE.	BOROUGH.	P.D.	MAP.
Heber Mansions	Fulham	W.	Cg
Heber Road	Camberwell	S.E.	Hj
Heber Road	Willesden U.D.	N.W.	Cb
Hebron Road	Hammersmith	W.	Cf
Heckfield Place	Fulham	S.W.	Cg
Heckford Street	Stepney	E.	Je
Heddon Street	Westminster	W.	Fe
Hedge Lane	Hendon U.D.	N.W.	Ca
Hedgers Grove	Hackney	E.	Jc
Hedgley Mews	Lewisham	S.E.	Kj
Hedgley Street	Lewisham	S.E.	Kj
Hedsor Buildings	Bethnal Green	E.	Hd
Hedworth Road	Wandsworth	S.W.	Dl
Heiron Street	Southwark	S.E.	Gg
Helen Street	Woolwich		Ng
Helena Road	Ealing	W.	Sd
Helena Road	Walthamstow U.D.	E.	Ja
Helena Road	West Ham	E.	Ld
Helen's Place	Bethnal Green	E.	Hd
Helenslea	Hendon U.D.	N.W.	Da
Helix Gardens	Lambeth	S.W.	Fj
Helix Road	Lambeth	S.W.	Fj
Helmet Court	City of London	E.C.	Ge
Helmet Court	Westminster	W.C.	Fe
Helmet Row	Finsbury	E.C.	Gd
Helmsley Cottages	Hackney	E.	Hc
Helmsley Place	Hackney	E.	Hc
Helmsley Street	Hackney	E.	Hc
Helmsley Terrace	Hackney	E.	Hc
Helvetia Street	Lewisham	S.E.	Jk
Heman's Street	Lambeth	S.W.	Fh
Hemberton Road	Lambeth	S.W.	Fh
Hemingford Road	Islington	N.	Fc
Hemming Street	Bethnal Green	E.	Hd
Hemp Row	Southwark	S.E.	Gg
Hemstall Road	Hampstead	N.W.	Dc
Hemsworth Street	Shoreditch	N.	Gd
Hemsworth Street	West Ham	E.	Le
Hen and Chickens Crt.	City of London	E.C.	Fe
Henderson Road	Wandsworth	S.W.	Dk
Henderson's Road	East Ham	E.	Lc
Hendham Road	Wandsworth	S.W.	Ek
Hendon House	Hendon U.D.	N.W.	Ca
Hendon Park Row	Hendon U.D.	N.W.	Da
Hendon Station	**Hendon U.D.**	N.W.	Ba
Hendre Cottages	Southwark	S.E.	Hg
Heneage Lane	City of London	E.C.	He
Heneage Street	Stepney	E.	He
Henfield Road	Merton U.D and Wimbledon	S.W.	Cm
Hengrave Road	Lewisham	S.E.	Hk
Henley Road	Willesden U.D.	N.W.	Cc
Henley Road	Woolwich	E.	Mf
Henley Square	Southwark	S.E.	Gf
Henley Street	Battersea	S.W.	Eh
Henley Terrace	Southwark	S.E.	Gf
Henniker Road	West Ham	E.	Kc
Henning Street	Battersea	S.W.	Dh
Henrietta Mansions	Marylebone	W.	Ee
Henrietta Passage	Marylebone	W.	Ee
Henrietta Road	Tottenham U.D.	N.	Ha
Henrietta Street	Bethnal Green	E.	Hd
Henrietta Street	Greenwich	S.E.	Jg
Henrietta Street	Marylebone	W.	Ee
Henrietta Street	Shoreditch	E.C.	Gd
Henrietta Street	Westminster	W.C.	Fe
Henry Place	Hammersmith	W.	Cf
Henry Place	Islington	N.	Fd
Henry Road	Stoke Newington	N.	Ga
Henry Road	Tottenham U.D.	N.	Ga
Henry Street	Battersea	S.W.	Dh
Henry Street	Bermondsey	S.E.	Gf
Henry Street	Bromley		Lm
Henry Street	Finsbury	E.C.	Gd
Henry Street	Holborn and St. Pancras	W.C.	Fd
Henry Street	Lambeth	S.E.	Fg
Henry Street	Marylebone	N.W.	Ed
Henry Street	Stepney	E.	Jd
Henry Street	Woolwich		Mg
Henryson Street	Lewisham	S.E.	Jj
Henshall Street	Islington	N.	Gc

STREET OR PLACE.	BOROUGH.	P.D.	MAP.
Holly Place	Hampstead	N.W.	Db
Holly Road	Chiswick U.D.	W.	Af
Holly Road	Leyton U.D.	E.	La
Holly Street	Hackney	E.	Hc
Holly Terrace	Hampstead	N.W.	Db
Holly Terrace	St. Pancras	N.	Eb
Holly Village	St. Pancras	N.	Eb
Hollybush	Wanstead U.D.	E.	La
Hollybush Gardens	Bethnal Green	E.	Hd
Hollybush Hill	Hampstead	N.W.	Db
Hollybush Hill	Leyton U.D. and Wanstead U.D.	E.	La
Hollybush Place	Bethnal Green	E.	Hd
Hollybush Road	West Ham	E.	Ld
Hollybush Street	West Ham	E.	Ld
Hollybush Vale	Hampstead	N.W.	Db
Hollycroft Avenue	Hampstead	N.W.	Db
Hollydale Road	Camberwell	S.E.	Hh
Hollyhedge House	Lewisham	S.E.	Kh
Hollyhedge Terrace	Lewisham	S.E.	Kj
Hollyoak Wood	Bexley		Nk
Hollytree Terrace	Lewisham	S.E.	Kj
Hollywood House	Wandsworth	S.W.	Ck
Hollywood Mews	Kensington	S.W.	Dg
Hollywood Road	Kensington	S.W.	Dg
Hollywood Villas	Wandsworth	S.W.	Cj
Holm Road	Willesden U.D.	N.W.	Cb
Holman Road	Battersea	S.W.	Dh
Holmbrook Street	Hackney	E.	Jc
Holmbush	Hendon U.D.	N.W.	Ca
Holmbush Road	Wandsworth	S.W.	Cj
Holmbury View	Hackney	E.	Ha
Holmby Street	Camberwell	S.E.	Gg
Holmdale Road	Hackney and Tottenham U.D.	N.	Ha
Holmdale Road	Hampstead	N.W.	Db
Holmdene Avenue	Camberwell	S.E.	Oj
Holme Road	East Ham	E.	Md
Holmes Place	Kensington	S.W.	Dg
Holmes Road	St. Pancras	N.W.	Ec
Holmes Terrace	Lambeth	S.E.	Ff
Holmesdale Road	Hornsey	N.	Ea
Holmesdale Road	Richmond		Ah
Holmes ey Street	Lewisham	S.E.	Jj
Holmewood Gardens	Wandsworth	S.W.	Fk
Holmewood Road	Wandsworth	S.W.	Fk
Holmewood Villas	Wandsworth	S.W.	Fk
Holmwood Terrace	Wandsworth	S.W.	Dj
Holms Street	Shoreditch	E.	Hd
Holmsdale Road	Teddington U.D.		Sl
Holmshaw Road	Lewisham	S.E.	Jl
Holmleigh Road	Hackney	N.	Ha
Holmwood	The Maldens and Coombe U.D.		Al
Holmwood	Wimbledon	S.W.	Bl
Holness Road	West Ham	E.	Lc
Holroyd Road	Wandsworth	S.W.	Cj
Holsworthy Square	Holborn	W.C.	Fd
Holt Place	Shoreditch	N.	Gd
Holtham Road	Hampstead	N.W.	Dc
Holwood Road	Bromley		Lm
Holyoak Road	Lambeth and Southwark	S.E.	Gg
Holyport Road	Fulham	S.W.	Cg
Holyrood Avenue	Croydon		Fm
Holywell Lane	Shoreditch	N.	Hd
Holywell Mount	Shoreditch	N.	Hd
Holywell Place	Shoreditch	N.	Hd
Holywell Row	Shoreditch	N.	Gd
Home Park Road	Wimbledon	S.W.	Cl
Home Road	Battersea	S.W.	Dh
Homecroft Road	Lewisham	S.E.	Hl
Homefield	Ealing	W.	Se
Homefield Road	Bromley		Lm
Homefield Road	Chiswick U.D.	W.	Bg
Homefield Road	Wimbledon	S.W.	Cl
Homefield Villas	Wandsworth	S.W.	Fl
Home for Incurables	Lambeth	S.W.	Gl
Homeleigh Road	Camberwell	S.E.	Jj
Homer Grange	Lewisham	S.E.	Hl
Homer Road	Hackney	E.	Jc
Homer Row	Marylebone	N.W. & W.	Ee

STREET OR PLACE.	BOROUGH.	P.D.	MAP.
Homer Street	Lambeth	S.E.	Ff
Homer Street	Marylebone	N.W. & W.	Ee
Homer St. Buildings	Marylebone	W.	Ee
Homerton Grove	Hackney	E.	Jc
Homerton Road	Hackney	E.	Jc
Homerton Row	Hackney	E.	Jc
Homerton Terrace	Hackney	E.	Jc
Homes Road	Kingston		Am
Homestall Road	Camberwell	S.E.	Hj
Homestead, The	Hendon U.D.	N.W.	Ca
Homestead Road	Fulham	S.W.	Cg
Homewood	Chislehurst U.D.		Nm
Homewood Road	Croydon R.D.		Dm
Homewood Terrace	Wandsworth	S.W.	Dj
Homfray Street	Hackney	E.	Jc
Honduras Street	Finsbury	E.C.	Gd
Honduras Wharf	Southwark	S.E.	Ge
Honey Lane	City of London	E.C.	Ge
Honey Lane Market	City of London	E.C.	Ge
Honeybourne Road	Hampstead	N.W.	Db
Honeybrook Road	Wandsworth	S.W.	Ek
Honeypot Lane	Wembley U.D.		Sc
Honeysuckle Court	City of London	E.C.	Ge
Honeywell Road	Battersea	S.W.	Ej
Honeywood Road	Willesden U.D.	N.W.	Bd
Honiton Court	Camberwell	S.E.	Hh
Honiton Road	Willesden U.D.	N.W.	Cd
Honley Road	Lewisham	S.E.	Kk
Honourable Artillery Co's. Hd. Qrs.	Finsbury	E.C.	Ge
Honour Oak Park	Camberwell and Lewisham	S.E.	Hk
Honour Oak Park Sta.	**Lewisham**	S.E.	Jj
Honour Oak Rise	Camberwell	S.E.	Hk
Honour Oak Road	Lewisham	S.E.	Hk
Honour Oak Station	**Lewisham**	S.E.	Hk
Honor Oak Villas	Camberwell	S.E.	Hk
Hoo, The	Camberwell	S.E.	Hl
Hood Street	Deptford	S.E.	Jg
Hook Lane	Bexley U.D.		Oj
Hooker's Court	City of London	E.C.	Ge
Hook's Road	Camberwell	S.E.	Hh
Hoop Lane	Hendon U.D.	N.W.	Da
Hooper Road	West Ham	E.	Le
Hooper Street	Stepney	E.	He
Hoopwick Street	Deptford	S.E.	Jg
Hop Gardens	Westminster	W.C.	Fe
Hope Cottage Road	Woolwich		Ng
Hope Park	Bromley		Lm
Hope Place	Islington	N.	Fc
Hope Place	Stepney	E.	Je
Hope Road	Islington	N.	Fc
Hope Street	Battersea	S.W.	Dh
Hope Street	Islington	N.	Fc
Hope Street	Stepney	E.	He
Hope Tavern	Battersea	S.W.	Ek
Hope Wharf	Bermondsey	S.E.	Hf
Hopedale Road	Greenwich		Lg
Hopeford Avenue	Willesden U.D.	N.W.	Cd
Hopewell Place	Camberwell	S.E.	Gh
Hopgood Street	Hammersmith	W.	Bf
Hopkin's Street	Westminster	W.	Fe
Hopton Road	Wandsworth	S.W.	Fl
Hopwood Street	Southwark	S.E.	Gg
Horace Road	West Ham	E.	Lb
Horace Street	Lambeth	S.W.	Fg
Horace Street	Marylebone	W.	Ee
Horace St. Buildings	Marylebone	W.	Ee
Horatio Street	Bethnal Green	E.	Hd
Horbury Crescent	Kensington	W.	De
Horbury Mews	Kensington	W.	De
Horder Road	Fulham	S.W.	Ch
Hormead Road	Paddington	W.	Cd
Horn Lane	Acton U.D.	W.	Ae
Horn Lane	Greenwich	S.E.	Lg
Horn Park	Woolwich	S.E.	Lk
Horn Park Lane	Woolwich	S.E.	Lj
Hornby Road	Camberwell	S.E.	Hh
Horney Lane	Bermondsey	S.E.	Hf
Horner Road	Hackney	E.	Hd
Horniman Gardens	Lewisham	S.E.	Hk
Horniman Museum	**Lewisham**	S.E.	Hk

STREET OR PLACE.	BOROUGH.	P.D.	MAP.
Horns, The	Lambeth	S.E.	Gg
Hornsey Chambers	Hackney	E.	Hb
Hornsey Lane	Hornsey and Islington	N.	Fa
Hornsey Lane Gardens	Hornsey	N.	Fa
Hornsey Place	Stoke Newington	N.	Gb
Hornsey Rise	Islington	N.	Fa
Hornsey Rise Gardens	Islington	N.	Fa
Hornsey Rise Mews	Islington	N.	Fa
Hornsey Road	Islington	N.	Fb
Hornsey Road Station	**Islington**	N.	Fa
Hornsey Street	Islington	N.	Fb
Hornsey Wood Tavern	Stoke Newington	N.	Ga
Hornshay Place	Deptford	S.E.	Hg
Hornshay Street	Deptford	S.E.	Hg
Hornton Mews	Kensington	W.	Df
Hornton Place	Kensington	W.	Df
Hornton Street	Kensington	W.	Df
Horse End	Woolwich		Ne
Horse Fair	Kingston		Sm
Horse & Dolphin Yard	Westminster	W.C.	Fe
Horse and Groom Yard	Southwark	S.E.	Gg
Horseferry Branch Rd.	Stepney	E.	Je
Horseferry Road	Greenwich	S.E.	Kg
Horseferry Road	Stepney	E.	Je
Horseferry Road	Westminster	S.W.	Ff
Horseford Road	Lambeth	S.W.	Fj
Horse Guards Avenue	Westminster	S.W.	Ff
Horse Guards Parade	Westminster	S.W.	Ff
Horsely Street	Southwark	S.E.	Gg
Horseleydown Lane	Bermondsey	S.E.	Hf
Horseleydown Stairs	Bermondsey	S.E.	Hf
Horseleydown Wharf	Stepney	S.E.	Hf
Horsell Road	Islington	N.	Gc
Horsendon House	Greenford U.D.		Sc
Horse Shoe Alley	Westminster	S.W.	Fe
Horseshoe Alley	Shoreditch	E.C.	Ge
Horseshoe Alley	Southwark	S.E.	Gf
Horseshoe Point	Hackney	E.	Ha
Horsford Road	Lambeth	S.W.	Fj
Horsley's Buildings	Stepney	E.	He
Horsman Street	Southwark	S.E.	Gg
Horsted Road	Deptford	S.E.	Jj
Hortensia Road	Chelsea	S.W.	Dg
Horton Road	Hackney	E.	Hc
Horton Street	Lewisham	S.E.	Kh
Hosack Road	Wandsworth	S.W.	Ek
Hosier Lane	City of London	E.C.	Ge
Hosier Street	Deptford	S.E.	Jh
Hoskins Court	Greenwich	S.E.	Kg
Hoskins Street	Greenwich	S.E.	Kg

HOSPITALS.

STREET OR PLACE.	BOROUGH.	P.D.	MAP.
Alexandra (Children)	Holborn	W.C.	Fe
Battersea General	Battersea	S.W.	Eh
Belgrave (Children)	Westminster	S.W.	Eg
Blackheath & Charlton Cottage	Greenwich	S.E.	Lh
British Lying-in	Holborn	W.C.	Fe
British Home for Incurables	Lambeth	S.W.	Gl
Brompton (Chest)	Kensington	S.W.	Dg
Cancer	Chelsea	S.W.	Dg
Central London Ophthalmic	St. Pancras	W.C.	Fd
Central London Throat and Ear	St. Pancras	W.C.	Fd
Charing Cross	Westminster	W.C.	Fe
Chelsea (Women)	Chelsea	S.W.	Dg
Cheyne (Incurable Children)	Chelsea	S.W.	Eg
City of London (Chest)	Bethnal Green	E.	Jd
City of London Lying-in	Finsbury	E.C.	Gd
City Orthopædic	Holborn	E.C.	Ge
Clapham Matercity	Lambeth	S.W.	Fh
East End Mothers' Home (Lying-in)	Stepney	E.	Je
East London (Children)	Stepney	E.	Je
Eastern Fever	Hackney	E.	Jc
Evelina (Children)	Southwark	S.E.	Gf
Fountain (Fever)	Wandsworth	S.W.	Dl
French	Holborn	W.C.	Fe
General Lying-in	Lambeth	S.W.	Ff
German	Hackney	E.	Hc
Gordon (Fistula)	Westminster	S.W.	Ef
Great Northern Central	Islington	N.	Fb
Grosvenor (Women & Children)	Westminster	S.W.	Fg
Grove (Fever)	Wandsworth	S.W.	Dl
Guy's	Southwark	S.E.	Gf
Hampstead	Hampstead	N.W.	Eb
Hospital for Diseases of Skin	Southwark	S.E.	Gf
Hospital for Epilepsy, Paralysis, etc.	Marylebone	N.W.	Dd
Hospital for Sick Children	Holborn	W.C.	Fe
Hospital for Women	Westminster	W.	Fe
Infants	Westminster	S.W.	Fg
Italian	Holborn	W.C.	Fe
Kensington (Children)	Kensington	W.	Df
Kensington (Jubilee)	Kensington	S.W.	Dg
King's College	Westminster	W.C.	Fe
Lady Gomm	Bermondsey	S.E.	Hg
Leyton, Walthamstow and Wanstead	Walthamstow U.D.	E.	Ka
London	Stepney	E.	He
London Fever	Islington	N.	Gd
London Homœopathic	Holborn	W.C.	Fe
London Lock	Paddington	W.	De
London Skin	St. Pancras	W.	Fd
London Temperance	St. Pancras	N.W.	Fd
London Throat	Marylebone	W.	Ee
Margaret St. (Consumption)	Marylebone	W.	Ee
Metropolitan	Hackney	E.	Hc
Metropolitan Throat and Ear	St. Pancras	W.	Fd
Middlesex	Marylebone	W.	Fe
Mildmay Cottage H.	Islington	N.	Gc
Mildmay Mission	Bethnal Green	E.	Hd
Military	Westminster	W.	Fg
Miller Hosp. for S.E. London	Greenwich	S.E.	Kh
Mount Vernon (Chest)	Hampstead	N.W.	Db
Municipal Infirmary (Throat)	Finsbury	E.C.	Gd
National Dental	Marylebone	W.	Ee
National (Epilepsy, &c.)	Holborn	W.C.	Fe
National (Heart)	Westminster	W.	Fe
New (Women)	St. Pancras	N.W.	Fd
North-Eastern (Child.)	Hackney	E.	Hd
North-Eastern Fever	Tottenham U.D.	N.	Ga
North-Western Fever	Hampstead	N.W.	Eb
North-West London	St. Pancras	N.W.	Ec
Paddington Green (Children)	Paddington	W.	De
Poplar	Poplar	E.	Ke
Queen Charlotte's Lying-in	Marylebone	W.	Ee
Royal (Chest)	Islington	N.	Gd
Royal Dental	Westminster	W.C.	Fe
Royal Ear	Westminster	W.	Fe
Royal Free	St. Pancras	W.C.	Fd
Royal (Incurables)	Wandsworth	S.W.	Cj
Royal London Ophthalmic	Finsbury	E.C.	Gd
Royal Maternity Charity	Finsbury	E.C.	Ge
Royal National Orthopædic	Marylebone	W.	Ed
Royal South London Ophthalmic	Southwark	S.E.	Gf
Royal Waterloo (Children and Women)	Lambeth	S.E.	Ff
Royal Westminster Ophthalmic	Westminster	W.C.	Fe
Samaritan Free (Women)	Marylebone	N.W.	Ee
St. Bartholomew's	City of London	E.C.	Ge
St. George's	Westminster	S.W.	Ef
St. John's	Lewisham	S.E.	Kh
St. John's (Skin)	Westminster	W.C.	Fe
St. John's (Skin)	Hammersmith	W.	Cf

STREET OR PLACE.	BOROUGH.	P.D.	MAP.
St. Mary's	Paddington	W.	De
St. Mark's (Fistula)	Finsbury	E.C.	Gd
St. Mary's (Children)	West Ham	E.	Ld
St. Paul's (Skin)	Holborn	W.C.	Fe
St. Peter's (Stone)	Westminster	W.C.	Fe
St. Saviour's (Women)	St. Pancras	N.W.	Ed
St. Thomas's	Lambeth	S.E.	Ff
South-Eastern Fever	Deptford	S.E.	Hg
South-Western Fever	Lambeth	S.W.	Fh
Throat	Westminster	W.	Fe
University College	St. Pancras	W.C.	Fd
Victoria (Children)	Chelsea	S.W.	Eg
West End (Nervous System)	Marylebone	W.	Ee
West Ham (Fever)	West Ham	E.	Ld
West Ham (General)	West Ham	E.	Kc
West London	Hammersmith	W.	Cg
Western (Fever)	Fulham	S.W.	Dg
Western Ophthalmic	Marylebone	W.	Ee
Western Skin	Marylebone	W.	Ee
Westminster	Westminster	S.W.	Ff
Woolwich Cottage	Woolwich		Mh

STREET OR PLACE.	BOROUGH.	P.D.	MAP.
Hospital Waterworks	Woolwich		Mh
Hotel Cecil	Westminster	W.C.	Fe
Hotham Road	Wandsworth	S.W.	Ch
Hotham Road	Wimbledon	S.W.	Dm
Hotham Street	West Ham	E.	Kd
Hothfield Place	Bermondsey	S.E.	Jf
Hotspur Street	Lambeth	S.E.	Fg
Hot Water Court	Finsbury	E.C.	Ge
Houblon Road	Richmond		Sj
Houghton Place	St. Pancras	N.W.	Fd
Houghton Place	Westminster	W.C.	Fe
Houghton Road	Tottenham U.D.	N.	Ha
Houndsditch	City of London	E.	He
Houses of Parliament	**Westminster**	S.W.	Ff
Houston Road	Lewisham	S.E.	Jl
Hove Avenue	Walthamstow U.D.	E.	Ja
Hove Street	Camberwell	S.E.	Hh
Hoveden Road	Willesden U.D.	N.W.	Cb
Howard Road	Barking U.D.		Nd
Howard Road	Bromley		Lm
Howard Road	East Ham	E.	Md
Howard Road	Leyton U.D.	E.	Kb
Howard Road	Penge U.D.	S.E.	Hm
Howard Road	Stoke Newington	N.	Gb
Howard Road	The Maldens & Coombe U.D.		Bm
Howard Road	Tottenham U.D.	N.	Ha
Howard Road	West Ham	E.	Ld
Howard Road	Willesden U.D.	N.W.	Cb
Howard Street	Poplar	E.	Ke
Howard Street	Wandsworth	S.W.	Fh
Howard's Buildings	Finsbury	E.C.	Gd
Howard's Buildings	Woolwich		Mg
Howard's Lane	Wandsworth	S.W.	Cj
Howard's Place	Finsbury	E.C.	Gd
Howard's Road	West Ham	E.	Lc
Howard's Street	Westminster	W.C.	Fe
Howarth Road	Woolwich		Og
Howberry Road	Croydon		Gm
Howbury Road	Camberwell	S.E.	Hh
Howden Road	Croydon	S.E.	Gm
Howden Street	Camberwell	S.E.	Hh
Howell Street	Paddington	W.	Dd
Howford Buildings	City of London	E.C.	Ge
Howgate Road	Barnes U.D.	S.W.	Ah
Howick Place	Westminster	S.W.	Ff
Howie Street	Battersea	S.W.	Eh
Howitt Road	Hampstead	N.W.	Ec
Howland Mews East	St. Pancras	W.	Fe
Howland Mews West	St. Pancras	W.	Fe
Howland Street	St. Pancras	W.	Fe
Howlett Road	Camberwell	S.E.	Gj
Howley Place	Lambeth	S.E.	Ef
Howley Place	Paddington	W.	De
Howley Street	Lambeth	S.E.	Ff
Howe Street	Shoreditch	E.	Hd
Howson Road	Deptford and Lewisham	S.E.	Jj
Hoxton Market	Shoreditch	N.	Gd
Hoxton Place	Shoreditch	N.	Hd
Hoxton Residences	Shoreditch	N.	Gd
Hoxton Square	Shoreditch	N.	Gd
Hoxton Street	Shoreditch	N.	Hd
Hoy Street	West Ham	E.	Le
Hoyle Road	Wandsworth	S.W.	Dl
Hubbard Street	West Ham	E.	Kd
Hubbuck's Wharf	Stepney	E.	Je
Hubert Grove	Lambeth	S.W.	Fh
Hubert Road	Wimbledon	S.W.	Dm
Huddart Street	Stepney	E.	Je
Huddleston Road	Islington	N.	Fb
Huddleston Road	Leyton U.D.	E.	Lb
Huddleston Road	Willesden U.D.	N.W.	Bc
Hudson Road	Kingston		Sm
Hudson Road	West Ham	E.	Le
Hudson Road	Woolwich		Ng
Hudson Terrace	Westminster	S.W.	Fg
Hudson's Place	Westminster	S.W.	Ef
Huggin Court	City of London	E.C.	Ge
Huggin Lane	City of London	E.C.	Ge
Hugh Place	Westminster	S.W.	Fg
Hugh Street	Westminster	S.W.	Eg
Hughan Road	West Ham	E.	Kc
Hughenden Avenue	Croydon		Fm
Hughenden Road	Hammersmith	W.	Bg
Hughes Fields	Greenwich	S.E.	Jg
Hughes Fields Cottages	Greenwich	S.E.	Jg
Hughes Recreation Ground	Greenwich	S.E.	Jg
Hughes Terrace	Bermondsey	S.E.	Jg
Hugo Road	Islington	N.	Fb
Hugon Road	Fulham	S.W.	Dh
Huguenot Place	Wandsworth	S.W.	Dj
Huguenot Road	Camberwell	S.E.	Hh
Hulsh's Court	City of London	E.C.	Ge
Hull Alley	Hackney	N.	Hb
Hull Place	Woolwich		Ng
Hull Street	Finsbury	E.C.	Ge
Hull's Place	Finsbury	E.C.	Ge
Humber Road	Greenwich	S.E.	Lg
Humber Road East	Greenwich	S.E.	Lg
Humbolt Road	Fulham	W.	Cg
Hume Road	Hammersmith	W.	Cf
Humphrey Street	Bermondsey and Camberwell	S.E.	Hg
Hungerford Bridge	Westminster	W.C.	Ff
Hungerford Road	Islington	N.	Fc
Hungerford Stables	Islington	N.	Fc
Hungerford Street	Stepney	E.	He
Hunsden Road	Deptford	S.E.	Jg
Hunt Court	Stepney	E.	He
Hunt Place	Stepney	E.	He
Hunt Street	Chiswick U.D.	W.	Bg
Hunt Street	Hammersmith	W.	Ce
Hunt Street	Lambeth	S.E.	Fg
Hunt Street	Stepney	E.	He
Hunter Mews	St. Pancras	W.C.	Fd
Hunter Road	Croydon		Sm
Hunter Street	St. Pancras	W.C.	Fd
Hunter's Place	Southwark	S.E.	Gf
Huntington Bldgs.	Bethnal Green	E.	Hd
Huntingdon Street	Islington	N.	Fc
Huntingdon Street	West Ham	E.	Le
Huntley Street	Bethnal Green	E.	Jd
Huntley Street	Holborn and St. Pancras	W.C.	Fd
Hunt's Court	Westminster	W.C.	Fe
Huntslett Street	Bethnal Green	E.	Jd
Huntsman Street	Southwark	S.E.	Gg
Huntsmoor Road	Wandsworth	S.W.	Dj
Huntspill Street	Wandsworth	S.W.	Dl
Huntsworth Mews	Marylebone	N.W.	Ed
Huntsworth Terrace	Marylebone	N.W.	Dd
Hurdwick Place	St. Pancras	N.W.	Fd
Hurlbutt Place	Southwark	S.E.	Gg
Hurley Bldgs.	Bethnal Green	E.	Hd
Hurley Road	Lambeth	S.E.	Gg
Hurlingham House	Fulham	S.W.	Ch
Hurlingham Mansions	Fulham	S.W.	Ch
Hurlingham Park	Fulham	S.W.	Ch
Hurlingham Road	Fulham	S.W.	Ch

STREET OR PLACE.	BOROUGH.	P.D.	MAP.
Huron Road	Wandsworth	S.W.	Ek
Hurst	Bexley U.D.		Ok
Hurst Bldgs.	Lewisham	S.E.	Kh
Hurst Cottages	Bexley U.D.		Ok
Hurst Cottages	Lewisham	S.E.	Kh
Hurst Road	Bexley U.D.		Ol
Hurst Street	Lambeth	S.E.	Gj
Hurstbourne Road	Lewisham	S.E.	Jk
Hurstmead Road	Tottenham U.D.	N.	Ha
Hurstway Street	Kensington	W.	Ce
Hurt House Lodge.	Erith U.D.		Og
Hut Barracks	Greenwich		Mh
Hutchings Street	Poplar	E.	Jf
Hutchison Street	City of London	E.	He
Hutchison's Avenue	City of London	E.	He
Hutley Place	Shoreditch	N.	Gd
Hutton Road	Lambeth	S.E.	Fg
Hutton Street	City of London	E.C.	Ge
Huxbear Street	Lewisham	S.E.	Jj
Huxley Road	Leyton U.D.	E.	Kb
Huxley Street	Paddington	W.	Cd
Hyacinth Road	Wandsworth	S.W.	Bk
Hyde, The	Hendon U.D.	N.W.	Ba
Hyde House	Kingsbury U.D.		Ba
Hyde Lane	Battersea	S.W.	Eh
Hyde Park	**Westminster**	W.	Ee
Hyde Park Barracks	**Westminster**	S.W.	Ef
Hyde Park Corner	Westminster	W.	Ef
Hyde Park Corner Sta.	**Westminster**	W.	Ef
Hyde Park Court	Westminster	S.W.	Ef
Hyde Park Gardens	Paddington	W.	De
Hyde Park Gardens Mews	Paddington	W.	De
Hyde Park Gate	Kensington	S.W.	Df
Hyde Park Gate South Mews	Kensington	S.W.	Df
Hyde Park Mansions	Marylebone.	N.W.	Ee
Hyde Park Place	Paddington	W.	De
Hyde Park Square	Paddington	W.	Ea
Hyde Park Square Mews	Paddington	W.	Ee
Hyde Park Street	Paddington	W.	Ee
Hyde Park Terrace	Paddington	W.	Ee
Hyde Place	Westminster	S.W.	Fg
Hyde Road	Shoreditch	N.	Gd
Hyde Street	Deptford	S.E.	Jg
Hyde Street	Holborn	W.C.	Fe
Hyde Vale	Greenwich	S.E.	Kh
Hyde's Place East	Islington	N.	Gc
Hyde's Place West	Islington.	N.	Gc
Hyde Thorpe Road	Wandsworth	S.W.	Ek
Hylett Crescent	Hammersmith	W.	Bf
Hylton Street	Woolwich		Ng
Hyndman Grove	Camberwell	S.E.	Hg
Hyndman Place	Camberwell	S.E.	Hg
Hyndman Street	Camberwell	S.E.	Hg
Hyson Road	Bermondsey	S.E.	Hg
Hythe Road	Hammersmith	N.W.	Bd
Hythe Road	Croydon		Gm
Ibsen Place	Greenwich	S.E.	Kg
Icehouse Wood	Chislehurst U.D.		Nm
Iceland Road	Poplar	E.	Jd
Ickburgh Road	Hackney	E.	Hb
Ida Road	Tottenham U.D.	N.	Ga
Ida Street	Poplar	E.	Ke
Iddesleigh Mansions	Westminster	S.W.	Ff
Idlecombe Road	Wandsworth	S.W.	El
Idmiston Road	Lambeth	S.E.	Gk
Idmiston Road	West Ham	E.	Lc
Idol Lane	City of London	E.C.	Ge
Idonia Street	Deptford	S.E.	Jh
Iffley Bldgs.	Bethnal Green	E.	Hd
Iffley Road	Hammersmith	W.	Cf
Ifield Road	Kensington	S.W.	Dg
Ilbert Street	Paddington	W.	Cd
Ilchester Gardens	Paddington	W.	De
Ilchester Mansions	Kensington	W.	Df
Ildersley Grove	Camberwell	S.E.	Gk
Ilderton Road	Bermondsey, Camberwell & Deptford	S.E.	Hg
Ilex Road	Willesden U.D.	N.W.	Bc
Iliffe Street	Southwark	S.E.	Gg
Iliffe Street Yard	Southwark	S.E.	Gg
Ilfracombe Bldgs.	Southwark	S.E.	Gf
Ilminster Gardens	Battersea	S.W.	Ej
Ilton Road	Walthamstow U.D.	E.	Ja
Ilva Place	Shoreditch	E.	Hd
Imperial Arcade	City of London	E.C.	Ge
Imperial Bldgs.	City of London	E.C.	Ge
Imperial Chambers	Southwark	S.E.	Gg
Imperial Institute	**Westminster**	S.W.	Df
Imperial Institute Rd.	Kensington and Westminster	S.W.	Df
Imperial Road	Fulham	S.W.	Dh
Imperial Square	Fulham	S.W.	Dh
Imperial Street	Poplar.	E.	Kd
Imperial Theatre	**Westminster**	S.W.	Ff
Inchmery Road	Lewisham	S.E.	Kk
Inderwick Road	Hornsey	N.	Fa
India Office	**Westminster**	S.W.	Ff
India Stores Depot	Lambeth	S.E.	Ff
Industry Terrace	Lambeth	S.W.	Fj
Infant Orphan Asylum	Wanstead U.D.	E.	La
Ingal Lane	West Ham	E.	Le
Ingarsby	Wimbledon	S.W.	Cm
Ingelheim Place	Poplar	E.	Jg
Ingelow Road	Battersea	S.W.	Eh
Ingersall Road	Hammersmith	W.	Bf
Ingestre Bldgs.	Westminster	W.	Fe
Ingestre Court	Westminster	W.	Fe
Ingestre Place	Westminster	W.	Fe
Ingestre Road	St. Pancras	N.W.	Eb
Ingestre Road	West Ham	E.	Lb
Ingham Road	Hampstead	N.W.	Db
Ingleborough Place	Lambeth	S.W.	Fh
Ingleborough Street	Lambeth	S.W.	Fh
Ingleby Road	Islington	N.	Fb
Ingledew Road	Woolwich		Ng
Ingleheim Place	Poplar	E.	Jg
Inglemere Road	Croydon R.D.		Em
Inglemere Road	Lewisham	S.E.	Jl
Ingleside	Wandsworth	S.W.	Ck
Ingleside Grove	Greenwich	S.E.	Lg
Inglethorpe Street	Fulham	S.W.	Ch
Ingleton Street	Lambeth	S.W.	Fh
Inglewood	Bromley		Mm
Inglewood	Lambeth	S.E.	Gl
Inglewood Road	Hampstead	N.W.	Dc
Inglis Road	Ealing	W.	Ae
Inglis Street	Lambeth	S.E.	Gh
Ingoldisthorpe Grove	Camberwell	S.E.	Hg
Ingram Court	City of London	E.C.	Ge
Ingram Road	Croydon		Gm
Ingrave Street	Battersea	S.W.	Dh
Inkerman Road	St. Pancras	N.W.	Ec
Inkerman Terrace	Kensington	W.	Df
Inman Road	Wandsworth	S.W.	Dk
Inman Road	Willesden U.D.	N.W.	Bc
Inner Circle	Marylebone	N.W.	Ed
Inner Park Road	Wandsworth	S.W.	Ck
Inner Temple, The	City of London	E.C.	Fe
Inner Temple Gardens	City of London	E.C.	Fe
Inner Temple Lane	City of London	E.C.	Fe
Innis Place	Southwark	S.E.	Gg
Invarine Road	Woolwich		Lg
Inver Road	Hackney	E.	Hb
Inver Terrace	Hackney	E.	Hb
Inverary Place	Woolwich		Ng
Inverine Road	Greenwich		Lg
Inverness Place	Paddington	W.	De
Inverness Place	Woolwich		Ng
Inverness Place Mews	Paddington	W.	De
Inverness Terrace	Battersea	S.W.	Ek
Inverness Terrace	Paddington	W.	De
Inverton Road	Camberwell	S.E.	Jj
Invicta Road	Greenwich		Lg
Invicta Road	West Ham	E.	Le
Inville Road	Southwark	S.E.	Gg
Inworth Street	Battersea	S.W.	Eh
Ion Square	Bethnal Green	E.	Hd
Iona Road	Hampstead	N.W.	Ec
Ipplepen Road	Tottenham U.D.	N.	Ha
Ipswich Road	Shoreditch	E.	Hd
Ireland Yard	City of London	E.C.	Ge

STREET OR PLACE.	BOROUGH.	P.D	MAP.
Irene Road	Fulham	S.W.	Dh
Ireton Street	Poplar	E.	Jd
Iris Place	Southwark	S.E.	Gf
Iron Bridge	Poplar and West Ham	E.	Ke
Iron Yard	Bermondsey	S.E.	Jf
Irongate Wharf	Stepney	E.	Hf
Irongate Wharf Road	Paddington	W.	De
Ironmill Place	Wandsworth	S.W.	Dj
Ironmill Road	Wandsworth	S.W.	Dj
Ironmonger Lane	City of London	E.C.	Ge
Ironmonger Passage	Finsbury	E.C.	Gd
Ironmonger Row	Finsbury	E.C.	Gd
Ironmonger Street	Finsbury	E.C.	Gd
Ironmonger's Almshos.	Shoreditch	E.	Hd
Irving Grove	Lambeth	S.W.	Fh
Irwell Place	Bermondsey	S.E.	Hf
Isabel Street	Lambeth	S.W.	Fh
Isabella Place	Wandsworth	S.W.	Ch
Isabella Plantation	Ham U.D.		Al
Isabella Road	Hackney	E.	Jc
Isabella Row	Lambeth	S.E.	Gg
Isabella Street	Southwark	S.E.	Gf
Isis Street	Wandsworth	S.W.	Dk
Isla Road	Woolwich		Ng
Island, The	Hackney	E.	Ha
Island Dock	Bermondsey	S.E.	Jf
Island Gardens	Poplar	E.	Kg
Island Row	Stepney	E.	Je
Isledon Road	Islington	N.	Fb
Isle of Dogs	Poplar	E.	Kf
Isleworth Ait	Heston and Isleworth U.D.		Sh
Islington Place	Islington	N.	Gc
Islington Generating Station	Islington	N.	Fc
Islington Green	Islington	N.	Gd
Islington Infirmary	Islington	N.	Fb
Islington Workhouse	Islington	N.	Fa
Islip Street	St. Pancras	N.W.	Fc
Ismailia Road	West Ham	E.	Lc
Ismalia Road	Fulham	S.W.	Dh
Italian Walk	Lambeth	S.E.	Fg
Ivan Place	Lambeth	S.E.	Gg
Ivanhoe Road	Camberwell	S.E.	Hh
Iveley Road	Wandsworth	S.W.	Eh
Iverna Gardens	Kensington	W.	Df
Iverson Road	Hampstead	N.W.	Dc
Iverson Yard	Hampstead	N.W.	Dc
Ives Street	Chelsea	S.W.	Ef
Ivimey Street	Bethnal Green	E.	Hd
Ivy Bridge Lane	Westminster	W.C.	Fe
Ivy Gardens	Hornsey	N.	Fa
Ivy House	Hendon U.D.	N.W.	Da
Ivy Lane	City of London	E.C.	Ge
Ivy Lane	Lewisham	S.E.	Jj
Ivy Lane	Shoreditch	N.	Gd
Ivy Place	Shoreditch	N.	Gd
Ivy Road	Willesden U.D.	N.W.	Cb
Ivy Street	Shoreditch	N.	Gd
Ivy Terrace	Hackney	E.	Ha
Ivy Terrace	Lewisham	S.E.	Kj
Ivydale Road	Camberwell	S.E.	Jh
Ivydene Road	Hackney	E.	Hc
Ivyhurst	Wimbledon	S.W.	Cm
Ivymount Road	Lambeth	S.E.& S.W.	Fl
Jackson Road	Islington	N.	Fb
Jackson Street	Woolwich		Mg
Jackson's Buildings	Hackney	E.	Hc
Jackson's Lane	Hornsey	N.	Ea
Jackson's Terrace	Islington	N.	Fa
Jack Straw's Castle	Hampstead	N.W.	Dh
Jacob Street	Bermondsey	S.E.	Hf
Jacobin Street	Shoreditch	E.	Hd
Jacob's Well Mews	Marylebone	W.	Ee
Jacob's Well Passage	City of London	E.C.	Ge
Jamaica Passage	Stepney	E.	Je
Jamaica Road	Bermondsey	S.E.	Hf
Jamaica Street	Stepney	E.	Je
Jamaica Wharf	Southwark	S.E.	Ge
James Court	City of London	E.C.	Ge
James Court	Stepney	E.	He

STREET OR PLACE.	BOROUGH.	P.D.	MAP.
James Grove	Battersea	S.W.	Eh
James Grove	Camberwell	S.E.	Hh
James Lane	Leyton U.D.	E.	Ka
James Place	Bethnal Green	E.	He
James Place	Lambeth	S.E.	Ff
James Place	Paddington	W.	De
James Place	Stepney	E.	Je
James Place	Stepney	E.	Hf
James Street	Bethnal Green	E.	Hd
James Street	Chiswick U.D.	W.	Bg
James Street	Finsbury	E.C.	Gd
James Street	Hammersmith	W.	Cg
James Street	Islington	N.	Gc
James Street	Islington	N.	Fc
James Street	Marylebone	W.	Ee
James Street	Paddington	W.	De
James Street	Poplar	E.	Kd
James Street	St. Pancras	N.W.	Ec
James Street	Stepney	E.	He
James Street	Stepney	E.	Je
James Street	West Ham	E.	Le
James Street	Westminster	W.C.	Fe
James Street	Westminster	S.W.	Fe
James Street	Woolwich		Mg
James Street Mansions	Westminster	S.W.	Ff
James Terrace	Wandsworth	S.W.	Dl
Jameson Street	Kensington	W.	De
Jane Court	Stepney	E.	He
Jane Street	Southwark	S.E.	Gf
Jane Street	Stepney	E.	He
Jane Shore Court	Shoreditch	E.	Hd
Janet Road	West Ham	E.	Le
Janet Street	Poplar	E.	Jf
Janeway Street	Bermondsey	S.E.	Hf
Janson Road	West Ham	E.	Kc
Japan Crescent	Islington	N.	Fa
Jardin Street	Camberwell	S.E.	Gg
Jarvis Road	Camberwell	S.E.	Hj
Jasmine Grove	Penge U.D.	S.E.	Hm
Jason Court	Marylebone	W.	Ee
Jasper Passage	Camberwell and Lambeth.	S.E.	Hl
Jasper Road	Camberwell and Lambeth	S.E.	Hl
Javen's Buildings	Bermondsey	S.E.	Gf
Jay's Buildings	Finsbury	N.	Fd
Jay's Mews	Westminster	S.W.	Df
Jebb Street	Poplar	E.	Kd
Jedburgh Road	West Ham	E.	Ld
Jedburgh Street	Battersea	S.W.	Ej
Jeddo Road	Hammersmith	W.	Bf
Jefferson Street	Poplar	E.	Kd
Jeffreys Place	St. Pancras	N.W.	Ec
Jeffreys Road	Lambeth	S.W.	Fh
Jeffrey's Square	City of London	E.C.	Ge
Jeffrey's Street	St. Pancras	N.W.	Ec
Jeffrie's Court	Stepney	E.	He
Jelf Road	Lambeth	S.W.	Fj
Jenkin's Court	Holborn	W.C.	Fe
Jenner Mews	Hackney	N.	Hb
Jenner Road	Hackney	N.	Hb
Jennings Road	Camberwell	S.E.	Hj
Jephson Road	East Ham	E.	Lc
Jeptha Avenue	Wandsworth	S.W.	Cj
Jerdan Place	Fulham	S.W.	Dg
Jeremiah Street	Poplar	E.	Ke
Jermyn Street	Westminster	S.W.	Fe
Jerningham Road	Deptford	S.E.	Jh
Jerome Place	Southwark	S.E.	Gg
Jerrard Street	Lewisham	S.E.	Kh
Jerrold Street	Shoreditch	N.	Hd
Jersey Road	Leyton U.D.	E.	Kb
Jersey Street	Bethnal Green	E.	Hd
Jervis Road	Fulham	S.W.	Cg
Jerusalem Court	Finsbury	E.C.	Gd
Jerusalem Passage	Finsbury	E.C.	Gd
Jerusalem Passage	Hackney	E.	Hc
Jerusalem Square	Hackney	E.	Hc
Jervis's Lane	East Ham	E.	Me
Jerviston House	Croydon	S.W.	Fl
Jesmond	Croydon R.D.		Em
Jesmond Street	Southwark	S.E.	Gg
Jesse Road	Leyton U.D.	E.	Kb

STREET OR PLACE.	BOROUGH.	P.D.	MAP.
Jessica Road	Wandsworth	S.W.	Dj
Jessop Road	Lambeth	S.E.	Gj
Jewel Street	Stepney	E.	Je
Jewett Street	Rotherhithe	S.E.	Jg
Jewin Crescent	City of London	E.C.	Ge
Jewin Street	City of London	E.C.	Ge
Jewry Street	City of London	E.C.	He
Jews Row	Wandsworth	S.W.	Dj
Jews Walk	Lewisham	S.E.	Hl
Jeypore Road	Wandsworth	S.W.	Dj
Joan Place	Lambeth	S.E.	Fg
Jocelyn Road	Richmond		Sh
Jocelyn Street	Camberwell	S.E.	Hh
Jockey Fields	Holborn	W.C.	Fe
Jodrell Road	Poplar	E.	Jc
Jodrell Terrace	Poplar	E.	Jc
Johanna Street	Lambeth	S.E.	Ff
John Place	Lewisham	S.E.	Kj
John Street	Battersea	S.W.	Fh
John Street	Battersea and Wandsworth	S.W.	Dj
John Street	Camberwell	S.E.	Gh
John Street	City of London	E.C.	He
John Street	Fulham	W.	Cg
John Street	Greenwich	S.E.	Kh
John Street	Hackney	E.	Jc
John Street	Hampstead	N.W.	Eb
John Street	Holborn	W.C.	Fe
John Street	Islington	N.	Fc
John Street	Marylebone	W.	Ee
John Street	Paddington	W.	Cd
John Street	Poplar	E.	Ke
John Street	St. Pancras	W.	Fe
John Street	Shoreditch	E.	Hd
John Street	Shoreditch	N.	Gd
John Street	Stepney	E.	Je
John Street	West Ham	E.	Ld
John Street	West Ham	E.	Le
John Street	Westminster	S.W.	Ff
John Street	Westminster	W.	Ee
John Street	Westminster	W.C.	Fe
John Street	Woolwich		Mg
John Street West	Marylebone	W.	Ee
John Street West	Southwark	S.E.	Gf
John Campbell Road	Stoke Newington	N.	Hc
John Bull Yard	Stepney	E.	He
John Carpenter Street	City of London	E.C.	Ge
John Penn Street	Greenwich	S.E.	Kh
John's Court	Southwark	S.E.	Gg
John's Court	Stepney	E.	He
John's Mews	Finsbury	E.C.	Gd
John's Place	Finsbury	E.C.	Ge
John's Place	Holborn	W.C.	Fd
John's Place	Islington	N.	Gd
John's Place	Stepney	E.	He
John's Place	Shoreditch	E.C.	Ge
John's Road	Hackney	N.	Hb
John's Yard	Marylebone	N.W.	Ed
Johnson Buildings	Poplar	E.	Ke
Johnson Mansions	Fulham	W.	Cg
Johnson Street	Kensington	W.	De
Johnson Street	Poplar	E.	Kg
Johnson Street	St. Pancras	N.W.	Fd
Johnson Street	Stepney	E.	He
Johnson Street	Stepney	E.	Je
Johnson Street	Westminster	S.W.	Ff
Johnson's Court	City of London	E.C.	Ge
Johnson's Court	Stepney	E.	He
Johnson's Mews	Paddington	W.	De
Johnson's Place	Westminster	S.W.	Fg
Johnstone Road	East Ham	E.	Md
Johnstone Terrace	Hendon U.D	N.W.	Cb
Joiner Street	Bermondsey	S.E.	Gf
Joiner's Arms Yard	Camberwell	S.E.	Gh
Joiner's Court	Bermondsey	S.E.	Hf
Jolliffe's Yard	Kensington	W.	Ce
Jolly Anglers	Hackney	E.	Jb
Jonathan Place	Southwark	S.E.	Gf
Jonathan Street	Lambeth	S.E.	Fg
Jones Street	Westminster	W.	Ee
Jones' Court	Bermondsey	S.E.	Hf
Joseph Street	Stepney	E.	Je
Joseph Street	Stepney	E.	Jd
Joseph Street	Woolwich		Mg
Josephine Avenue	Lambeth	S.W.	Fj
Joshua Street	Poplar	E.	Ke
Joubert Street	Battersea	S.W.	Eh
Joyce Road	Lewisham	S.E.	Kj
Jubilee Almshouses	Greenwich	S.E.	Kh
Jubilee Chambers	Hammersmith	W.	Cf
Jubilee Clump	Wimbledon	S.W.	Bl
Jubilee Hall	Kensington	W.	Ce
Jubilee Place	Chelsea	S.W.	Eg
Jubilee Plantation	Richmond and Barnes U.D.		Ak
Jubilee Street	Stepney	E.	He
Jubilee Terrace	Shoreditch	N.	Gd
Jubilee Wharf	Stepney	E.	Hf
Judd Street	St. Pancras	W.C.	Fd
Judge's Walk	Hampstead	N.W.	Db
Judkin's Street	Poplar	E.	Kf
Juer Street	Battersea	S.W.	Eh
Julia Street	St. Pancras	N.W.	Eb
Julian Avenue	Acton U.D.	W.	Ae
Junction Dock	Poplar	E.	Kf
Junction Mews	Paddington	W.	Ee
Junction Place	Hackney	E.	Hb
Junction Road	Acton U.D.	W.	Af
Junction Road	Deptford	S.E.	Jg
Junction Road	Ealing	W.	Sf
Junction Road	Islington	N.	Fb
Junction Road	Leyton U.D.	E.	Kb
Junction Road	West Ham	E.	Kc
Junction Road	West Ham	E.	Ld
Junction Road Station	**Islington**	N.	Fb
Junction Street	West Ham	E.	Ke
Junction Arms	Willesden U.D.	N.W.	Bd
Juniper Bldgs.	Stepney	E.	He
Juniper Court	Camberwell	S.E.	Hh
Juniper Court	Stepney	E.	He
Juniper Street	Stepney	E.	He
Jupp Road	West Ham	E.	Kc
Jupp's Road	Stepney	E.	Jd
Justice Walk	Chelsea	S.W.	Dg
Jutland Road	Lewisham	S.E.	Kk
Juxon Street	Lambeth	S.E.	Ff
Kambala Road	Battersea	S.W.	Dh
Kangley Bridge Road	Lewisham	S.E.	Jl
Karl Place	Lambeth	S.E.	Ff
Kashgar Road	Woolwich		Ng
Kassala Road	Battersea	S.W.	Eh
Kate Street	Wandsworth	S.W.	Ek
Kate's Place	Woolwich	E.	Mf
Katharine Road	East Ham	E.	Lc
Katherine Road	East Ham	E.	Md
Kathleen Road	Battersea	S.W.	Eh
Kay Road	Lambeth	S.W.	Fg
Kay Street	Bethnal Green	E.	Hd
Kean Street	Westminster	W.C.	Fe
Keble Street	Wandsworth	S.W.	Dl
Keeley Street	Holborn	W.C.	Fe
Keen's Yard	Islington	N.	Ge
Keemor Street	Woolwich		Mg
Keeton Road	Bermondsey	S.E.	Hf
Keildon Road	Battersea	S.W.	Ej
Keith Gardens	Hammersmith	W.	Bf
Keith Grove	Hammersmith	W.	Bf
Keith Road	Barking U.D.		Nd
Kelday Road	Poplar	E.	Jc
Kelfield Gardens	Kensington	W.	Ce
Kell Street	Southwark	S.E.	Gf
Kelland Road	West Ham	E.	Ld
Kellerton Road	Lewisham	S.E.	Kj
Kellett Houses	St. Pancras	W.C.	Fd
Kellett Road	Lambeth	S.W.	Fj
Kellino Street	Wandsworth	S.W.	El
Kelland Street	Finsbury	E.C.	Fd
Kelloway Street	Lambeth	S.W.	Fj
Kelly Road	West Ham	E.	Le
Kelly Street	St. Pancras	N.W.	Ec
Kelmore Grove	Camberwell	S.E.	Hj
Kelmore Villas	Camberwell	S.E.	Hj
Kelmscott Road	Battersea	S.W.	Ej
Kelross Road	Islington	N.	Gb
Kelsey Manor	Beckenham U.D.		Jm
Kelsey Park	Beckenham U.D		Jm

STREET OR PLACE.	BOROUGH.	P.D.	MAP.
Kelsey Park Road	Beckenham U.D.		Jm
Kelsey Street	Bethnal Green	E.	Hd
Kelso	Wandsworth	S.W.	Fl
Kelso Place	Kensington	W.	Df
Kelson Street	Hampstead	N.W.	Dc
Kelvedon Road	Fulham	S.W.	Ch
Kelvin Grove	Lewisham	S.E.	Hl
Kelvin Road	Islington	N.	Gb
Kelvin Terrace	Lewisham	S.E.	Hk
Kelvington Road	Camberwell	S.E.	Jj
Kelway Place	Fulham	S.W.	Cg
Kemble Road	Lewisham	S.E.	Jk
Kemble Street	Westminster	W.C.	Fe
Kemerton Road	Beckenham U.D.		Km
Kemerton Road	Lambeth	S.E.	Gh
Kemey Street	Hackney	E.	Jc
Kemnal	Chislehurst U.D.		Nl
Kemnal Road	Chislehurst U.D.		Nm
Kemp Cottages	Woolwich		Mg
Kempe Road	Willesden U.D.	N.W.	Cd
Kemplay Road	Hampstead	N.W.	Db
Kemp's Court	Westminster	W.	Fe
Kemp's Place	Finsbury	E.C.	Fd
Kempsford Gardens	Kensington	S.W.	Dg
Kempsford Road	Lambeth	S.E.	Gg
Kempshed Road	Camberwell	S.E.	Hg
Kempshott Road	Wandsworth	S.W.	Fm
Kempslade Street	Deptford	S.E.	Jg
Kempson Road	Fulham	S.W.	Dh
Kempton Road	East Ham	E.	Md
Kemsing Road	Greenwich	S.E.	Lg
Ken Wood	Hampstead	N.W.	Ea
Kenbury Street	Lambeth	S.E.	Gh
Kenchester Street	Lambeth	S.W.	Fh
Kendal Road	Willesden U.D.	N.W.	Bb
Kendall Mews	Marylebone	W.	Ee
Kendell Terrace	Greenwich	S.E.	Kg
Kender Grove	Deptford	S.E.	Hh
Kender Place	Deptford	S.E.	Hh
Kender Street	Deptford	S.E.	Hh
Kendoa Road	Wandsworth	S.W.	Fj
Kendrick Mews	Kensington	S.W.	Dg
Kenilford Road	Wandsworth	S.W.	Ek
Kenilworth Avenue	Wimbledon	S.W.	Cl
Kenilworth Road	Bethnal Green	E.	Jd
Kenilworth Road	Ealing	W.	Sf
Kenilworth Road	Willesden U.D.	N.W.	Cc
Kenilworth Street	Kensington	W.	Ce
Kenilworth Villas	Camberwell	S.E.	Hj
Kenley Road	Brentford U.D.		Sg
Kenley Street	Kensington	W.	Ce
Kenlor Road	Wandsworth	S.W.	Dl
Kenmont Gardens	Hammersmith	N.W.	Bd
Kenmont Terrace	Hammersmith	W.	Cd
Kenmure Avenue	Hackney	E.	Hc
Kenmure Road	Hackney	E.	Hc
Kennard Road	West Ham	E.	Kc
Kennard Street	Battersea	S.W.	Eh
Kennard Street	East Ham	E.	Mf
Kennedy Road	Barking U.D.		Nd
Kennedy's Court	Holborn	W.C.	Fe
Kennet Wharf	City of London	E.C.	Ge
Kennet Wharf Lane	City of London	E.C.	Ge
Kenneth Road	Fulham	W.	Cg
Kenneth Terrace	Woolwich		Nk
Kennet Road	Paddington	W.	Cd
Kenninghall Road	Hackney	E.	Hb
Kenning's Street	Bermondsey	S.E.	Hf
Kennington Grove	Lambeth	S.E.	Fg
Kennington Mansions	Lambeth	S.E.	Fg
Kennington Oval	Lambeth	S.E.	Fg
Kennington Park	Lambeth and Southwark	S.E.	Gg
Kennington Park Gdns.	Southwark	S.E.	Gg
Kennington Park Road	Lambeth and Southwark	S.E.	Gg
Kennington Road (part)	Lambeth	S.E.	Fg
Kennington Road (part)	Southwark	S.E.	Ff
Kennington Station	**Southwark**	S.E.	Gg
Kennington Terrace	Lambeth	S.E.	Gg
Kennington Theatre	**Southwark**	S.E.	Gg
Kenry House	The Maldens and Coombe U.D.		Al
Kensal Green Cemetery	Kensington	W.	Cd
Kensal Rise Station	**Willesden U.D.**	N.W.	Cd
Kensal Road	Kensington and Paddington	W.	Cd
Kensington Avenue	Croydon		Fm
Kensington Buildings	Kensington	W.	Df
Kensington Court	Kensington	W.	Df
Kensington Court Gdns.	Kensington	W.	Df
Kensington Court Mansions	Kensington	W.	Df
Kensington Court Place	Kensington	W.	Df
Kensington Court Stables	Kensington	W.	Df
Kensington Crescent	Kensington	W.	Cf
Kensington Gardens	Kensington Paddington & Westminster	W.	Df
Kensington Gards. Sq.	Paddington	W.	De
Kensington Gate	Kensington	W.	Df
Kensington Gore	Westminster	S.W.	Df
Kensington Mansions	Kensington	W.	Dg
Kensington Palace	**Kensington**	W.	Df
Kensington Palace Gardens	Kensington	W.	Df
Kensington Palace Mansions	Kensington	W.	Df
Kensington Park Gdns.	Kensington	W.	Ce
Kensington Park Mews	Kensington	W.	Ce
Kensington Park Road	Kensington	W.	Ce
Kensington Place	Kensington	W.	Df
Kensington Place	Westminster	W.	Ff
Kensington Road	Kensington and Westminster	S.W. & W.	Df
Kensington Square	Kensington	W.	Df
Kensington Square Mansions	Kensington	W.	Df
Kensington (High St.) Station	**Kensington**	W.	Df
Kensington Union Workhouse	Kensington	W.	Df
Kent Avenue	Ealing	W.	Sd
Kent Gardens	Ealing	W.	Sd
Kent House	Ealing	W.	Sd
Kent House Lane	Beckenham U.D. and Lewisham	S.E.	Jl
Kent House Pleasure Grounds	Beckenham U.D.		Jm
Kent House Road	Beckenham U.D. Lewisham and Penge U.D.	(part) S.E.	Jl-m
Kent House Station	**Beckenham U.D.**		Jm
Kent Road	Richmond		Ag
Kent Road	Tottenham U.D.	N.	Ha
Kent Street	Shoreditch	E.	Hd
Kent Street	West Ham	E.	Ld
Kent Terrace	Deptford	S.E.	Jg
Kent Terrace	Marylebone	N.W.	Ed
Kentfield Place	Deptford	S.E.	Jh
Kentish Buildings	Southwark	S.E.	Gf
Kentish Town Almshouses	St. Pancras	N.W.	Fc
Kentish Town Road	St. Pancras	N.W.	Ec
Kentish Town Sta. (N.L.)	**St. Pancras**	N.W.	Ec
Kentish Town Sta. (Mid.)	**St. Pancras**	N.W.	Fc
Kentish Town (Tube)	**St. Pancras**	N.W.	Fc
Kentish Town S. (Tube)	**St. Pancras**	N.W.	Fc
Kentmere Road	Woolwich		Ng
Kenton Lane	Kingsbury U.D.		Aa
Kenton Road	Hackney	E.	Jc
Kenton Road	Wembley U.D.		Sa
Kenton Street	Holborn and St. Pancras	W.C.	Fd
Kent's Place	Paddington	W.	De
Kenway Road	Kensington	S.W.	Df
Kenwood Avenue	Deptford	S.E.	Jh
Kenwood Road	Hornsey	N.	Ea
Kenwyn Road	Wandsworth	S.W.	Fj
Kenyon Mansions	Fulham	S.W.	Cg
Kenyon Mansions	Lambeth	S.W.	Fj
Kenyon Street	Fulham	S.W.	Ch
Keogh Road	West Ham	E.	Kc
Kepler Road	Lambeth	S.W.	Fj
Keppel Mews North	Holborn	W.C.	Fe

STREET OR PLACE.	BOROUGH.	P.D.	MAP.
Keppel Mews South	Holborn	W.C.	Fe
Keppel Road	East Ham	E.	Md
Keppel Street	Chelsea	S.W.	Eg
Keppel Street	Holborn	W.C.	Fe
Keppel Street	Southwark	S.E.	Gf
Kerbela Street	Bethnal Green	E.	Hd
Kerby Street	Poplar	E.	Ke
Kerfield Crescent	Camberwell	S.E.	Gh
Kerfield Place	Camberwell	S.E.	Gh
Kerrison Place	Ealing	W.	Se
Kerrison Road	Battersea	S.W.	Eh
Kerrison Road	Ealing	W.	Sf
Kerry Road	Deptford	S.E.	Jg
Kersfield Road	Wandsworth	S.W.	Cj
Kersley Mews	Battersea	S.W.	Eh
Kersley Road	Stoke Newington	N.	Hb
Kersley Street	Battersea	S.W.	Eh
Keslake Road	Willesden U.D.	N.W.	Cd
Keston Street	Camberwell	S.E.	Hj
Kestrell Avenue	Lambeth	S.E.	Gj
Keswick Road	Wandsworth	S.W.	Cj
Kettering Street	Wandsworth	S.W.	Em
Kettle Court	Southwark	S.E.	Gg
Kew Bridge	Brentford U.D. and Richmond		Ag
Kew Bridge Road	Brentford U.D.		Sg
Kew Bridge Station	**Brentford U.D.**		Ag
Kew Gardens	**Richmond**		Ag
Kew Gardens Road	Richmond		Sg-h
Kew Gardens Station	**Richmond**		Ah
Kew Green	Richmond		Ag
Kew Lane	Richmond	S.W.	Ah
Kew Observatory	Richmond		Sh
Kew Palace	Richmond		Ag
Kew Pier	Richmond		Ag
Kew Road	Richmond		Sh
Kew Foot Lane	Richmond		Sh
Kew Foot Road	Richmond		Sh
Key Street	Stepney	E.	He
Keyes Road	Willesden U.D.	N.W.	Cb
Keymer Road	Wandsworth	S.W.	Fk
Kezia Street	Deptford	S.E.	Jg
Khama Street	Wandsworth	S.W.	Dl
Khartoum Road	Wandsworth	S.W.	Dl
Khartoum Road	West Ham	E.	Ld
Khedive Road	West Ham	E.	Lc
Khyber Road	Battersea	S.W.	Eh
Kibworth St.	Lambeth	S.W.	Fh
Kid Brook	Greenwich	S.E.	Lh
Kidbrooke Gardens	Greenwich	S.E.	Lh
Kidbrooke Green	Greenwich	S.E.	Lh
Kidbrooke Grove	Greenwich	S.E.	Lh
Kidbrooke Lane	Greenwich	S.E.	Lh
Kidbrooke Lane	Lewisham	S.E.	Lj
Kidbrooke Park Road	Greenwich	S.E.	Lj
Kidbrooke Station	**Greenwich**	S.E.	Lj
Kidd Street	Woolwich		Mg
Kidderpore Avenue	Hampstead	N.W.	Db
Kidderpore Gardens	Hampstead	N.W.	Db
Kidney Wood	Richmond		Sk
Kiel Street	Greenwich	S.E.	Jg
Kifford Court	Westminster	S.W.	Ff
Kilburn Lane	Paddington and Willesden U.D.	W.	Cd
Kilburn Park Road	Fulham and Paddington	W.	Dd
Kilburn Priory	Hampstead	N.W.	Dd
Kilburn Road	Paddington	N.W.	Dd
Kilburn Square	Willesden U.D.	N.W.	Dc
Kilburn Station	**Hampstead**	N.W.	Dd
Kilburn Vale	Hampstead	N.W.	Dc
Kilburn and Brondesbury Station	**Willesden U.D.**	N.W.	Cc
Kildare Gardens	Paddington	W.	De
Kildare Terrace	Paddington	W.	De
Kildoran Road	Lambeth	S.W.	Fj
Kilgour Street	Lewisham	S.E.	Jj
Kilkie Street	Fulham	S.W.	Dh
Killarney Road	Wandsworth	S.W.	Dj
Killcat Corner	Barnes U.D.		Bk
Killearn Road	Lewisham	S.E.	Kk
Killick's Yard	Wandsworth	S.W.	Eh
Killieser Avenue	Wandsworth	E.	Fk
Killowen Road	Hackney	S.W.	Jc
Killyon Road	Wandsworth	S.W.	Fh
Kilmaine Road	Fulham	S.W.	Ch
Kilmarsh Road	Hammersmith	W.	Cf
Kilmartin Avenue	Croydon	S.W.	Fm
Kilmorie Road	Lewisham	S.E.	Jk
Kilravock Street	Paddington	W.	Cd
Kilton Street	Battersea	S.W.	Eh
Kimbell Gardens	Fulham	S.W.	Dg
Kimberley Gardens	Fulham	S.W.	Ch
Kimberley Gardens	Tottenham U.D.	N.	Ga
Kimberley Road	Beckenham U.D.		Jm
Kimberley Road	Camberwell	S.E.	Hh
Kimberley Road	Lambeth	S.W.	Fg
Kimberley Road	Leyton U.D.	E.	Kb
Kimberley Road	West Ham	E.	Ld
Kimberley Road	Willesden U.D.	N.W.	Cc
Kimbolton Cottages	Chelsea	S.W.	Dg
Kimbolton Row	Chelsea	S.W.	Dg
Kimpton Road	Camberwell	S.E.	Gh
Kinburn Street	Bermondsey	S.E.	Jf
Kincaid Road	Camberwell	S.E.	Hh
Kincardine Street	Wandsworth	S.W.	Ck
Kinder Street	Stepney	E.	He
Kinfauns Road	Lambeth and Wandsworth	S.W.	Fk
King Place	St. Pancras	N.W.	Fd
King Square	Finsbury	E.C.	Gd
King Street	Acton U.D.	W.	Ae
King Street	Chelsea	S.W.	Eg
King Street, Cheapside	City of London	E.C.	Ge
King Street, Snow Hill	City of London	E.C.	Ge
King Street	Hammersmith	W.	Bg
King Street	Kensington	W.	Df
King Street	Leyton U.D.	E.	Ka
King Street	Marylebone	W.	Ee
King Street	Poplar	E.	Je
King Street	Richmond		Sj
King Street	St. Pancras	N.W.	Fd
King Street	Stepney	E.	He
King Street	Walthamstow U.D.	E.	Ja
King Street	West Ham	E.	Le
King Street	Westminster	S.W.	Ff
King Street	Westminster	W.C.	Fe
King Street Mews	Westminster	W.	Ee
King Street Road	West Ham	E.	Le
King and Queen Stairs	Bermondsey	S.E.	Jf
King and Queen Street	Southwark	S.E.	Gg
King Arthur Street	Camberwell	S.E.	Hh
King David Lane	Stepney	E.	He
Kingdon Road	Hampstead	N.W.	Dc
King Edward Gardens	Acton U.D.	W.	Ae
King Edward's Grove	Teddington U.D.		Sl
King Edward Road	Leyton U.D.	E.	Ka
King Edward School	Southwark	S.E.	Gf
King Edward Street	City of London	E.C.	Ge
King Edward Street	Islington	N.	Gd
King Edward Street	Lambeth and Southwark	S.E.	Gf
King Edward Street	Stepney	E.	He
King Edward Street	Stepney	E.	Hf
King Edward's Road	Barking U.D.		Nd
King Edward's Road	Hackney	E.	Hc
Kingfield Street	Poplar	E.	Kg
King George Street	Greenwich	S.E.	Kh
King Henry Street	Islington	N.	Hc
King Henry's Road	Hampstead and St. Pancras	N.W.	Ec
King Henry's Stairs	Stepney	E.	Hf
King Henry's Walk	Islington	N.	Gc
King Henry's Wharf	Stepney	E.	Hf
Kinghorn Street	City of London	E.C.	Ge
King James Street	Southwark	S.E.	Gf
King John Street	Stepney	E.	Je
King John's Court	Shoreditch	E.C.	Hd
King John's Court	Stepney	E.	Je
King John's Lane	Woolwich		Mk
Kinglake Street	Southwark	S.E.	Gg
Kingly Street	Westminster	W.	Fe
King's Avenue	Ealing	W.	Se
King's Avenue	The Maldens and Coombe U.D.		Bm

STREET OR PLACE.	BOROUGH.	P.D.	MAP.
King's Avenue . .	Wandsworth .	S.W.	Fj
King's Cottages . .	Hampstead . .	N.W.	Db
King's Court . . .	Southwark . .	S.E.	Gf
King's Garden . .	Woolwich . .		Mj
King's Highway . .	Woolwich . .		Ng
King's Mews . . .	Holborn . .	W.C.	Fe
King's Mews . . .	Westminster .	W.	Ee
King's Mere . . .	Wandsworth .	S.W.	Ck
King's Place . . .	Bethnal Green .	E.	Hd
King's Place . . .	Marylebone . .	W.	Ee
King's Place . . .	Southwark . .	S.E.	Gf
King's Road . . .	Barnes U.D. .	S.W.	Bh
King's Road . . .	Camberwell .	S.E.	Hh
King's Road . . .	Chelsea&Fulham	S.W.	Dg
King's Road . . .	Croydon . .		Hm
King's Road . . .	Croydon R.D. .		Em
King's Road . . .	Ealing . . .	W.	Se
King's Road . . .	East Ham . .	E.	Ld
King's Road . . .	Kingston . .		Sm
King's Road . . .	Leyton U.D. .	E.	Ka
King's Road . . .	Richmond . .		Sj
King's Road . . .	St. Pancras . .	N.W.	Fc
King's Road . . .	Stoke Newington	N.	Gb
King's Road . .	Twickenham U.D.		Sj
King's Road . . .	Westminster .	S.W.	Ef
King's Road . . .	Willesden U.D. .	N.W.	Bc
King's Road . . .	Wimbledon . .	S.W.	Cm
King's Stairs . . .	Bermondsey .	S.E.	Jf
King's Yard . . .	St. Pancras . .	N.W.	Fc
King's Arch Place .	Southwark . .	S.E.	Gg
King's Arms . . .	Acton U.D. . .	W.	Af
King's Arms . . .	Kingsbury U.D. .	N.W.	Ba
King's Arms Court .	Stepney . .	E.	He
King's Arms Place .	Bermondsey .	S.E.	Hf
King's Arms Yard .	City of London .	E.C.	Ge
King's Arms Yard .	Westminster .	W.	Fe
King's Bench Walk .	City of London .	E.C.	Ge
Kingsbridge Road .	Kensington . .	W.	Ce
King's Bench Walk .	Southwark . .	S.E.	Gf
Kingsbury Bridge .	Willesden U.D. .	N.W.	Bb
Kingsbury Green . .	Kingsbury U.D. .		Ba
Kingsbury House . .	Kingsbury U.D. .		Aa
Kingsbury Road . .	Hendon U.D. and Kingsbury U.D.	N.W.	Ba
Kingsbury Road . .	Islington . .	N.	Hc
Kingsbury Villa . .	Kingsbury U.D. .		Aa
Kingsbury & Neasden Station	**Willesden U.D. .**	N.W.	Bb
Kingscliff Gardens .	Wandsworth .	S.W.	Ck
Kings College . .	Westminster .	W.C.	Fe
Kings College Mews East	Hampstead . .	N.W.	Ec
King's College Mews West	Hampstead . .	N.W.	Ec
King's College Road .	Hampstead . .	N.W.	Ec
King's College School .	Wimbledon . .	S.W.	Cl
Kingscote Road . .	Acton U.D. . .	W.	Af
Kingscote Street . .	City of London .	E.C.	Ge
King's Court Mansions	Fulham . .	S.W.	Ch
Kingscourt Road . .	Wandsworth .	S.W.	Fl
Kingscroft Street . .	Islington . .	N.	Gc
King's Cross . . .	Islington and St. Pancras . .	N.	Fd
King's Cross Road .	Finsbury and St. Pancras . .	W.C.	Fd
King'sCrossSta.(Metn.)	**St. Pancras . .**	W.C.	Fd
King's CrossSta.(Tube)	**St. Pancras . .**	W.C.	Fd
King's Cross Terminus	**St. Pancras . .**	N.	Fd
Kingsdale Road . .	Beckenham U.D.	S.E.	Jn
Kingsdale Road . .	Woolwich . .		Nh
Kingsdown Road . .	Islington . .	N.	Fb
Kingsdown Road . .	Leyton U.D. .	E.	Kb
Kingsdown Villas .	Camberwell .	S.E.	Hj
Kingsford Street . .	Hampstead . .	N.W.	Eb
Kingsgate . . .	Hampstead . .	N.W.	Cc
Kingsgate Place . .	Hampstead . .	N.W.	Dc
Kingsgate Road . .	Hampstead . .	N.W.	Dc
Kingsgate Street . .	Holborn . .	W.C.	Ge
Kingshall Road . .	Beckenham U.D.		Jm
King's Head Court .	Bermondsey .	S.E.	Gf
King's Hd.Ct.,Beech St.	City of London .	E.C.	Ge
King's Hd. Ct., Holborn	City of London .	E.C.	Ge
King's HeadCt.ShoeLa.	City of London .	E.C.	Ge

STREET OR PLACE.	BOROUGH.	P.D.	MAP.
King's Head Court .	Westminster .	S.W.	Ff
King's Head Yard .	Holborn . .	W.C.	Fe
King's Head Yard .	Westminster .	W.C.	Fg
Kingslyn . . .	Croydon . .	S.E.	Gm
Kingsland Mews . .	Shoreditch . .	N.	Hc
Kingsland Passage .	Hackney and Islington . .	E. & N.	Hc
Kingsland Road . .	Hackney and Shoreditch .	E.	Hd
Kingsland Road . .	West Ham . .	E.	Ld
Kingsley Buildings .	Southwark . .	S.E.	Hg
Kingsley Mansions .	Fulham . .	W.	Cg
Kingsley Mews . .	Kensington . .	W.	Df
Kingsley Road . .	West Ham . .	E.	Lc
Kingsley Road . .	Willesden U.D. .	N.W.	Cc
Kingsley Road . .	Wimbledon . .	S.W.	Dl
Kingsley Street . .	Battersea . .	S.W.	Eh
Kingsman Street . .	Woolwich . .		Mg
Kingsmead Road . .	Wandsworth .	S.W.	Fk
Kingsnorth Place . .	Shoreditch . .	N.	Gd
Kingsthorpe Road .	Lewisham . .	S.E.	Jl
Kingston Bridge . .	Hampton Wick . U.D. & Kingston		Sm
Kingston Cemetery .	Kingston . .		Am
Kingston Hall Road .	Kingston . .		Sm
Kingston Hill . .	Kingston & The Maldens and Coombe U.D. .		Al-m
Kingston Hill Place .	The Maldens and Coombe U.D.		Al
Kingston House . .	Westminster .	S.W.	Df
Kingston Road . .	Chelsea . .	S.W.	Dg
Kingston Road . .	Leyton U.D. .	E.	Kb
Kingston Road . .	Merton U.D. and Wimbledon .	S.W.	Cm
Kingston Road . .	Teddington U.D.		Sm
Kingston Road . .	Wandsworth .	S.W.	Ck
Kingston Sewage Works	Kingston . .		Sm
Kingston Station . .	**Kingston . .**		Sm
Kingston Street . .	St. Pancras . .	N.W.	Ec
Kingston Street . .	Southwark . .	S.E.	Gg
Kingston Vale . .	The Maldens and Coombe U.D.	S.W.	Bl
King's Villas . . .	Wandsworth .	S.W.	El
Kingsway . . .	Holborn and Westminster .	N.W.	Fe
Kingswood Avenue .	Willesden U.D. .	W.C.	Cc
Kingswood House .	Camberwell .	S.E.	Gl
Kingswood Place . .	Lewisham . .	S.E.	Kj
Kingswood Road . .	Acton U.D. . .	W.	Af
Kingswood Road . .	Beckenham U.D.		Km
Kingswood Road . .	Beckenham U.D. & Penge U.D.	S.E.	Hm
Kingswood Road . .	Camberwell .	S.E.	Gl
Kingswood Road . .	Wandsworth .	S.W.	Fj
Kingwell Yard . .	Holborn . .	W.C.	Fe
Kingswood Road . .	Wimbledon . .	S.W.	Cm
King William Lane .	Greenwich . .	S.E.	Kg
King William Street .	City of London .	E.C.	Ge
King William Street .	Greenwich . .	S.E.	Kg
King William Street .	Westminster .	W.C.	Fe
K. William IV. Asylum	Beckenham U.D. & Penge U.D.	S.E.	Hm
Kingwood Road . .	Fulham . .	S.W.	Ch
Kinlock Street . .	Islington . .	N.	Fb
Kinnaird Avenue . .	Bromley . .		Km
Kinnear Road . .	Hammersmith .	W.	Bf
Kinnerton Mews . .	Westminster .	S.W.	Ef
Kinnerton Place North	Westminster .	S.W.	Ef
Kinnerton Place South	Westminster .	S.W.	Ef
Kinnerton Street . .	Westminster .	S.W.	Ef
Kinnerton Yard . .	Westminster .	S.W.	Ef
Kinnoul Road . .	Fulham . .	W.	Cg
Kinross Street . .	Bermondsey .	S.E.	Hf
Kinsale Road . .	Camberwell .	S.E.	Hj
Kinstock Road . .	Woolwich . .		Nh
Kintore Street . .	Bermondsey .	S.E.	Hg
Kinveachy Gardens .	Greenwich . .		Mg
Kinver Road North .	Lewisham . .	S.E.	Hl
Kinver Road South .	Lewisham . .	S.E.	Hl
Kipling Street . .	Bermondsey .	S.E.	Gf
Kirby Street . . .	Bermondsey .	S.E.	Gf
Kirby Street . . .	Holborn . .	E.C.	Ge

STREET OR PLACE.	BOROUGH.	P.D.	MAP.
Leabourne Road	Hackney	N.	Ha
Leacroft	Lewisham	S.E.	Hl
Lead Mill Point	Leyton U.D.	E.	Jb
Lead Mill Stream	Hackney, Leyton U.D. and West Ham	E.	Jb
Lead Street	Poplar	E.	Kg
Leadenhall Market	**City of London**	E.C.	Ge
Leadenhall Place	City of London	E.C.	Ge
Leadenhall Street	City of London	E.C.	Ge
Leader Street	Chelsea	S.W.	Eg
Leaders Gardens	Wandsworth	S.W.	Ch
Leading Street	Stepney	E.	Je
Leage Street	Finsbury	E.C.	Gd
Leagrave Street	Hackney	E.	Jb
Leahurst Road	Lewisham	S.E.	Kj
Lealand Road	Tottenham U.D.	N.	Ha
Leamington Avenue	Walthamstow U.D.	E.	Ka
Leamington Park	Acton U.D.	W.	Ae
Leamington Road	Paddington	W.	Ce
Leamington Rd. Villas	Paddington	W.	Ce
Leamington Street	Westminster	W.	Fe
Leamington Villas	Lewisham	S.E.	Kj
Leander Road	Lambeth	S.W.	Fj
Leaside Road	Hackney	E.	Jb
Leasowes Road	Leyton U.D.	E.	Ka
Leather Bottle, The	Wandsworth	S.W.	Dk
Leather Lane	City of London and Holborn	E.C.	Ge
Leather Market	Bermondsey	S.E.	Gf
Leatherdale Street	Stepney	E.	Jd
Leathersellers' Almsh.	Lewisham	S.E.	Kj
Leathwaite Road	Battersea	S.W.	Ej
Leathwell Road	Greenwich and Lewisham	S.E.	Kh
Lebanon Gardens	Wandsworth	S.W.	Cj
Lebanon Road	Wandsworth	S.W.	Cj
Lebanon Street	Southwark	S.E.	Gg
Le Blond's Buildings	Shoreditch	E.C.	Gd
Leconfield Road	Islington & Stoke Newington	N.	Gb
Ledbury Mews, East	Kensington	W.	De
Ledbury Mews, West	Kensington	W.	De
Ledbury Road	Kensington and Paddington	W.	De
Ledbury Street	Camberwell	S.E.	Hg
Ledbury Villas	Camberwell	S.E.	Hj
Ledrington Road	Penge U.D.	S.E.	Hm
Lee Cemetery	Lewisham	S.E.	Kk
Lee Green	Lewisham and Woolwich	S.E.	Lj
Lee High Road	Lewisham	S.E.	Kj
Lee Manor House	Lewisham	S.E.	Kj
Lee Park	Lewisham	S.E.	Kj
Lee Road	Greenwich, Lewisham and Woolwich	S.E.	Lj
Lee Station	**Lewisham**	S.E.	Lk
Lee Street	Shoreditch	E.	Hc
Lee Street	Stepney	E.	Je
Lee Street	Woolwich		Ng
Lee Terrace	Deptford	S.E.	Jg
Lee Terrace	Lewisham	S.E.	Kj
Lee Terrace	Lewisham	S.E.	Lj
Leeds Place	Islington	N.	Fb
Leeke Street	St. Pancras	W.C.	Fd
Lee's Mews	Westminster	W.	Ee
Lee's Yard	Westminster	W.	Fe
Leeson Road	Lambeth	S.E.	Gh
Leesons	Bromley R.D.		Nm
Lefern Road	Hammersmith	W.	Bf
Lefevre Road	Poplar	E.	Jd
Lefevre Terrace	Poplar	E.	Jd
Leffler Road	Woolwich		Ng
Lefroy Road	Hammersmith	W.	Bf
Legard Road	Islington	N.	Gb
Legg's Court	Poplar	E.	Jd
Leghorn Road	Willesden U.D.	N.W.	Bd
Leghorn Road	Woolwich		Ng
Leg of Mutton Pond	Hampstead	N.W.	Db
Leg of Mutton Pond	Richmond		Ak
Le Grand Place	Lambeth	S.E.	Ff
Leicester Place	Finsbury	E.C.	Gd
Leicester Place	Wandsworth	S.W.	El
Leicester Place	Westminster	W.C.	Fe
Leicester Road	Wanstead U.D.	E.	La
Leicester Road	Willesden U.D.	N.W.	Bc
Leicester Square	Westminster	W.C.	Fe
Leicester Sq. Station	**Westminster**	W.C.	Fe
Leicester Street	Poplar	E.	Kf
Leicester Street	Westminster	W.C.	Fe
Leigh Place	Holborn	E.C.	Fe
Leigh Road	Islington	N.	Gb
Leigh Road	Leyton U.D.	E.	Ka
Leigh Street	Holborn	W.C.	Fe
Leigh Street	St. Pancras	W.C.	Fd
Leigham Avenue	Wandsworth	S.W.	Fl
Leigham Court Road	Lambeth and Wandsworth	S.W.	Fl
Leigham House	Wandsworth	S.W.	Fl
Leigham Vale	Lambeth and Wandsworth	S.W.	Fl
Leighton Buildings	Westminster	S.W.	Fg
Leighton Crescent	St. Pancras	N.W.	Fb
Leighton Gardens	Willesden U.D.	N.W.	Cc
Leighton Grove	St. Pancras	N.W.	Fb
Leighton Mansions	Fulham	W.	Cg
Leighton Road	St. Pancras	N.W.	Fc
Leinster Avenue	Barnes U.D.	S.W.	Ah
Leinster Gardens	Paddington	W.	De
Leinster Mews	Paddington	W.	De
Leinster Place	Paddington	W.	De
Leinster Road	Willesden U.D.	N.W.	Dd
Leinster Square	Paddington	W.	De
Leinster Street	Paddington	W.	De
Leinster Terrace	Paddington	W.	De
Leinster Yard	Paddington	W.	De
Leipsic Road	Camberwell	S.E.	Gh
Leith Yard	Hampstead	N.W.	De
Leman Street	Stepney	E.	He
Leman Street Station	**Stepney**	E.	He
Lemna Road	Leyton U.D.	E.	La
Lemon Place	Lambeth	S.E.	Fg
Lemon Tree Yard	Westminster	S.W.	Fe
Lemon Tree Yard	Westminster	W.C.	Fe
Lemon's Terrace	Stepney	E.	Je
Lemonwell	Woolwich		Nj
Lemuel Street	Wandsworth	S.W.	Dj
Lena Gardens	Hammersmith	W.	Cf
Lenham Road	Croydon		Gm
Lenham Road	Lewisham	S.E.	Kj
Lennard Place	Marylebone	N.W.	Dd
Lennard Road	Beckenham U.D.	S.E.	Jm
Lennox Avenue	Walthamstow U.D.	E.	Ja
Lennox Gardens	Chelsea	S.W.	Ef
Lennox Gardens Mews	Chelsea	S.W.	Ef
Lennox Road	Islington	N.	Fb
Lennox Road	Walthamstow U.D.	E.	Ja
Lensden Place	Finsbury	E.C.	Gd
Lenthall Mews	Kensington	S.W.	Df
Lenthall Road	Hackney	E.	Hc
Lenthorp Road	Greenwich	S.E.	Kg
Leo Street	Camberwell	S.E.	Hh
Leonard Place	Kensington	W.	Df
Leonard Place	Stoke Newington	N.	Gb
Leonard Road	Croydon R.D. & Wandsworth	S.W.	Em
Leonard Road	West Ham	E.	Lb
Leonard Street	Finsbury and Shoreditch	E.C.	Gd
Leonard Street	West Ham	E.	Mf
Leonard Wharf	Poplar	E.	Ke
Leonidas Street	Deptford	S.E.	Jh
Leopard Court	Bermondsey	S.E.	Jf
Leopold Road	Ealing	W.	Se
Leopold Road	Walthamstow U.D.	E.	Ja
Leopold Road	Willesden U.D.	N.W.	Bc
Leopold Road	Wimbledon	S.W.	Cl
Leopold Street	Lambeth	S.E.	Fg
Leopold Street	Stepney	E.	Je
Leppoc Road	Wandsworth	S.W.	Fj
Leroy Street	Bermondsey	S.E.	Gf

STREET OR PLACE.	BOROUGH.	P.D.	MAP.
Limetree Terrace	Camberwell	S.E.	Hh
Limfield Gardens	Hampstead	N.W.	Db
Linacre Road	Willesden U.D.	N.W.	Cc
Lincoln Road	Kingston		Am
Lincoln Street	Chelsea	S.W.	Eg
Lincoln Street	Leyton U.D.	E.	Kb
Lincoln Street	Stepney	E.	Jd
Lincoln's Inn	Holborn & Westminster	W.C.	Fe
Lincoln's Inn Fields	Holborn & Westminster	W.C.	Fe
Lind Street	Deptford	S.E.	Jh
Linda Street	Battersea	S.W.	Dj
Landale Street	Poplar	E.	Ke
Lindais Road	Lewisham	S.E.	Jj
Linden Avenue	Willesden U.D.	N.W.	Cd
Linden Buildings	Bethnal Green	E.	Hd
Linden Gardens	Chiswick U.D.	W.	Bg
Linden Gardens	Kensington	W.	Da
Linden Grove	Beckenham U.D.	S.E.	Hm
Linden Grove	Camberwell	S.E.	Hj
Linden Mews	Kensington	W.	De
Linden Road	Walthamstow U.D.	E.	Ja
Lindfield Gardens	Hampstead	N.W.	Db
Lindia Street	Battersea	S.W.	Dj
Lindley Road	Leyton U.D.	E.	Kb
Lindley Street	Stepney	E.	He
Lindo Street	Camberwell and Deptford	S.E.	Hh
Lindore Road	Battersea	S.W.	Ej
Lindrop Road	Fulham	S.W.	Dh
Lindsell Street	Greenwich	S.E.	Kh
Lindsey Cottages	Islington	N.	Gc
Lindsey Grove	Bermondsey	S.E.	Hf
Lindsey Road	Bermondsey	S.E.	Hf
Lindsey Street	City of London & Finsbury	E.C.	Ge
Lindum Road	Teddington U.D.		Sm
Lindum Terrace	Wandsworth	S.W.	El
Linford Street	Battersea	S.W.	Eh
Lingen Street	Poplar	E.	Kd
Lingfield Road	Wimbledon	S.W.	Cl
Lingham Street	Lambeth	S.W.	Fh
Lingo Road	West Ham	E.	Le
Linguard Road	Lewisham	S.E.	Kj
Linhope Street	Marylebone	N.W.	Ed
Link Street	Hackney	E.	Jc
Linnel Road	Camberwell	S.E.	Gh
Linom Road	Lambeth	S.W.	Fj
Linscott Road	Hackney	E.	Hb
Linstead Street	Hampstead	N.W.	Dc
Lintaine Flats	Fulham	W.	Cg
Lintaine Grove	Fulham	W.	Cg
Linthorpe Road	Hackney	N.	Ha
Linton Grove	Lambeth	S.E.	Gl
Linton Road	Leyton U.D.	E.	Kb
Linton Street	Islington	N.	Gd
Linton Street	Marylebone	N.W.	Ee
Linver Road	Fulham	S.W.	Ch
Linwood Place	Southwark	S.E.	Gf
Lion Brewery	Lambeth	S.E.	Ff
Lion Buildings	Southwark	S.E.	Gf
Lion Passage	Southwark	S.E.	Gf
Lion Row	Finsbury	E.C.	Gd
Lion Street	Poplar	E.	Ke
Lion Street	Southwark	S.E.	Gf
Lion Terrace	Wandsworth	S.W.	Ch
Lion Wharf	Poplar	E.	Jf
Lionel Mews	Kensington	W.	Ce
Lionel Road	Brentford U.D.		Sg
Lion Gate Gardens	Richmond		Sh
Lion's Corner	City of London	E.	He
Lisbon Buildings	Bethnal Green	E.	He
Lisbon Street	Bethnal Green	E.	He
Lisburne Road	Hampstead	N.W.	Eb
Lisford Street	Camberwell	S.E.	Hh
Lisgar Terrace	Fulham	W.	Cf
Lisle Street	Westminster	W.	Fe
Lismore Circus	St. Pancras	N.W.	Eb
Lismore Gardens	St. Pancras	N.W.	Eb
Lismore Road	St. Pancras	N.W.	Eb
Lissenden Gardens	St. Pancras	N.W.	Eb
Lisson Buildings	Marylebone	N.W.	Ed
Lisson Grove	Marylebone	N.W.	Ed
Lisson Row	Marylebone	N.W.	Ed
Lisson Street	Marylebone	N.W.	Ee
Lisson Street Buildings	Marylebone	N.W.	Ee
Lisson St. Residences	Marylebone	N.W.	Ee
Lister Mews	Islington	N.	Fb
Liston Road	Wandsworth	S.W.	Fh
Listowel Street	Lambeth	S.W.	Gh
Listria Park	Stoke Newington	N.	Hb
Litcham Street	St. Pancras	N.W.	Ec
Litchfield Street	Holborn and Westminster	W.C.	Fe
Lithgow Street	Battersea	S.W.	Dh
Lithos Road	Hampstead	N.W.	Dc
Litlington Road	Bermondsey	S.E.	Hg
Little Abbey Street	Bermondsey	S.E.	Hf
Little Acre Court	Wandsworth	S.W.	Fj
Little Albany Street	St. Pancras	N.W.	Ed
Little Alie Street	Stepney	E.	He
Little Ann Street	Stepney	E.	He
Little Argyll Street	Westminster	W.	Ee
Little Arthur Street	Finsbury	E.C.	Ge
Little Baltic Street	Finsbury	E.C.	Gd
Little Barlow Street	Marylebone	W.	Ee
Little Bath Street	Holborn	E.C.	Gd
Little Bennett Street	Greenwich	S.E.	Kg
Little Birch Wood	Bexley U.D.		Ol
Little Birches	Bexley U.D.		Nl
Little Blake Hall	Wanstead U.D.	E.	La
Little Blenheim Street	Chelsea	S.W.	Dg
Little Boston	Ealing	W.	Sf
Little Bridge Street	Stepney	E.	Jd
Little Britain	City of London	E.C.	Ge
Little Bruton Street	Westminster	W.	Ee
Little Cadogan Place	Chelsea	S.W.	Ef
Little Camden Street	St. Pancras	N.W.	Fc
Little Camera Street	Chelsea	S.W.	Dg
Little Canterbury Place	Lambeth	S.E.	Ff
Little Carlisle Street	Marylebone	N.W.	Ed
Little Catherine Street	Westminster	W.C.	Fe
Little Chapel Street	Westminster	W.	Fe
Little Charles Place	St. Pancras	N.W.	Ec
Little Chatham Place	Southwark	S.E.	Gg
Little Chatham Street	Southwark	S.E.	Gg
Little Chester Street	Westminster	S.W.	Ef
Little Church Lane	Hammersmith	W.	Cg
Little Church Street	Marylebone	N.W.	Ed
Little Clarendon Street	St. Pancras	N.W.	Fd
Little Clayton Street	Lambeth	S.E.	Fg
Little College Street	City of London	E.C.	Ge
Little College Street	St. Pancras	N.W.	Fc
Little College Street	Westminster	S.W.	Ff
Little Collingwood St.	Bethnal Green	E.	Hd
Little Combe	Woolwich		Lg
Little Cross Street	Islington	N.	Gc
Little Crown Court	Shoreditch	E.C.	Hd
Little Cumming Street	Islington	N.	Fd
Little Dean Street	Westminster	W.	Fe
Little Denmark Street	Holborn	W.C.	Fe
Little Dorrit Playgd.	Southwark	S.E.	Gf
Little Drummond St.	St. Pancras	N.W.	Fd
Little Duke Street	Lambeth	S.E.	Gf
Little Durweston St.	Marylebone	W.	Ee
Little Ealing Lane	Ealing	W.	Sf
Little Earl Street	Holborn	W.C.	Fe
Little Earl Street	Marylebone	N.W.	Ee
Little Ebury Street	Westminster	S.W.	Eg
Little Edward Street	St. Pancras	N.W.	Ed
Little Elm Place	Kensington	S.W.	Dg
Little Essex Street	Shoreditch	N.	Hd
Little Essex Street	Westminster	W.C.	Fe
Little Europa Place	Battersea	S.W.	Dh
Little Exeter Street	Chelsea	S.W.	Ef
Little Exeter Street	Marylebone	N.W.	Ed
Little Exmouth Street	St. Pancras	N.W.	Fd
Little George Street	City of London	E.C.	Ge
Little George Street	St. Pancras	N.W.	Fd
Little George Street	Westminster	S.W.	Ff
Little Gloucester St.	Lambeth	S.W.	Fh
Little Goodge Street	St. Pancras	W.	Fd
Little Gray's Inn Lane	Holborn	E.C.	Fe

STREET OR PLACE.	BOROUGH.	P.D.	MAP.
Logan Mews	Kensington	W.	Df
Logan Place	Kensington	W.	Df
Logs, The	Hampstead	N.W.	Db
Logs Hill	Bromley		Mm
Logton Street	Poplar	E.	Jd
Lolesworth Buildings	Stepney	E.	He
Lolesworth Street	Stepney	E.	He
Lollard Street	Lambeth	S.E.	Fg
Loman Street	Southwark	S.E.	Gf
Lomas Buildings	Stepney	E.	Je
Lomas Street	Stepney	E.	Je
Lombard Buildings	Chelsea	S.W.	Dg
Lombard Court	City of London	E.C.	Ge
Lombard Dwellings	Battersea	S.W.	Dh
Lombard Road	Battersea	S.W.	Dh
Lombard Wall	Greenwich		Lg
Lombard Street	City of London	E.C.	Ge
Lombard Street	Southwark	S.E.	Gf
Londesborough Road	Stoke Newington	N.	Gb
London Bridge	City of London & Southwark	E.C. S.E.	Gf Gf
London Bridge Station	**Bermondsey**	S.E.	Gf
London Bridge Wharf	City of London	E.C.	Ge
London & Continental Wharf	Stepney	E.	Hf
London College of Divinity	Islington	N.	Gb
London Conservatoire of Music	Paddington	W.	De
London County Athletic Grounds	Camberwell	S.E.	Gj
London C.C. County Hall	Westminster	S.W.	Ff
London C.C. Edncn. Dept.	Westminster	W.C.	Fe
London C.C. Generating Station	Greenwich	S.E.	Kg
London C.C. Pumping Station	Chelsea	S.W.	Dh
London C. Woolwich Ferry	Woolwich	E.	Mf
London Docks	Stepney	E.	Hf
London Fields	Hackney	E.	Hc
London Fields Station	**Hackney**	E.	Hc
London Institution	City of London	E.C.	Ga
London Lane	Bromley		Lm
London Lane	Hackney	E.	Hc
London Mews	St. Pancras	W.	Fe
London Place	Hackney	E.	Hc
London Road	Bermondsey and Deptford	S.E.	Jg
London Road	Bromley		Km
London Road	Croydon R.D.		Em
London Road	Croydon	S.W.	Fm
London Road	Hackney	E.	Hb
London Road	Heston & Isleworth U.D.		Ag
London Road	Kingston		Am
London Road	Lewisham	S.E.	Hk
London Road	Southwark	S.E.	Gf
London Road	Wembley U.D.		Sc
London Road	West Ham	E.	Ld
London Street	Bethnal Green	E.	Hd
London Street	City of London	E.C.	He
London Street	Greenwich	S.E.	Jg
London Street	Islington	N.	Fd
London Street	Kingston		Sm
London Street	Paddington	W.	De
London Street	Stepney	E.	Je
London Terrace	Stepney	E.	Je
London University	**Westminster**	S.W.	Df
London Wall	City of London	E.C.	Ga
London Wall Avenue	City of London	E.C.	Ge
London Wharf	Poplar	E.	Jf
London Wharf	Stepney	E.	Je
London Yard	Poplar	E.	Kg
Lonesome Lane	Croydon R.D.	S.W.	Em
Long Acre	Westminster	W.C.	Fe
Long Alley	Richmond		Sj
Long Lane	Bermondsey and Southwark	S.E.	Gf
Long Lane	Bexley U.D. and Woolwich		Oh
Long Lane	City of London	E.C.	Ge
Long Mark	Barking U.D.		Nd
Long Pond	Wandsworth	S.W.	Ej
Long Street	Shoreditch	E.	Hd
Long Walk	Barnes U.D.	S.W.	Bh
Long Walk	Bermondsey	S.E.	Hf
Long Walk	Brentford U.D.	W.	Af
Longbeach Road	Battersea	S.W.	Ej
Longbridge Road	Lewisham	S.E.	Kj
Longcroft Road	Camberwell	S.E.	Hg
Longfellow Road	Stepney	E.	Jd
Longfellow Road	Walthamstow U.D.	E.	Ja
Longfield	Bromley		Lm
Longfield Avenue	Ealing	W.	Se
Longfield Avenue	Walthamstow U.D.	E.	Ja
Longfield Road	Ealing	W.	Se
Longfield Street	Wandsworth	S.W.	Ck
Longford Street	St. Pancras	N.W.	Ed
Longhedge Street	Battersea	S.W.	Eh
Longhope Place	Lambeth	S.E.	Gf
Longhurst Road	Lewisham	S.E.	Kj
Longlands Park Cres.	Foots Cray U.D.		Nl
Longlands Nursery	Foots Cray U.D.		Nl
Longlands Park Road	Foots Cray U.D.		Nl
Longlands Road	Bexley U.D. and Foots Cray U.D.		Nl
Longlead Road	Wandsworth	S.W.	El
Longley Road	Wandsworth	S.W.	El
Longley Street	Bermondsey	S.E.	Hg
Longnor Road	Stepney	E.	Jd
Longpond Road	Lewisham	S.E.	Kh
Longridge Road	Kensington	S.W.	Dg
Long's Court	Westminster	W.C.	Fe
Long's Wharf	Woolwich		Mf
Long's Yard	Holborn	W.C.	Fe
Longsdale Road	Barnes U.D.	S.W.	Bg
Longstone Road	Wandsworth	S.W.	El
Longton Avenue	Lewisham	S.E.	Hl
Longton Grove	Lewisham	S.E.	Hl
Longton Grove Cots.	Lewisham	S.E.	Hl
Longton Grove Ter.	Lewisham	S.E.	Hl
Longville Road	Southwark	S.E.	Gg
Lonsdale Mews	Kensington	W.	Ce
Lonsdale Road	Chiswick U.D.	W.	Bf
Lonsdale Road	Kensington	W.	Ce
Lonsdale Road	Leyton U.D.	E.	La
Lonsdale Road	Willesden U.D.	N.W.	Cd
Lonsdale Square	Islington	N.	Gc
Lonsdale Yard	Kensington	W.	Ce
Loo Road	Croydon		Sm
Loraine Mews	Islington	N.	Fc
Loraine Place	Islington	N.	Fb
Loraine Road	Islington	N.	Fb
Lord Street	West Ham	E.	Mf
Lord Holland Lane	Kensington	W.	Cf
Lord's Cricket Ground	Marylebone	N.W.	Dd
Lordship Grove	Stoke Newington	N.	Gb
Lordship Lane	Camberwell	S.E.	Hj
Lordship Lane Station	**Camberwell**	S.E.	Hk
Lordship Mews	Stoke Newington	N.	Gb
Lordship Park	Stoke Newington	N.	Gb
Lordship Park Mews	Stoke Newington	N.	Gb
Lordship Place	Chelsea	S.W.	Dg
Lordship Road	Stoke Newington	N.	Ga
Lordship Terrace	Stoke Newington	N.	Gb
Lorenzo Street	Finsbury	W.C.	Fd
Loris Road	Hammersmith	W.	Cf
Lorn Road	Lambeth	S.W.	Fh
Lorna Road	Fulham	S.W.	Cg
Lorne Gardens	Kensington	W.	Cf
Lorne Gardens	Marylebone	N.W.	Ed
Lorne Road	Hornsey	N.	Fa
Lorne Road	Walthamstow U.D.	E.	Ja
Lorne Road	West Ham	E.	Lb
Lorrimore Buildings	Southwark	S.E.	Gg
Lorrimore Road	Southwark	S.E.	Gg
Lorrimore Square	Southwark	S.E.	Gg
Lorrimore Street	Southwark	S.E.	Gg
Lorton Terrace	Kensington	W.	Ce

STREET OR PLACE.	BOROUGH.	P.D.	MAP.
Lothair Road North	Tottenham U.D. and Hornsey	N.	Ga
Lothair Road South	Hornsey	N.	Ga
Lothair Street	Battersea	S.W.	Dh
Lothair Villas	Ealing	W.	Sf
Lothbury	City of London	E.C.	Ge
Lothian Road	Lambeth	S.W.	Gh
Lothrot Street	Paddington	W.	Cd
Lots Road	Chelsea	S.W.	Dg
Lotus Court	Stepney	E.	Ha
Loubet Street	Wandsworth	S.W.	El
Loudoun Road	Hampstead and Marylebone	N.W.	Dc
Loudoun Road Mews	Hampstead	N.W.	Dc
Loudoun Rd. Station	**Hampstead**	N.W.	Dc
Loughborough Junction Station	**Lambeth**	S.W.	Gh
Loughborough Mans.	Lambeth	S.W.	Gj
Loughborough Park	Lambeth	S.W.	Gj
Loughborough Park Buildings	Lambeth	S.W.	Gh
Loughborough Road	Lambeth	S.W.	Gh
Loughborough Street	Lambeth	S.E.	Fg
Louisa Gardens	Stepney	E.	Je
Louisa Square	Shoreditch	N.	Gd
Louisa Street	Shoreditch	N.	Gd
Louisa Street	Stepney	E.	Je
Louise Road	West Ham	E.	Kc
Louisville Road	Wandsworth	S.W.	El
Lousehall	Barking U.D.		Od
Louvain Road	Battersea	S.W.	Dj
Louvain Street	Stepney	E.	Je
Lovat Place	Bermondsey	S.E.	Jf
Love Court	Stepney	E.	Je
Love Lane	Bromley		Lm
Love Lane	City of London	E.C.	Ge
Love Lane	Croydon R.D.		Dm
Love Lane	Fulham	S.W.	Ch
Love Lane	Hackney	E.	Hb
Love Lane	Lambeth	S.W.	Fh
Love Lane	Lewisham	S.E.	Kh
Love Lane	Poplar	E.	Kd
Love Lane	Southwark	S.E.	Gf
Love Lane	Stepney	E.	Hf
Love Lane	Stepney	E.	Je
Love Lane	Woolwich		Mg
Love Lane Square	Stepney	E.	Je
Love Walk	Camberwell	S.E.	Gh
Loveday Road	Ealing	W.	Se
Lovegrove Cottages	Camberwell	S.E.	Hg
Lovegrove Street	Camberwell	S.E.	Hg
Lovelace Road	Lambeth	S.E.	Gk
Lovelinch Street	Deptford	S.E.	Jg
Lovell's Court	City of London	E.C.	Ge
Loveridge Mews	Hampstead	N.W.	Cc
Loveridge Road	Hampstead	N.W.	Cc
Lovers Walk	Greenwich	S.E.	Kg
Lovers Walk	Westminster	W.	Ef
Lovett's Place	Wandsworth	S.W.	Dj
Low hall Lane	Walthamstow U.D.	E.	Ja
Lowden Road	Lambeth	S.E.	Gj
Lowder Street	Stepney	E.	Hf
Lowell Street	Stepney	E.	Je
Lower Ashby Street	Finsbury	W.C.	Gd
Lower Belgrave Street	Westminster	S.W.	Ef
Lower Berkeley Street	Marylebone	W.	Ee
Lower Bland Street	Southwark	S.E.	Gf
Lower Brunswick Yard	Bermondsey	S.E.	Jf
Lower Camden	Bromley and Chistlehurst U.D.		Mm
Lower Cedars Mews	Wandsworth	S.W.	Ej
Lower Chapman Street	Stepney	E.	He
Lower Charles Street	Finsbury	W.C.	Gd
Lower Clapton Road	Hackney	E.	Hb
Lower Common	Wandsworth	S.W.	Ch
Lower Conduit Mews	Paddington	W.	De
Lower Cross Road	Hampstead	N.W.	Eb
Lower Downs Road	Wimbledon	S.W.	Cm
Lower East Lane Whf.	Bermondsey	S.E.	Hf
Lower East Smithfield	Stepney	E.	Hf
Lower Elmers End	Beckenham U.D.		Jm
Lower Fenton Street	Stepney	E.	He
Lower Green	Croydon R.D.		Em
Lower Grosvenor Place	Westminster	S.W.	Ef
Lower Grove	Wandsworth	S.W.	Dj
Lower Grove House	Wandsworth	S.W.	Bj
Lower Gutter Hedge Lane	Hendon U.D.	N.W.	Ba
Lower Ham Road	Ham U.D. and Kingston		Sl
Lower Harden Street	Woolwich		Mg
Lower Heath	Hampstead	N.W.	Eb
Lower James Street	Westminster	W.	Fe
Lower John Street	Shoreditch	N.	Gd
Lower John Street	Stepney	E.	Je
Lower John Street	Westminster	W.	Fe
Lower Kennington Lane	Lambeth	S.E.	Gg
Lower Kid Brook	Greenwich	S.E.	Mh
Lower Mall	Hammersmith	W.	Cg
Lower Market Street	Woolwich		Mg
Lower Marsh	Lambeth	S.E.	Ff
Lower Marylebone St.	Marylebone	W.	Ee
Lower Maryon Road	Greenwich		Mg
Lower Mortlake Road	Richmond		Sh
Lower North Street	Stepney	E.	Je
Lower Orchard Street	Wandsworth	S.W.	Fj
Lower Park Fields	Wandsworth	S.W.	Cj
Lower Park Rd. (part)	Camberwell	S.E.	Hg
Lower Park Rd. (part)	Camberwell	S.E.	Hh
Lower Pellipar Road	Woolwich		Mg
Lower Phillimore Place	Kensington	W.	Df
Lower Pool	Bermondsey and Stepney	E. & S.E.	Jf
Lower Porchester St.	Paddington	W.	Ee
Lower Quebec Yard	Bermondsey	S.E.	Jf
Lower Ranelagh Road	Westminster	S.W.	Eg
Lower Richmond Road	Barnes U.D. and Richmond	S.W.	Ah
Lower Richmond Road	Wandsworth	S.W.	Ch
Lower Road	Bermondsey and Deptford	S.E.	Hf
Lower Road	Richmond		Sj
Lower Road	West Ham	E.	Ld
Lower Seymour Street	Marylebone	W.	Ee
Lower Shadwell	Stepney	E.	Je
Lower Sloane Street	Chelsea	S.W.	Eg
Lower Smith Street	Finsbury	W.C.	Gd
Lower Sydenham Stn.	**Lewisham**	S.E.	Jl
Lower Teddington Rd.	Hampton Wick U.D.		Sm
Lower Terrace	Hampstead	N.W.	Db
Lower Terrace	Penge U.D.	S.E.	Hl
Lower Thames Street	City of London	E.C.	Ge
Lower Watergate	Greenwich	S.E.	Jg
Lower William Street	Marylebone	N.W.	Ed
Lower Winchester Rd.	Lewisham	S.E.	Jk
Lower Wood Street	Woolwich		Mg
Lowfield Road	Hampstead	N.W.	Dc
Lowman Road	Islington	N.	Fb
Lowndes Court	Westminster	W.	Fe
Lowndes Mews	Westminster	S.W.	Ef
Lowndes Place	Westminster	S.W.	Ef
Lowndes Square	Chelsea and Westminster	S.W.	Ef
Lowndes Street	Chelsea and Westminster	S.W.	Ef
Lowndes Terrace	Chelsea and Westminster	S.W.	Ef
Lownds Avenue	Bromley		Lm
Lowood	Camberwell	S.E.	Hl
Lowood	Woolwich		Nh
Lowood Street	Stepney	E.	He
Lowth Road	Camberwell	S.E.	Gh
Lowther Gardens	Westminster	S.W.	Df
Lowther Hill	Lewisham	S.E.	Jk
Lowther Road	Islington	N.	Fc
Lowther Road	Kingston		Sm
Loxford Avenue	East Ham	E.	Md
Loxham Street	St. Pancras	N.W.	Fd
Loxley Road	Wandsworth	S.W.	Dk
Loxton Road	Lewisham	S.E.	Jk
Loxwood Place	St. Pancras	N.W.	Ed
Luard Street	Islington	N.	Fc

STREET OR PLACE.	BOROUGH.	P.D	MAP.
Lubbock Road	Chislehurst U.D.		Mm
Lubbock Street	Deptford	S.E.	Jh
Lubeck Street	Battersea	S.W.	Eh
Lucas Place	St. Pancras	W.C.	Fd
Lucas Road	Beckenham U.D.	S.E.	Hm
Lucas Road	Lambeth and Southwark	S.E.	Gg
Lucas Road	West Ham	E.	Kd
Lucas Street	Bethnal Green	E.	Hd
Lucas Street	Deptford	S.E.	Jh
Lucas Street	Stepney	E.	He
Lucerne Road	Islington	N.	Gb
Lucey Road	Bermondsey	S.E.	Hf
Lucien Road	Wandsworth	S.W.	El
Lucknow Street	Woolwich		Nh
Ludon's Place	Stepney	E.	He
Ludgate Arcade	City of London	E.C.	Ge
Ludgate Circus	City of London	E.C.	Ge
Ludgate Circus Bldgs.	City of London	E.C.	Ge
Ludgate Hill	City of London	E.C.	Ge
Ludgate Hill Station	**City of London**	E.C.	Ge
Ludgate Square	City of London	E.C.	Ge
Ludlam Road	Lewisham	S.E.	Jj
Ludlow Street	Finsbury	E.C.	Gd
Ludwick Road	Deptford	S.E.	Jg
Lufton Place	Stepney	E.	Je
Lugard Road	Camberwell	S.E.	Hh
Luke Street	Shoreditch	N.	Ga
Lullington Road	Penge U.D.	S.E.	Hm
Lulot Street	St. Pancras	N.	Eb
Lulworth Road	Camberwell	S.E.	Hh
Lumber Court	Holborn	W.C.	Fe
Lumley Court	Westminster	W.C.	Fe
Lumley Street	Westminster	W.	Ee
Lumley's Almhouses	Shoreditch	N.	Gd
Lumsden Street	Poplar	E.	Kf
Luna Road	Croydon		Sm
Luna Street	Chelsea	S.W.	Dg
Lunham Road	Lambeth	S.E.	Gm
Luntley Place	Stepney	E.	He
Lupton Street	St Pancras	N.W.	Fb
Lupus Street	Westminster	S.W.	Eg
Lurgan Avenue	Fulham	W.	Cg
Lurgan Street	Hammersmith	W.	Cg
Lurline Gardens	Battersea	S.W.	Eh
Luscombe Street	Lambeth	S.W.	Fg
Lushington Road	Willesden U.D.	N.W.	Bd
Lush's Buildings	St. Pancras	N.W.	Fd
Lutheran Place	Wandsworth	S.W.	Fk
Luton Place	Greenwich	S.E.	Kh
Luton Road	West Ham	E.	Le
Luton Street	Marylebone	N.W.	Dd
Lutton Terrace	Hampstead	N.W.	Db
Luttrell Avenue	Wandsworth	S.W.	Bj
Lutwych Road	Lewisham	S.E.	Jk
Luxemburg Gardens	Hammersmith	W.	Cf
Luxford Street	Bermondsey	S.E.	Jg
Luxmore Street	Deptford and Lewisham	S.E.	Jh
Luxor Street	Lambeth	S.E.	Gh
Lyall Mews East	Westminster	S.W.	Ef
Lyall Mews West	Westminster	S.W.	Ef
Lyall Place	Chelsea	S.W.	Ef
Lyall Road	Bethnal Green, Poplar and Stepney	E.	Jd
Lyall Street	Westminster	S.W.	Ef
Lycett Mews	Hammersmith	W.	Bf
Lyceum Theatre	**Westminster**	W.C.	Fe
Lydbrook Street	Stepney	E.	Je
Lydden Grove	Wandsworth	S.W.	Dk
Lydden Road	Wandsworth	S.W.	Dk
Lydenberg Street	Greenwich		Lf
Lydford Road	Paddington	W.	Cd
Lydford Road	Willesden U.D.	N.W.	Co
Lydhurst Avenue	Wandsworth	S.W.	Fk
Lydia Buildings	Lambeth	S.E.	Ff
Lydia Street	Stepney	E.	Je
Lydon Road	Wandsworth	S.W.	Eh
Lydyard Road	Islington	N.	Fa
Lyford Road	Wandsworth	S.W.	Dk
Lyford Street	Greenwich and Woolwich		Mg
Lyham Road	Lambeth and Wandsworth	S.W.	Fj
Lyme Street	St. Pancras	N.W.	Fc
Lyme Terrace	St. Pancras	N.W.	Fc
Lymington Road	Hampstead	N.W.	Dc
Lynch Road	Wandsworth	S.W.	Dj
Lyncroft Gardens.	Ealing	W.	Sf
Lyncroft Gardens	Hampstead	N.W.	Db
Lyncroft Mansions	Hampstead	N.W.	Db
Lyndale	Hendon U.D.	N.W.	Db
Lyndhurst	Ealing	W.	Sc
Lyndhurst Drive	Leyton U.D.	E.	Ka
Lyndhurst Gardens	Hampstead	N.W.	Dc
Lyndhurst Grove	Camberwell	S.E.	Ha
Lyndhurst Road	Camberwell	S.E.	Hh
Lyndhurst Road	Hampstead	N.W.	Db
Lyndhurst Square	Camberwell	S.E.	Hh
Lynedock Street	Shoreditch	N.	Hd
Lynette Avenue	Wandsworth	S.W.	Ej
Lynmouth Road	Hackney	N.	Ha
Lynn Street	Wandsworth	S.W.	Ek
Lynton Mews	Bermondsey	S.E.	Hg
Lynton Road	Acton U.D.	W.	Ae
Lynton Road	Bermondsey	S.E.	Hg
Lynton Road	Hornsey	N.	Fa
Lynton Road	Willesden U.D.	N.W.	Cd
Lyon Place	Marylebone	N.W.	Dd
Lyon Street	Islington	N.	Fc
Lyon's Yard	Woolwich		Mg
Lyric Place	Lambeth	S.E.	Gg
Lyric Road	Barnes U.D.	S.W.	Bh
Lyric Theatre	**Westminster**	W.	Fe
Lysander Grove	Islington	N.	Fa
Lysia Street	Fulham	S.W.	Cg
Lysias Road	Wandsworth	S.W.	Ej
Lytchett Road	Bromley		Lm
Lytcott Grove	Camberwell	S.E.	Hj
Lyte Street	Bethnal Green	E.	Hd
Lyttelton Road	Leyton U.D.	E.	Kb
Lytton Grove	Wandsworth	S.W.	Cj
Lytton Road	Leyton U.D.	E.	Ka
Lyveden Road	Greenwich	S.E.	Lh
Lyvedon Road	Croydon R.D.	S.W.	Em
Maberley Road	Croydon & Penge U.D.		Hm
Mabledon Place	St. Pancras	W.C.	Fd
Mablethorpe Road	Fulham	S.W.	Cg
Mabley Street	Hackney	E.	Jc
Mabyn Street	Woolwich		Ng
Macartney House	Greenwich	S.E.	Kh
Macaulay Road	East Ham	E.	Md
Macaulay Road	Wandsworth	S.W.	Eh
McBean's Buildings	Southwark	S.E.	Gf
Macclesfield Gate	Marylebone	N.W.	Ed
Macclesfield Place	Finsbury	E.C.	Gd
Macclesfield Street	Finsbury	E.C.	Gd
Macclesfield Street	Westminster	W.	Fe
Macclesfield Terrace	Finsbury	E.C.	Gd
McDermott Road	Camberwell	S.E.	Hh
Macdonald Road	West Ham	E.	Lb
McDowall Road	Camberwell	S.E.	Gh
Macduff Road	Battersea	S.W.	Eh
Mace Street	Bethnal Green	E.	Jd
Macfarlane Place	Hammersmith	W.	Cf
Macfarlane Road	Hammersmith	W.	Cf
McGrath Place	Shoreditch	N.	Hd
Machell Road	Camberwell	S.E.	Hh
Mackay Road	Wandsworth	S.W.	Eh
Mackenzie Road	Beckenham U.D.		Hm
McKerrell Road	Camberwell	S.E.	Hh
Mackeson Road	Hampstead	N.W.	Eb
Mackintosh Lane	Hackney	E.	Jc
Mackley Road	Poplar	E.	Kg
Macklin Street	Holborn	W.C.	Fe
Mack's Road	Bermondsey	S.E.	Hg
Maclaren Street	Hackney	E.	Jb
Maclean Street	Lewisham	S.E.	Jj
McLeod Mews	Kensington	S.W.	Df
McLeod Road	Woolwich		Og
Maclise Mansions	Hammersmith	W.	Cf
Maclise Road	Hammersmith	W.	Cf
McNeil Road	Camberwell	S.E.	Gh
Macoma Road	Woolwich		Ng

STREET OR PLACE.	BOROUGH.	P.D.	MAP.
Macoma Terrace	Woolwich		Ng
Macready Place	Islington	N.	Fb
Macroom Road	Paddington	W.	Dd
Madden Road	Lewisham	S.E.	Jl
Maddox Street	Westminster	W.	Ee
Madeira Avenue	Bromley		Km
Madeira Avenue	Lewisham	S.E.	Km
Madeira Road	Leyton U.D.	E.	Kb
Madeira Road	Wandsworth	S.W.	Fl
Madeley Road	Ealing	W.	Se
Madeline Road	Penge U.D.	S.E.	Hm
Madras Place	Islington	N.	Gc
Madrid Place.	Lambeth	S.W.	Fh
Madron Street	Southwark	S.E.	Hg
Madron Villas	Southwark	S.E.	Hg
Mafeking Avenue	Brentford U.D.		Sg
Mafeking Road	West Ham	E.	Ld
Magdala Road	Islington	N.	Fb
Magdalen Passage	Stepney	E.	He
Magdalen Road	Wandsworth	S.W.	Dk
Magdalen Street	Bermondsey	S.E.	Gf
Magee Street	Lambeth	S.E.	Fg
Magnolia Road	Chiswick U.D.	W.	Ag
Magpie Alley	City of London	E.C.	Ge
Magpie Alley.	Shoreditch	E.	He
Maida Hill West	Paddington	W.	Dd
Maida Vale	Hampstead, Marylebone and Paddington	W.	Dd
Maiden Lane.	City of London	E.C.	Ge
Maiden Lane.	Westminster	W.C.	Fe
Maiden Lane Station	**Islington**	N.	Fc
Maiden Road	West Ham	E.	Kc
Maidenhead Court	City of London	E.C.	Ge
Maidenhead Passage	Westminster	W.	Fe
Maidenstone Hill.	Greenwich	S.E.	Kh
Maidenstone Terrace	Greenwich	S.E.	Kh
Maidman Street	Stepney	E.	Jd
Maid of Honour Road.	Richmond		Sj
Maidstone Buildings	Southwark	S.E.	Gf
Maidstone Street	Shoreditch	E.	Hd
Maine Road	Lewisham	S.E.	Jk
Maine Street	Stepney	E.	Je
Maiwand Street	Hackney	E.	Jb
Maishman's Place	Shoreditch	E.C.	Ge
Maismore Terrace	Camberwell.	S.E.	Hg
Maitland Park Road	St. Pancras	N.W.	Ec
Maitland Park Villas	St. Pancras	N.W.	Ec
Maitland Road	Beckenham U.D.	S.E.	Hm
Maitland Road	West Ham	E.	Lc
Maitland Terrace.	Battersea	S.W.	Ej
Maize Row	Stepney	E.	Je
Majendie Road	Woolwich		Ng
Major Road	Bermondsey	S.E.	Hf
Major Road	West Ham	E.	Kc
Malabar Street	Poplar	E.	Jf
Malay Street	Stepney	E.	Hf
Malbrook Road	Wandsworth	S.W.	Bj
Malcolm Road	Penge U.D.	S.E.	Hm
Malcolm Road	Wimbledon	S.W.	Cm
Malden Crescent	St. Pancras	N.W.	Ec
Malden Hill	The Maldens and Coombe U.D.		Bm
Malden Hill Gardens	The Maldens and Coombe U.D.		Bm
Malden Place	St. Pancras	N.W.	Ec
Malden Road.	The Maldens and Coombe U.D.		Am-Bm
Malden Road	St. Pancras	N.W.	Ec
Maldon Road.	Acton U.D.	W.	Ae
Malet Street	Holborn	W.C.	Fe
Maley Avenue	Lambeth	S.E.	Gk
Malfors Road.	Camberwell.	S.E.	Hh
Malham Road	Lewisham	S.E.	Jk
Mall, The	Ealing	W.	Se
Mall, The	Kensington	W.	De
Mall, The	Westminster	S.W.	Ff
Mall Chambers	Kensington	W.	De
Mall Cottages	Kensington	W.	De
Mall Road	Hammersmith	W.	Bg
Mall Villas	Hammersmith	W.	Bg
Mall West, The	Kensington	W.	Df
Mall West Mews	Kensington	W.	Df
Mallard Street	Hackney	E.	Jc
Mallet Road	Lewisham	S.E.	Kj
Malling Place	Southwark	S.E.	Hg
Mallinson Road	Battersea	S.W.	Ej
Malmesbury Road	Poplar	E.	Jd
Malmesbury Road	West Ham	E.	Ke
Malmsey Place	Lambeth	S.E.	Fg
Malpas Road	Deptford	S.E.	Jh
Malt House Mews.	Marylebone	N.W.	Dd
Malt House Yard	Wandsworth	S.W.	Dj
Malt Street	Camberwell	S.E.	Hg
Malt Street Mews	Camberwell	S.E.	Hg
Malta Road	Leyton U.D.	E.	Ja
Malta Street	Finsbury	E.C.	Gd
Maltby Street	Bermondsey	S.E.	Hf
Malton Street	Woolwich		Nh
Malva Road	Wandsworth	S.W.	Dj
Malvern Road	East Ham	E.	Md
Malvern Road	Hackney	E.	Hc
Malvern Road	Leyton U.D.	E.	Lb
Malvern Road	Willesden U.D.	N.W.	Dd
Malvern Terrace	Islington	N.	Fc
Malveston Road	Bromley		Mm
Malvinia Road	Lewisham	S.E.	Jj
Malvolio Road	Fulham	S.W.	Dg
Malwood Road	Wandsworth	S.W.	Ej
Malyon's Road	Lewisham	S.E.	Jj
Managers Street	Poplar	E.	Kf
Manaton Road	Camberwell	S.E.	Hh
Manbey Grove	West Ham	E.	Kc
Manbey Park	West Ham	E.	Kc
Manbey Road	West Ham	E.	Kc
Manbey Street	West Ham	E.	Kc
Manby Road	Leyton U.D.	E.	Kb
Manby Road	Walthamstow U.D.	E.	Ja
Manchester Avenue	City of London	E.C.	Ge
Manchester Buildings	Southwark	S.E.	Gg
Manchester Mews	Kensington	W.	Ce
Manchester Mews south	Marylebone	W.	Ee
Manchester Place	Betnnal Green	E.	Hd
Manchester Road	Croydon		Gm
Manchester Road	Hammersmith & Kensington.	W.	Ce
Manchester Road.	Poplar	E.	Kg
Manchester Road.	Tottenham U.D.	N.	Ga
Manchester Square	Marylebone.	W.	Ee
Manchester Sq. Bldgs.	Marylebone.	W.	Ee
Manchester Street	Marylebone.	W.	Ee
Manchester Street	St. Pancras	W.C.	Fd
Manchester Terrace	Hammersmith	W.	Ce
Manchuria Road	Battersea	S.W.	Ej
Manciple Place	Southwark	S.E.	Gf
Mandalay Road	Wandsworth	S.W.	Ej
Mander Place	Southwark	S.E.	Gf
Mandeville Place	Marylebone.	W.	Ee
Mandeville Street.	Hackney	E.	Jb
Mandeville's Yard	Southwark.	S.E.	Gf
Mandrake Road	Wandsworth	S.W.	Ek
Mandrell Road	Lambeth	S.W.	Fj
Manette Street	Westminster	W.	Fe
Manfred Road	Wandsworth	S.W.	Cj
Manilla Street	Poplar	E.	Jf
Manisty Street	Poplar	E.	Ke
Manley Street	St. Pancras	N.W.	Ec
Mann Street	Southwark	S.E.	Gg
Manners Street	Lambeth	S.E.	Ff
Manning Place	Lambeth	S.E.	Fg
Manning Place	Marylebone.	N.W.	Ee
Manning Place	Stepney	E.	Je
Manning Road	Walthamstow U.D.	E.	Ja
Manning Street	Stepney	E.	Je
Mann's Brewery	Bethnal Green	E.	He
Manor, The	Lewisham	S.E.	Hk
Manor Buildings	Chelsea	S.W.	Dg
Manor Club	Merton U.D.	S.W.	Dm
Manor Court	Stepney	E.	Je
Manor Gardens	Chelsea	S.W.	Dg
Manor Gardens	Islington	N.	Fb
Manor Grove.	Beckenham U.D.		Jm
Manor Grove.	Camberwell	S.E.	Hg
Manor Grove.	Richmond		Ah

STREET OR PLACE.	BOROUGH.	P.D.	MAP.
Market Hill	Woolwich		Mf
Market Mews	Westminster	W.	Ef
Market Place	Acton U.D.	W.	Af
Market Place	Kingston		Sm
Market Place	Marylebone	W.	Ee
Market Place	West Ham	E.	Mf
Market Road	Islington	N.	Fc
Market Road	Richmond		Ah
Market Row	Hackney	E.	Hc
Market Row	Southwark	S.E.	Gf
Market Square	Bromley		Lm
Market Street	Bermondsey	S.E.	Gf
Market Street	East Ham	E.	Md
Market Street	Islington	N.	Fc
Market Street	Paddington	W.	De
Market Street	Poplar	E.	Ke
Market Street	Shoreditch	E.C.	Ge
Market Street	Stepney	E.	Je
Market Street	Westminster	W.	Ef
Market Street	Woolwich		Mg
Market Street Mews	Paddington	W.	De
Markfield Road	Tottenham U.D.	N.	Ha
Markham Place	Chelsea	S.W.	Eg
Markham Square	Chelsea	S.W.	Eg
Markham Street	Chelsea	S.W.	Eg
Markham Terrace	Chelsea	S.W.	Eg
Markhouse Avenue	Walthamstow U.D.	E.	Ja
Markhouse Place	Walthamstow U.D.	E.	Ja
Markhouse Road	Leyton U.D. and Walthamstow U.D.	E.	Ja
Mark's Place	Shoreditch	E.	Hc
Marl Street	Wandsworth	S.W.	Dj
Marlborough Court	Westminster	W.	Fe
Marlborough Gate	Paddington	W.	De
Marlborough Grove	Acton U.D.	W.	Bf
Marlborough Hill	Marylebone	N.W.	Dc
Marlborough House	**Westminster**	S.W.	Ff
Marlborough Lane	Greenwich		Lh
Marlborough Mansions	Westminster	S.W.	Ff
Marlborough Mews	Marylebone	N.W.	Dd
Marlborough Mews	Paddington	W.	De
Marlborough Place	Chelsea	S.W.	Eg
Marlborough Place	Marylebone	N.W.	Dd
Marlborough Road	Acton U.D.	W.	Bf
Marlborough Road	Camberwell	S.E.	Hg
Marlborough Road	Chiswick U.D.	W.	Ag
Marlborough Road	Croydon R.D.	S.W.	Dm
Marlborough Road	Ealing	W.	Sf
Marlborough Road	East Ham	E.	Lc
Marlborough Road	Hackney and Shoreditch	E.	Hc
Marlborough Road	Islington	N.	Fb
Marlborough Road	Lewisham	S.E.	Kj
Marlborough Road	Leyton U.D.	E.	Kb
Marlborough Road	Marylebone	N.W.	Dd
Marlborough Road	Richmond		Sj
Marlborough Road Station	**Marylebone**	N.W.	Dd
Marlborough Row	Westminster	W.	Fe
Marlborough Square	Chelsea	S.W.	Eg
Marlborough Street	Chelsea	S.W.	Eg
Marlborough Street	Greenwich	S.E.	Kg
Marlborough Street	Lambeth and Southwark	S.E.	Gf
Marlborough Street	Paddington	W.	De
Marlborough Theatre	**Islington**	N.	Fb
Marlborough Villas	Camberwell	S.E.	Hj
Marler Road	Lewisham	S.E.	Jk
Marloes Road	Kensington	W.	Df
Marlow Buildings	Bethnal Green	E.	Hd
Marlow Road	Beckenham U.D.	S.E.	Hm
Marlow Road	Hackney	E.	Jc
Marlton Street	Greenwich	S.E.	Lg
Marmadon Street	Woolwich		Ng
Marmaduke Court	Stepney	E.	He
Marmaduke Place	Stepney	E.	He
Marmion Mews	Battersea	S.W.	Ej
Marmion Road	Battersea	S.W.	Ej
Marmont Road	Camberwell	S.E.	Hh
Marmora Road	Camberwell	S.E.	Hj

STREET OR PLACE.	BOROUGH.	P.D.	MAP.
Marne Lodge	Wandsworth	S.W.	Bk
Marne Street	Paddington	W.	Cd
Marner Street	Poplar	E.	Kd
Marney Road	Battersea	S.W	Ej
Marnock Road	Lewisham	S.E.	Jj
Maroon Street	Stepney	E.	Je
Marquess Grove	Islington	N.	Gc
Marquess Road	Islington	N.	Gc
Marquis Road	Hornsey	N.	Fa
Marquis Road	St. Pancras	N.W.	Fc
Marriott Road	Islington	N.	Fa
Marryatt Road	Wimbledon	S.W.	Cl
Marsala Road	Lewisham	S.E.	Kj
Marsden Road	Camberwell	S.E.	Hj
Marsden Street	St. Pancras	N.W.	Ec
Marsh Hill	Hackney	E.	Jc
Marsh Lane	Greenwich	S.E.	Kg
Marsh Lane	Leyton U.D.	E.	Jb
Marsh Lane	West Ham	E.	Kd
Marsh Street	Poplar	E.	Kg
Marshall Street	Southwark	S.E.	Gf
Marshall Street	Westminster	W.	Fe
Marshall's Grove	Woolwich		Mg
Marshall's Road	Bermondsey	S.E.	Hf
Marshalsea Road	Southwark	S.E.	Gf
Marsham Street	Shoreditch	N.	Gd
Marsham Street	Westminster	S.W.	Ff
Marshfield Street	Poplar	E.	Kf
Marshgate Bridge	Hackney	E.	Jc
Marsland Road	Southwark	S.E.	Gg
Marsom Street	Shoreditch	N.	Gd
Marston Road	Greenwich		Lj
Mart, The	City of London	E.C.	Ge
Mart Street	Lambeth	S.E.	Gg
Martaban Road	Stoke Newington	N.	Hb
Martagon Place	Southwark	S.E.	Gf
Martell Road	Lambeth	S.E.	Gl
Martello Terrace	Hackney	E.	Hc
Martha Place	Deptford	S.E.	Hh
Martha Street	Stepney	E.	He
Martha's Buildings	Finsbury	E.C.	Gd
Martin Road	West Ham	E.	Le
Martin Street	Bermondsey	S.E.	Hf
Martin Street	Kensington	W.	Ce
Martin Street	West Ham	E.	Kc
Martindale Road	Wandsworth	S.W.	Ek
Martindale Road	West Ham	E.	Le
Martineau Road	Islington	N.	Gb
Martin's Buildings	Lambeth	S.E.	Ff
Martin's Hill	Bromley		Lm
Martin's Lane	City of London	E.C.	Ge
Martin's Road	Bromley		Km
Martlett's Court	Westminster	W.C.	Fe
Marton Road	Stoke Newington	N.	Hb
Marvel's Lane	Lewisham	S.E.	Ll
Marvellous Terrace	Lewisham	S.E.	Jk
Marville Road	Fulham	S.W.	Ch
Marvin Street	Hackney	E.	He
Mary Place	Kensington	W.	Ce
Mary Place	Poplar	E.	Ke
Mary Place	Stepney	E.	Hf
Mary Place	Woolwich		Mg
Mary Street	Islington	N.	Gd
Mary Street	Poplar	E.	Jd
Mary Street	Stepney	E.	He
Mary Street	West Ham	E.	Le
Mary Terrace	St. Pancras	N.W.	Ed
Mary Ann Buildings	Deptford	S.E.	Jg
Mary Ann Place	Greenwich	S.E.	Kg
Mary Ann Place	Shoreditch	E.C.	Ge
Mary Ann Place	Southwark	S.E.	Gf
Mary Ann Street	Stepney	E.	He
Mary Ann Street	Woolwich	E.	Mf
Maryland Park	West Ham	E.	Kc
Maryland Point	West Ham	E.	Kc
Maryland Point Sta.	**West Ham**	E.	Kc
Maryland Road	West Ham	E.	Kc
Maryland Square	West Ham	E.	Kc
Maryland Street	West Ham	E.	Kc
Marylands Road	Paddington	W.	Dd
Marylebone Lane	Marylebone	W.	Ee
Marylebone Mews	Marylebone	W.	Ee
Marylebone Passage	Marylebone	W	Fe

STREET OR PLACE.	BOROUGH.	P.D.	MAP.
Meeting House Court	City of London	E.C.	Ge
Meeting House Lane	Camberwell	S.E.	Hh
Meeting House Lane	Woolwich		Mf
Mehetabel Road	Hackney	E.	Hc
Mehetabel Villas	Hackney	E.	Hc
Meikle Road	Greenwich	S.E.	Jh
Melbourne Grove	Camberwell	S.E.	Hj
Melbourne Mansions	Fulham	W.	Cg
Melbourne Place	Lambeth	S.W.	Fh
Melbourne Place	Westminster	W.C.	Fe
Melbourne Road	East Ham	E.	Md
Melbourne Road	Leyton U.D.	E.	Ka
Melbourne Square	Lambeth	S.W.	Fh
Melbury Road	Kensington	W.	Cf
Melcombe Place	Marylebone	N.W.	Ed
Melford Road	Camberwell	S.E.	Hk
Melford Road	Leyton U.D.	E.	Kb
Melford Road	Walthamstow U.D.	E.	Ja
Melfort Road	Croydon	S.W.	Fm
Melgund Road	Islington	N.	Gc
Melina Place	Marylebone	N.W.	Dd
Melina Road	Hammersmith	W.	Bf
Melina Terrace	Hammersmith	W.	Bf
Melior Place	Bermondsey	S.E.	Gf
Melior Street	Bermondsey	S.E.	Gf
Mellicks Place	Bermondsey	S.E.	Hf
Melling Street	Woolwich		Nh
Mellish Street	Poplar	E.	Jf
Mellish's Wharf	Poplar	E.	Jf
Mellison Road	Wandsworth	S.W.	El
Melmouth Place	Fulham	S.W.	Dg
Melody Road	Wandsworth	S.W.	Dj
Melon Road	Camberwell	S.E.	Hh
Melrose Avenue	Croydon	S.W.	E&Fm
Melrose Avenue	Willesden U.D.	N.W.	Cb
Melrose Gardens	Hammersmith	W.	Cf
Melrose Gardens	Wandsworth	S.W.	Cj
Melrose Road	Acton U.D.	W.	Af
Melrose Road	Barnes U.D.	S.W.	Bh
Melrose Road	Merton U.D.	S.W.	Cm
Melrose Road	Wandsworth	S.W.	Cj
Melrose Terrace	Hammersmith	W.	Cf
Melton Mews	St. Pancras	N.W.	Fd
Melton Road	Tottenham U.D.	N.	Ha
Melton Street	Fulham	W.	Cg
Melton Street	St. Pancras	N.W.	Fa
Melverton Street	Lambeth	S.E.	Fg
Melville Road	Willesden U.D.	N.W.	Ac
Melvin Road	Penge U.D.	S.E.	Hm
Memel Place	Wandsworth	S.W.	Cj
Memel Street	Finsbury	E.C.	Gd
Memorial Avenue	West Ham	E.	Kd
Memorial Ground	West Ham	E.	Ld
Memorial Hall	City of London	E.C.	Ge
Memorial Hall	Islington	N.	Gc
Memorial Hall	Southwark	S.E.	Gg
Mendelssohn Gardens	Chelsea	S.W.	Eg
Mendip Road	Battersea	S.W	Dh
Mendora Road	Fulham	S.W.	Cg
Menelik Street	Hampstead	N.W.	Cb
Menotti Street	Bethnal Green	E.	Hd
Mentmore Terrace	Hackney	E.	Hc
Mentone Mansions	Kensington	S.W.	Dg
Mentone Road	Islington	N.	Gb
Meon Road	Acton U.D.	W.	Af
Mepham Street	Lambeth	S.E.	Ff
Mercer Avenue	Westminster	W.C.	Fe
Mercer Street	Westminster	W.C.	Fe
Mercer Street	Stepney	E.	He
Mercer's Court	Southwark	S.E.	Gf
Mercer's Road	Islington	N.	Fb
Mercer's Terrace	Islington	N.	Fa
Merchant Street	Poplar	E.	Jd
Merchant Seamen's Orphan Asylum	Wanstead U.D.	E.	La
Merchant Taylors' Institute	Lewisham	S.E.	Kj
Merchant's Row	Stepney	E.	Je
Merchiston Road	Lewisham	S.E.	Kk
Merchland Road	Foots Cray U.D.		Nk
Merchland Road West	Woolwich		Nk
Mercury Road	Brentford U.D.		Sg
Mercy Terrace	Lewisham	S.E.	Jj
Meredith Buildings	Stepney	E.	He
Meredith Road	Barnes U.D.	S.W.	Bh
Meredith Street	Finsbury	E.C.	Gd
Meredith Street	West Ham	E.	Ld
Merevale	Bromley		Mm
Merivale Road	Wandsworth	S.W.	Cj
Merlewood	Bromley		Mm
Merlin's Place	Finsbury	W.C.	Fd
Mermaid Court	Southwark	S.E.	Gf
Merredene Street	Lambeth	S.W.	Fj
Merritt Road	Lewisham	S.E.	Jj
Merrick Square	Southwark	S.E.	Gf
Merrington Road	Fulham	S.W.	Dg
Merritt's Buildings	Shoreditch	E.C.	Ge
Merrow Buildings	Southwark	S.E.	Gg
Merrow Street	Southwark	S.E.	Gg
Mersey Street	Hammersmith and Kensington	W.	Ce
Mersham Road	Croydon		Gm
Merthyr Road	Barnes U.D.	S.W.	Bg
Merton Abbey Station	**Merton U.D.**	S.W.	Dm
Merton Avenue	Chiswick U.D.	W.	Bf
Merton Bridge	Croydon R.D.	S.W.	Dm
Merton College	Wimbledon	S.W.	Cm
Merton Hall Road	Merton U.D. and Wimbledon	S.W.	Cm
Merton Lane	St. Pancras	N.	Ea
Merton Lodge	St. Pancras	N.	Ea
Merton Mills	Wimbledon	S.W.	Dm
Merton Park Station	**Merton U.D.**	S.W.	Dm
Merton Place	Greenwich and Lewisham	S.E.	Kh
Merton Road	Hampstead	N.W.	Ec
Merton Road	Kensington	S.W.	Df
Merton Road	Leyton U.D. and Walthamstow U.D.	E.	Ka
Merton Road	Wandsworth	S.W.	Dj
Merton Road	Wimbledon	S.W.	Dm
Merton Rush	Merton U.D.	S.W.	Cm
Merton's Court	Stepney	E.	He
Merttins Road	Camberwell	S.E.	Jj
Mervan Road	Lambeth	S.W.	Fj
Mesnard's Wharf	Stepney	S.E.	Hf
Messaline Road	Acton U.D.	W.	Ae
Messiter Place	Woolwich		Mk
Messina Avenue	Hampstead	N.W.	Dc
Meteor Street	Battersea	S.W.	Ej
Methlily Street	Lambeth	S.E.	Gg
Metropolitan Bldgs.	City of London	E.C.	Ge
Metropolitan Electric Supply Co.	Willesden U.D.	N.W.	Bd
Metropolitan Market	City of London	E.C.	Ge
Metropolitan Market	Islington	N.	Fc
Metropolitan Music Hall	**Paddington**	W.	De
Metropolitan Railway Works	Willesden U.D.	N.W.	Bb
Metropolitan Tabernacle	Southwark	S.E.	Gf
Metropolitan Theatre	**Lambeth**	S.E.	Gh
Metropolitan Water Bd.	Battersea	S.W.	Eg
Metropolitan Water Bd.	Lambeth	S.W.	Fj
Metropolitan Water Board Reservoirs	Deptford	S.E.	Kh
Metropolitan Wharf	Stepney	E.	Hf
Mews, The	Camberwell	S.E.	Hg
Mews, The	Camberwell	S.E.	Hh
Mews, The	Islington	N.	Fb
Mews, The	Lambeth	S.W.	Fg
Mews, The	Stoke Newington	N.	Gb
Mexfield Road	Wandsworth	S.W.	Cj
Meymott Street	Southwark	S.E.	Gf
Meynell Crescent	Hackney	E.	Jc
Meynell Road	Hackney	E.	Jc
Meyrick Road	Battersea	S.W.	Dh
Meyrick Road	Willesden U.D.	N.W.	Bc
Miall Road	Lewisham	S.E.	Jl
Micheldever Road	Lewisham	S.E.	Kj
Middle Court	City of London	E.C.	Ge
Middle Lane	Hornsey	N.	Fa
Middle New Street	City of London	E.C.	Ge

STREET OR PLACE.	BOROUGH.	P.D.	MAP.
Nadine Street	Greenwich		Lg
Nag's Head	Islington	N.	Fb
Nag's Head Court	Finsbury	E.C.	Gd
Nag's Head Yard	Southwark	S.E.	Gf
Nag's Head Yard	Stepney	E.	He
Nailour Street	Islington	N.	Fc
Nairn Street	Poplar	E.	He
Namur Terrace	Southwark	S.E.	Gg
Nancy Street	Shoreditch	N.	Gd
Nankin Street	Poplar	E.	Je
Nansen Road	Battersea	S.W.	Ej
Nant Road	Hendon U.D.	N.W.	Ca
Nant Road	Hendon U.D.	N.W.	Db
Nant Street	Beth al Green	E.	Hd
Napier Avenue	Fulham	S.W.	Ch
Napier Road	Deptford	S.E.	Jh
Napier Road	East Ham	E.	Md
Napier Road	Kensington	W.	Cf
Napier Road	Leyton U.D.	E.	Kb
Napier Road	Wembley U.D.		Sb
Napier Road	West Ham	E.	Kd
Napier Road	Willesden U.D.	N.W.	Bd
Napier Street	Deptford	S.E.	Jh
Napier Street	Shoreditch	N.	Gd
Napier Terrace	Islington	N.	Gc
Napier Yard	Poplar	E.	Jg
Napoleon Court	Bermondsey	S.E.	Hf
Napoleon Passage	Bermondsey	S.E.	Hf
Napoleon Road	Twickenh m U.D.		Sj
Narbonne Avenue	Wandsworth	S.W.	Ej
Narborough Street	Fulham	S.W.	Dh
Narcissus Road	Hampstead	N.W.	Db
Narford Road	Hackney	E.	Hb
Nargor Road	Woolwich		Ng
Narrow Street	Stepney	E.	Je
Nascot Street	Hammersmith	W.	Ce
Naseby Road	Croydon	S.E.	Gm
Naseby Road	Hammersmith	W.	Cf
Nasmyth Road	Hammersmith	W.	Bf
Nassau Road	Barnes U.D.	S.W.	Bh
Nassau Road	Tottenham U.D.	N.	Ha
Nassau Street	Marylebone	W.	Fe
Nassau Street	Westminster	W.	Fe
Nassington Street	Hampstead	N.W.	Eb
Natal Road	Croydon		Sm
Natal Road	Wandsworth	S.W.	Fl
National Athletic Grounds	Willesden U.D.	N.W.	Cd
National Gallery	**Westminster**	W.C.	Fe
National Portrait Gallery	**Westminster**	W.C.	Fe
National Society Training College	Battersea	S.W.	Dh
Natural History Museum	**Kensington**	S.W.	Df
Naval Row	Poplar	E.	Ke
Naval Row South	Poplar	E.	Ke
Navarino Grove	Hackney	E.	Hc
Navarino Road	Hackney	E.	Hc
Navarre Road	East Ham	E.	Md
Navarre Street	Bethnal Green	E.	Hd
Navy Street	Wandsworth	S.W.	Fh
Naylor Road	Camberwell	S.E.	Hg
Nazareth House	Hammersmith	W.	Cg
Neal Street	Holborn and Westminster	W.C.	Fe
Nealdon Street	Lambeth	S.W.	Fh
Neasden House	Willesden U.D.	N.W.	Bb
Neasden Lane	Willesden U.D.	N.W.	Bb
Neat Street	Camberwell	S.E.	Gg
Neath Place	Bethnal Green	E.	Hd
Nebraska Street	Southwark	S.E.	Gf
Neckinger, The	Bermondsey	S.E.	Hf
Neckinger Street	Bermondsey	S.E.	Hf
Needham Lane	Hendon U.D.	N.W.	Cb
Nelgarde Road	Lewisham	S.E.	Jk
Nelldale Road	Bermondsey	S.E.	Hg
Nelson Court	Poplar	E.	Kd
Nelson Court	Stepney	E.	He
Nelson Dock	Bermondsey	S.E.	Jf
Nelson Grove	Merton U.D.	S.W.	Dm
Nelson Grove	Wandsworth	S.W.	Fh
Nelson Grove Road	Merton U.D.	S.W.	Dm
Nelson Place	Acton U.D.	W.	Ae
Nelson Place	Bethnal Green	E.	Hd
Nelson Place	Foots Cray U.D.		Ol
Nelson Place	Islington	N.	Gc
Nelson Place	Islington	N.	Gd
Nelson Place	Southwark	S.E.	Gf
Nelson Place	Stepn y	E.	He
Nelson Road	Foots Cray U.D.		Ol
Nelson Road	H rnsey	N.	Fa
Nelson Road	Wanstead U.D.	E.	La
Nelson Road	Wimbledon	S.W.	Dm
Nelson Square	Camberwell	S.E.	Hh
Nelson Square	Southwark	S.E.	Gf
Nelson Street	Bethnal Green	E.	Hd
Nelson Street	East Ham	E.	Md
Nelson Street	Finsbury	E.C.	Gd
Nelson Street	Greenwich	S.E.	Kg
Nelson Street	Stepney	E.	He
Nelson Street	West Ham	E.	Le
Nelson Street	Woolwich		Mf
Nelson Terrace	Islington	N.	Gd
Nelson Terrace	Woolwich		Mg
Nelson Wharf	Poplar	E.	Kg
Nelson's Column	**Westminster**	W.C.	Ff
Nelson's Grove	Wandsworth	S.W.	Fj
Nelson's Row	Wandsworth	S.W.	Fj
Nelson's Wharf	Southwark	S.E.	Ge
Nemoure Road	Acton U.D.	W.	Ae
Nepaul Road	Battersea	S.W.	Eh
Nepean Street	Wandsworth	S.W.	Bj
Neptune Cottages	Lambeth	S.W.	Fh
Neptune Street	Bermondsey	S.E.	Hf
Neptune Street	Lambeth	S.W.	Fh
Neptune Street	Step ey	E.	He
Ness Street	Bermondsey	S.E.	Hf
Neston Street	Bermondsey	S.E.	Jf
Netheravon Road	Chiswick U.D.	W.	Bg
Netherby Road	Lewisham	S.E.	Hk
Netherfield Road	Wandsworth	S.W.	El
Netherford Road	Wandsworth	S.W.	Fh
Netherhall Gardens	Hampstead	N.W.	Dc
Netherland Place	Islington	N.	Fc
Netherton Grove	Chelsea	S.W.	Dg
Netherton Road	Tottenham U.D.	N.	Ga
Netherwood Road	Hammersmith	W.	Cf
Netherwood Street	Hampstead	N.W.	Dc
Netherwood Terrace	Wandsworth	S.W.	Fl
Netley Road	Bre tford U.D.		Sg
Netley Road	Walthamstow U.D.	E.	Ja
Netley Street	Paddington	W.	Dd
Netley Street	St. Pancras	N.W.	Fd
Nettleton Court	City of London	E.C.	Ge
Nettleton Road	Deptford	S.E.	Jh
Neuchatel Road	Lewisham	S.E.	Jk
Nevada Street	Greenwich	S.E.	Kg
Neve Street	Poplar	E.	Kd
Nevern Place	Kensington	S.W.	Dg
Nevern Road	Kensington	S.W.	Dg
Nevern Square	Kensington	S.W.	Dg
Nevill Road	Stoke Newington	N.	Hb
Neville Court	Marylebone	N.W.	Dd
Neville Road	West Ham	E.	Lc
Neville Road	Willesden U.D.	N.W.	Dd
Neville Street	Kensi gton	S.W.	Dg
Neville Street	Lambeth	S.E.	Fg
Neville Terrace	Kensington	S.W.	Dg
Nevill's Court	City of London	E.C.	Ge
Nevis Road	Wa dsworth	S.W.	Ek
New Alley	Southwark	S.E.	Gf
New Buildings	Hampstead	N.W.	Db
New Close	Croydon R.D.	S.W.	Dm
New Court	City of London	E.C.	Ge
New Court	Westminster	W.C.	Fe
New Court	Stepney	E.	He
New Cut	Lambeth and Southwark	S.E.	Gf
New Dock	Bermondsey	S.E.	Jf
New Inn	Ham U.D.		Sk
New Lane	Bermondsey	S.E.	Hf
New Place	Bermondsey	S.E.	Hf
New Ride	Westminster	S.W.	Ef
New River			Ga

STREET OR PLACE.	BOROUGH.	P.D.	MAP.
New Road	Battersea, Lambeth & Wandsworth	S.W.	Fh
New Road	Brentford U.D.		Sg
New Road	Bromley		Ll
New Road	Bromley R.D.		Mk
New Road	Chiswick U.D.	W.	Bg
New Road	Erith U.D.		Og
New Road	Ham U.D.		Sl
New Road	Heston and Isleworth U.D.		Sh
New Road	Hornsey	N.	Fa
New Road	Kensington	W.	Cf
New Road	Kingston		Am
New Road	Stepney	E.	Je
New Road	West Ham	E.	Ld
New Road	Woolwich		Mg
New Square	Bethnal Green	E.	He
New Square	City of London	E.C.	He
New Square	Bermondsey	S.E.	Hf
New Square	Holborn	W.C.	Fe
New Square	Stepney	E.	He
New Street	Camberwell	S.E.	Hg
New Street	City of London	E.	He
New Street	City of London	E.C.	Ge
New Street	City of London	E.C.	He
New Street	Finsbury	E.C.	Gd
New Street	Hackney	E.	Hc
New Street	Hammersmith	W.	Bg
New Street	Kensington	S.W.	Ef
New Street	Marylebone	N.W.	Ed
New Street	Southwark	S.E.	Gg
New Street	West Ham	E.	Kd
New Street	West Ham	E.	Ke
New Street	Westminster	S.W.	Ff
New Street	Westminster	W.	Fe
New Street	Westminster	W.C.	Fe
New Street Hill	City of London	E.C.	Ge
New Street Mews	Marylebone	N.W.	Ed
New Street Mews	Southwark	S.E.	Gg
New Street Square	City of London	E.C.	Ge
New Theatre	**Westminster**	W.C.	Fe
Newark Street	Stepney	E.	He
Newbald's Wharf	Bermondsey	S.E.	Hf
New Barn Street	West Ham	E.	Le
New Basinghall Street	City of London	E.C.	Ge
New Beckenham Sta.	**Beckenham U.D.**		Jm
Newbold Road	Camberwell	S.E.	Hh
Newbold Street	Stepney	E.	He
New Bond Street	Westminster	W.	Ee
New Bridge Street	City of London	E.C.	Ge
New Broad Street	City of London	E.C.	Ge
Newburgh Road	Acton U.D.	W.	Ae
New Burlington Mews	Westminster	W.	Fe
New Burlington Place	Westminster	W.	Fe
New Burlington Street	Westminster	W.	Fe
Newburn Buildings	Lambeth	S.E.	Fg
Newburn Street	Lambeth	S.E.	Fg
Newbury Road	Willesden U.D.	N.W.	Bc
Newbury Street	City of London	E.C.	Ge
Newby Place	Poplar	E.	Ke
Newby Street	Wandsworth	S.W.	Eh
Newcastle Place	Finsbury	E.C.	Gd
Newcastle Place	Stepney	E.	He
Newcastle Street	City of London	E.C.	Ge
Newcastle Street	Greenwich	S.E.	Kg
Newcastle Street	Poplar	E.	Kg
New Cavendish Street	Marylebone	W.	Ee
New Charles Street	Finsbury	E.C.	Gd
New Chesterfield St.	Marylebone	W.	Ee
New Church Court	Westminster	W.C.	Fe
New Church Road	Camberwell	S.E.	Gg
New Church Street	Bermondsey	S.E.	Hf
New Church Street	Bethnal Green & Stepney	E.	He
New City Road	West Ham	E.	Ld
Newcombe Street	Kensington	W.	De
Newcomen Road	Battersea	S.W.	Dh
Newcomen Road	Leyton U.D.	E.	Lb
Newcomen Street	Southwark	S.E.	Gf
New Compton Street	Holborn	W.C.	Fe
New Cottage Row	Bermondsey	S.E.	Hf
New Coventry Street	Westminster	W.	Fe
New Cross Empire	**Deptford**	S.E.	Jh
New Cross Gate	Deptford	S.E.	Jh
New Cross Hall	Deptford	S.E.	Jh
New Cross Road	Deptford	S.E.	Jh
New Cross Stations	**Deptford**	S.E.	Jh
New Cross Theatre	**Deptford**	S.E.	Jh
New Crane Wharf	Stepney	E.	Hf
New Eltham & Pope St. Station	**Woolwich**		Mk
New Eltringham Road	Wandsworth	S.W.	Dj
New End	Hampstead	N.W.	Db
New End Road	Hampstead	N.W.	Db
Newgate Street	City of London	E.C.	Ge
New Goulston Street	Stepney	E.	He
New Gravel Lane	Stepney	E.	He
Newhall Street	Islington	N.	Gc
Newham Row	Bermondsey	S.E.	Gf
Newick Road	Hackney	E.	Hb
Newington Butts	Lambeth and Southwark	S.E.	Gg
Newington Causeway	Southwark	S.E.	Gf
Newington Crescent	Southwark	S.E.	Gg
Newington Green	Islington & Stoke Newington	N.	Gb
Newington Green Road	Islington	N.	Gc
Newington Terrace	Lewisham	S.E.	Kj
New Inn Broadway	Shoreditch	E.C.	Hd
New Inn Gateway	Shoreditch	E.C.	Hd
New Inn Square	Shoreditch	E.C.	Hd
New Inn Street	Shoreditch	E.C.	Hd
New Inn Yard	Shoreditch	E.C.	Hd
New James Street	Camberwell	S.E.	Hh
New Kent Road	Southwark	S.E.	Gf
New King Street	Deptford	S.E.	Jg
New King's Road	Fulham	S.W.	Ch
Newlan Row	Paddington	W.	Ee
Newland Place	Kensington	W.	Df
Newlands Road	Croydon	S.W.	Fm
Newland Street	East Ham and West Ham	E.	Mf
Newlands House	Wandsworth	S.W.	El
Newlands Park	Beckenham U.D. and Lewisham	S.E.	Hl
Newlands Park Villas	Lewisham	S.E.	Hl
Newling Street	Bethnal Green	E.	Hd
New London Street	City of London	E.C.	Ge
Newman Mews	Marylebone	W.	Fe
Newman Road	Walthamstow U.D.	E.	Ja
Newman Road	West Ham	E.	Ld
Newman Street	Battersea	S.W.	Eh
Newman Street	Marylebone	W.	Fe
Newman's Row	Holborn	W.C.	Fe
New Martan Street	Stepney	E.	He
Newnham Street	Marylebone	W.	Ee
Newnham Street	Stepney	E.	He
Newnham Terrace	Lambeth	S.E.	Ff
New Norfolk Street	Shoreditch	E.C.	Gd
New North Buildings	Shoreditch	E.C.	Gd
New North Place	Shoreditch	E.C.	Gd
New North Road (part)	Islington	N.	Gc
New North Road (part)	Shoreditch	N.	Gd
New North Street	Holborn	W.C.	Fe
New Oxford Street	Holborn	W.C.	Fe
New Palace Yard	Westminster	S.W.	Ff
New Park Road	Lambeth and Wandsworth	S.W.	Fk
New Park Road	Wimbledon	S.W.	Ck
Newport Court	Westminster	W.C.	Fe
Newport Road	Leyton U.D.	E.	Kb
Newport Street	Lambeth	S.E.	Fg
Newport Street	Westminster	W.C.	Fe
New Quebec Street	Marylebone	W.	Ee
New River District Wks.	Stoke Newington	N.	Gb
New River Head	Finsbury	E.C.	Gd
New River Reservoirs	Stoke Newington	N.	Ga
New Rowsell Street	Stepney	E.	Je
New Scotland Yard	Westminster	S.W.	Ff
Newstead	Wimbledon	S.W.	Cl
Newstead Lodge	Lambeth	S.W.	Fl
Newstead Road	Lewisham	S.E.	Kk
Newton Avenue	Acton U.D.	W.	Af
Newton Grove	Acton U.D.	W.	Bf

STREET OR PLACE.	BOROUGH.	P.D.	MAP.
Notting Hill Station	**Kensington**	W.	Ce
Novar Road	Woolwich		Nk
Novello Street	Fulham	S.W.	Dh
Noyna Road	Wandsworth	S.W.	Ek
Nuding Road	Lewisham	S.E.	Jj
Nugent Road	Islington	N.	Fa
Nugent Terrace	Marylebone	N.W.	Dd
Nunhead Cemetery	Camberwell	S.E.	Hj
Nunhead Crescent	Camberwell	S.E.	Hj
Nunhead Green	Camberwell	S.E.	Hh
Nunhead Grove	Camberwell	S.E.	Hj
Nunhead Lane	Camberwell	S.E.	Hj
Nunhead Passage	Camberwell	S.E.	Hh
Nunhead Station	**Camberwell**	S.E.	Hh
Nunkeeling Road	Kensington	W.	Ce
Nursery Place	Leyton U.D.	E.	Ja
Nursery Road	Croydon R.D.		Dm
Nursery Road	Merton U.D.	S.W.	Dm
Nursery Road	Lambeth	S.W.	Fj
Nursery Street	Southwark	S.E.	Gg
Nursery Street	Wandsworth	S.W.	Eh
Nutbourne Street	Paddington	W.	Cd
Nutbrook Street	Camberwell	S.E.	Hj
Nutfield Road	Camberwell	S.E.	Hj
Nutfield Road	Leyton U.D.	E.	Kb
Nutford Place	Marylebone	W.	Ee
Nuthurst Avenue	Hornsey	N.	Fa
Nuthurst Avenue	Wandsworth	S.W.	Fk
Nutley Terrace	Hampstead	N.W.	Dc
Nutt Street	Camberwell	S.E.	Hg
Nuttall Street	Shoreditch	N.	Hd
Nutwell Street	Wandsworth	S.W.	El
Nyanza Street	Woolwich		Nh
Nye Street	Poplar	E.	Ke
Nynehead Street	Deptford	S.E.	Jg
Nyon Street	Lewisham	S.E.	Jk
Oak Crescent	West Ham	E.	Le
Oak Grove	Hampstead	N.W.	Cb
Oak Grove	Hendon U.D.	N.W.	Cb
Oak Grove	Penge U.D.	S.E.	Hm
Oak Grove Road	Penge U.D.	S.E.	Hm
Oak Hill	Hampstead	N.W.	Db
Oak Hill Place	Lambeth	S.E.	Fl
Oak Lane	Stepney	E.	Je
Oak Lawn	Lewisham	S.E.	Kh
Oak Lodge	Wandsworth	S.W.	Ck
Oak Place	Bermondsey	S.E.	Jf
Oak Road	West Ham	E.	Le
Oak Road	Willesden U.D.	N.W.	Bc
Oak Street	Ealing	W.	Se
Oak Street	Woolwich		Mg
Oak Village	St. Pancras	N.W.	Eb
Oak Wharf	Poplar	E.	Jf
Oakbank	Chislehurst U.D.		Mm
Oakbank Road	Lambeth	S.E.	Gj
Oakbury Road	Fulham	S.W.	Dh
Oakcroft Road	Lewisham	S.E.	Kh
Oakdale Road	Leyton U.D.	E.	Kb
Oakdale Road	Wandsworth	S.W.	Fl
Oakdale Road	West Ham	E.	Lc
Oakden Street	Lambeth	S.E.	Gg
Oakfield	Camberwell	S.E.	Gl
Oakfield	Hornsey	N.	Fa
Oakfield	The Maldens and Coombe U.D.		Al
Oakfield	Wandsworth	S.W.	Fk
Oakfield	Wimbledon	S.W.	Cl
Oakfield Crescent	St. Pancras	N.W.	Eb
Oakfield Road	East Ham	E.	Md
Oakfield Road	Hackney	E.	Hb
Oakfield Road	Hornsey	N.	Ga
Oakfield Road	Penge U.D.	S.E.	Hm
Oakfield Road	Tottenham U.D.	N.	Ga
Oakfield Street	Kensington	S.W.	Dg
Oakford Place	Shoreditch	E.	Hd
Oakford Road	St. Pancras	N.W.	Fb
Oakham Street	Chelsea	S.W.	Eg
Oakhill House	Hampstead	N.W.	Db
Oakhill Park	Hampstead	N.W.	Db
Oakhill Park Mews	Hampstead	N.W.	Db
Oakhill Road	Beckenham U.D.		Km
Oakhill Road	Wandsworth	S.W.	Cj
Oakhurst	Ealing	W.	Se

STREET OR PLACE.	BOROUGH.	P.D.	MAP.
Oakhurst Grove	Camberwell	S.E.	Hj
Oakhurst Road	West Ham	E.	Lc
Oakington Road	Paddington	W.	Dd
Oaklands	Wandsworth	S.W.	Ck
Oaklands	Woolwich		Ng
Oaklands Grove	Hammersmith	W.	Bf
Oaklands Road	Bromley		Km
Oaklands Road	Willesden U.D.	N.W.	Cb
Oaklawn	Wimbledon	S.W.	Cl
Oaklea	Wandsworth	S.W.	Ck
Oaklea Passage	Kingston		Sm
Oakleigh	Beckenham U.D.		Jm
Oakleigh	Chislehurst U.D.		Mm
Oakley Avenue	Ealing	W.	Ae
Oakley Court	Bethnal Green	E.	Hd
Oakley Crescent	Chelsea	S.W.	Eg
Oakley Crescent	Islington	E.C.	Gd
Oakley Gardens	Hornsey	N.	Fa
Oakley Place	Camberwell	S.E.	Hg
Oakley Road	Islington	N.	Gc
Oakley Square	St. Pancras	N.W.	Fd
Oakley Street	Chelsea	S.W.	Eg
Oakley Street	Lambeth	S.E.	Ff
Oakmead Road	Wandsworth	S.W.	Ek
Oaks, The	Beckenham U.D.	S.E.	Gl
Oaks, The	Lambeth	E.	La
Oaks, The	Wanstead U.D.	S.W.	Ck
Oaks, The	Wimbledon		Ng
Oaksford Avenue	Woolwich	S.E.	Hl
Oak Tree Place	Lewisham	N.W.	Dd
Oak Tree Road	Marylebone	N.W.	Dd
Oakwood	Marylebone		Km
Oakwood	Beckenham U.D.		Mm
Oakwood Avenue	Chislehurst U.D.		Km
Oakwood Court	Kensington	W.	Cf
Oareborough Road	Deptford	S.E.	Jg
Oat Lane	City of London	E.C.	Ge
Oban Road	Barking U.D.		Nd
Oban Street	Poplar	E.	He
Oberstein Road	Battersea	S.W.	Dj
Observatory, The	Barnes U.D.	S.W.	Aj
Observatory Gardens	Kensington	W.	Df
Observatory House	Lewisham	S.E.	Hk
Occupation Road	Ealing	W.	Sf
Occupation Road	Southwark	S.E.	Gg
Occupation Road	Woolwich		Mh
Ocean Street	Stepney	E.	Je
Ockendon Road	Islington	N.	Gc
Ockley Road	Wandsworth	S.W.	Fl
Octagon Street	Bethnal Green	E.	He
Octavia Street	Battersea	S.W.	Eh
Octavius Street	Deptford	S.E.	Jh
Odell Street	Camberwell	S.E.	Hg
Odessa Road	Leyton U.D. and West Ham	E.	Lb
Odessa Road	Woolwich		Ng
Odessa Street	Bermondsey	S.E.	Jf
Odessa Wharf	Bermondsey	S.E.	Jf
Odger Street	Battersea	S.W.	Eh
Offerton Road	Wandsworth	S.W.	Fh
Offley Road	Lambeth	S.W.	Fg
Offord Road	Islington	N.	Fc
Offord Street	Islington	N.	Fc
Offord Terrace	Islington	N.	Fc
Ogilby Street	Woolwich		Mg
Oglander Road	Camberwell	S.E.	Hj
Ogle Mews	Marylebone	W.	Fe
Ohio Road	West Ham	E.	Le
Ohren's Court	Stepney	E.	Je
Oil Wharf	Poplar	E.	Kg
Oil Mill Lane	Kingston		Sm
Okeburn Road	Wandsworth	S.W.	El
Okehampton Road	Willesden U.D.	N.W.	Cc
Olaf Street	Hammersmith	W.	Ce
Old Court	Shoreditch	E.	Hd
Old Court Place	Kensington	W.	Df
Old Park	Woolwich		Oh
Old Road	Bermondsey	S.E.	Hf
Old Road	Lewisham	S.E.	Kj
Old Square	Holborn	W.C.	Fe
Old Street	Finsbury and Shoreditch	E.C.	Gd
Old Street	West Ham	E.	Ld

STREET OR PLACE.	BOROUGH.	P.D.	MAP.
Packer's Court	City of London	E.C.	Ge
Packington Road	Acton U.D.	W.	Af
Packington Street	Islington	N.	Gc
Paddenswick Road	Hammersmith	W.	Bf
Paddington Basin	Paddington	W.	De
Paddington Cemetery	Willesden U.D.	N.W.	Cc
Paddington Green	Paddington	W.	De
Paddington Recreation Ground	Paddington	W.	Dd
Paddington Street	Islington	N.	Fb
Paddington Street	Marylebone	W.	Ee
Paddington Station	**Paddington**	W.	De
Paddington Workhouse	Paddington	W.	De
Paddock Street	Greenwich	S.E.	Kg
Padfield Street	Lambeth	S.E.	Gj
Padua Road	Penge U.D.	S.E.	Hm
Pagden Street	Battersea	S.W.	Eh
Page Green Common	Tottenham U.D.	N.	Ha
Page Green Road	Tottenham U.D.	N.	Ha
Page Street	Westminster	S.W.	Ff
Pageant's Stairs	Bermondsey	S.E.	Jf
Pageant's Wharf	Bermondsey	S.E.	Jf
Page's Walk	Bermondsey	S.E.	Gf
Paget Road	Stoke Newington	N.	Ga
Paget Road	Woolwich		Mh
Paget Terrace	Woolwich		Mh
Pagnell Street	Deptford	S.E.	Jh
Pagoda Avenue	Richmond		Sh
Paignton Road	Tottenham U.D.	N.	Ha
Pain's Road	Croydon R.D.		Em
Painsthorpe Road	Stoke Newington	N.	Hb
Pakeman Street	Islington	N.	Fb
Pakenham Street	St. Pancras	W.C.	Fd
Palace Avenue	Kensington	W.	Df
Palace Court	Paddington	W.	De
Palace Gardens Mans.	Kensington	W.	Df
Palace Gardens Mews	Kensington	W.	Df
Palace Gardens Terrace	Kensington	W.	Df
Palace Gate	Kensington	W.	Df
Palace Gate	Westminster	S.W.	Df
Palace Grove	Bromley		Lm
Palace Grove	Penge U.D.	S.E.	Hm
Palace Hotel	Penge U.D.	S.E.	Hm
Palace Road	Bromley		Lm
Palace Road	Hackney	E.	Jc
Palace Road	Hornsey	N.	Fa
Palace Road	Lambeth and Wandsworth	S.W.	Fk
Palace Road	Penge U.D.	S.E.	Hm
Palace Square	Penge U.D.	S.E.	Hm
Palace Street	St. Pancras	N.W.	Ec
Palace Street	Westminster	S.W.	Ef
Palace Yard	Richmond		Sj
Palamos Road	Leyton U.D.	E.	Ja
Palatine Avenue	Stoke Newington	N.	Hb
Palatine Road	Stoke Newington	N.	Hb
Palestine Crescent	Croydon R.D.	S.W.	Dm
Pale Well	Barnes U.D.	S.W.	Aj
Palewell Common	Barnes U.D.	S.W.	Aj
Palewell Park	Barnes U.D.	S.W.	Aj
Palgrave Road	Hammersmith	W.	Bf
Palissy Street	Bethnal Green	E.	Hd
Pall Mall	Westminster	S.W.	Ff
Pall Mall East	Westminster	S.W.	Ff
Pall Mall Place	Westminster	S.W.	Ff
Pallador Place	Camberwell	S.E.	Gh
Palliser Road	Fulham and Hammersmith	W.	Cg
Palm Street	Bethnal Green & Stepney	E.	Jd
Palmer Place	Islington	N.	Gc
Palmer Road	Poplar	E.	Jd
Palmer Street	Lambeth	S.E.	Gf
Palmer Street	Stepney	E.	He
Palmer Street	Westminster	S.W.	Ff
Palmer's Green	Kingston		Sm
Palmer's Road	Bethnal Green	E.	Jd
Palmer's Road	Lambeth and Southwark	S.E.	Gg
Palmerston, The	Hampstead	N.W.	Dc
Palmerston Buildings	City of London	E.C.	Ge
Palmerston Crescent	Wimbledon	S.W.	Dm
Palmerston Road	Acton U.D.	W.	Af
Palmerston Road	Hampstead	N.W.	Dc
Palmerston Road	Islington	N.	Fb
Palmerston Road	Southwark	S.E.	Gg
Palmerston Road	Walthamstow U.D.	E.	Ja
Palmerston Road	Wandsworth	S.W.	Dj
Palmerston Road	West Ham	E.	Lc
Palmerston Road	Wimbledon	S.W.	Dm
Palmerston Road	Woolwich		Ng
Palmerston Street	Battersea	S.W.	Eh
Pamber Street	Kensington	W.	Ce
Pancras Lane	City of London	E.C.	Ge
Pancras Road	St. Pancras	N.W.	Fd
Pancras Square	St. Pancras	N.W.	Fd
Pancras Street	St. Pancras	W.C.	Fd
Pancras Terrace	Camberwell	S.E.	Hj
Pandora Road	Hampstead	N.W.	Dc
Pandrel Street	Woolwich		Nh
Panmure Road	Lewisham	S.E.	Hl
Panton Square	Westminster	W.	Fe
Panton Street	Westminster	S.W& W.C.	Fe
Paper Street	City of London	E.C.	Ge
Parade, The	Hampstead	N.W.	Dd
Parade, The	Islington	N.	Fa
Paradise Road	Islington	N.	Gb
Paradise Road	Lambeth	S.W.	Fh
Paradise Road	Richmond		Sj
Paradise Street	Bermondsey	S.E.	Hf
Paradise Street	Finsbury and Shoreditch	E.C.	Gd
Paradise Street	Marylebone	W.	Ee
Paradise Walk	Chelsea	S.W.	Eg
Paragon, The	Greenwich	S.E.	Lh
Paragon, The	Southwark	S.E.	Gf
Paragon Buildings	Southwark	S.E.	Gg
Paragon Place	Lewisham	S.E.	Lh
Paragon Road	Hackney	E.	Hc
Paragon Row	Southwark	S.E.	Gg
Paragon Theatre	**Stepney**	E.	Je
Parbury Street	Lewisham	S.E.	Jj
Parcels Post Depôt	Finsbury	E.C.	Fd
Parchmore Road	Croydon		Sm
Pardoner Street	Southwark	S.E.	Gf
Parfett Street	Stepney	E.	He
Parfitt Road	Bermondsey	S.E.	Hg
Parfrey Street	Fulham	W.	Cg
Parian Street	Poplar	E.	He
Paris Street	Lambeth	S.E.	Ff
Paris Terrace	Poplar	E.	Je
Parish Lane	Beckenham U.D. & Penge U.D.	S.E.	Jm
Parish Road	Beckenham U.D.	S.E.	Hm
Parish Street	Bermondsey	S.E.	Hf
Parish Wood	Bexley U.D.		Nk
Park Avenue	Bromley		Lm
Park Avenue	Croydon R.D.		Em
Park Avenue	East Ham	E.	Md
Park Avenue	Hornsey	N.	Fa
Park Avenue	West Ham	E.	Kc
Park Avenue	Willesden U.D.	N.W.	Cc
Park Buildings	Camberwell	S.E.	Hh
Park Crescent	Marylebone	N.W. & W.	Ed
Park Crescent	Stoke Newington	N.	Gb
Park Crescent	Wandsworth	S.W.	Fj
Park Crescent Mews	Wandsworth	S.W.	Fj
Park Crescent Mews E.	Marylebone	W.	Ed
Park Crescent Mews W.	Marylebone	W.	Ed
Park End	Bromley		Km
Park End	Lewisham	S.E.	Hl
Park Farm Road	Bromley		Lm
Park Gate	Wanstead U.D.	E.	La
Park Grove	Battersea	S.W.	Eh
Park Grove	Bromley		Lm
Park Grove	West Ham	E.	Ld
Park Hall	Lewisham	S.E.	Hl
Park Hill	Wandsworth	S.W.	El
Park Hill	Wandsworth	S.W.	Fj
Park Hill Road	Beckenham U.D.		Km
Park Hill Road	Foots Cray U.D.		Nl
Park House	Hampton Wick U.D.		Sm
Park House	Hornsey	N.	Ea

STREET OR PLACE.	BOROUGH.	P.D.	MAP.
Patience Road	Battersea	S.W.	Eh
Patmore Road	Battersea	S.W.	Fh
Patmos Road	Lambeth	S.W.	Gh
Paton Street	Finsbury	E.C.	Gd
Patrick Road	West Ham	E.	Ld
Patriot Court	Bethnal Green	E.	Hd
Patriot Square	Bethnal Green	E.	Hd
Patshull Place	St. Pancras	N.W.	Fc
Patshull Road	St. Pancras	N.W.	Fc
Patten Alley	Richmond		Sj
Patten Road	Wandsworth	S.W.	Dk
Pattenden Road	Lewisham	S.E.	Jk
Patterson Road	Croydon R.D.		Em
Patterson Street	Stepney	E.	He
Pattison Road	Hendon U.D.	N.W.	Db
Pattison Road	Woolwich		Ng
Paul Street	Finsbury and Shoreditch	E.C.	Gd
Paul Street	West Ham	E.	Kc
Paulet Road	Lambeth	S.E.	Gh
Paulin Place	Bermondsey	S.E.	Hf
Paulin Street	Bermondsey	S.E.	Hf
Paul's Alley	City of London	E.C.	Ge
Paul's Passage	Southwark	S.E.	Gf
Paulton Square	Chelsea	S.W.	Dg
Paulton Street	Chelsea	S.W.	Dg
Pavement, The	City of London	E.C.	Ge
Pavilion, The London	**Westminster**	W.	Fe
Pavilion Hotel	Hammersmith	W.	Ce
Pavilion Road	Chelsea	S.W.	Ef
Pavilion Street	Chelsea	S.W.	Ef
Pavilion Theatre	**Stepney**	E.	He
Pawnbrokers Almsho.	West Ham	E.	Lc
Paxton Road	Chiswick U.D.	W.	Bg
Pay Wharf	West Ham	E.	Mf
Payn Road	Poplar	E.	Kd
Payne Street	Deptford	S.E.	Jg
Payne Street	Islington	N.	Fd
Peabody Avenue	Westminster	S.W.	Eg
Peabody Buildings	Finsbury	E.C.	Fd
Peabody Buildings	Finsbury	E.C.	Gd
Peabody Court	Islington	N.	Gc
Peabody Square	Islington	N.	Gc
Peabody Square	Southwark	S.E.	Gf
Peabody Square	Westminster	S.W.	Fe
Peach Street	Paddington	W.	Cd
Peacock Square	Southwark	S.E.	Gg
Peacock Street	Southwark	S.E.	Gg
Peak Hill	Lewisham	S.E.	Hl
Peak Hill Avenue	Lewisham	S.E.	Hl
Peak Hill Gardens	Lewisham	S.E.	Hl
Peak Hill Lodge	Lewisham	S.E.	Hl
Peak Hill Road	Lewisham	S.E.	Hl
Pear Street	Westminster	S.W.	Ff
Pearcefield Avenue	Lewisham	S.E.	Hk
Pearcroft Road	Leyton U.D.	E.	Kb
Peardon Street	Wandsworth	S.W.	Eh
Pearfield Road	Lewisham	S.E.	Jl
Pearl Street	Stepney	E.	Hf
Pearman Street	Lambeth	S.E.	Gf
Pearson Street	Battersea	S.W.	En
Pearson Street	Greenwich	S.E.	Kg
Pearson Street	Shoreditch	E.	Hd
Pear Tree Court	Finsbury	E.C.	Fd
Pear Tree Road	Bromley		Lm
Pear Tree Street	Finsbury	E.C.	Gd
Peary Place	Bethnal Green	E.	Hd
Peckham Wood	Camberwell	S.E.	Hl
Peckham Grove	Camberwell	S.E.	Gg
Peckham House	Camberwell	S.E.	Hh
Peckham Park Road	Camberwell	S.E.	Hh
Peckham Road	Camberwell	S.E.	Gh
Peckham Rye	Camberwell	S.E.	Hj
Peckham Rye Common	Camberwell	S.E.	Hj
Peckham Rye Park	Camberwell	S.E.	Hj
Peckham Rye Station	**Camberwell**	S.E.	Hh
Peckwater Street	St. Pancras	N.W.	Fc
Pedley Street	Bethnal Green	E.	Hd
Pedro Street	Hackney	E.	Jb
Pedworth Road	Bermondsey	S.E.	Hg
Peek Crescent	Wimbledon	S.W.	Cl
Peel Court	Finsbury	E.C.	Ge
Peel Grove	Bethnal Green	E.	Hd
Peel Grove Mews	Bethnal Green	E.	Hd
Peel Road	Wembley U.D.		Sb
Peel Road	Willesden U.D.	N.W.	Dd
Peel Street	Kensington	W.	Df
Peerless Buildings	Finsbury	E.C.	Gd
Peerless Street	Finsbury	E.C.	Gd
Pekin Street	Poplar	E.	Je
Peldon Avenue	Richmond		Sj
Pelham Crescent	Kensington	S.W.	Dg
Pelham Mews	Kensington	S.W.	Df
Pelham Mews	Kensington	W.	Ce
Pelham Place	Kensington	S.W.	Df
Pelham Road	Beckenham U.D.		Hm
Pelham Road	Tottenham U.D.	N.	Ha
Pelham Road	Wimbledon	S.W.	Dm
Pelham Street	Kensington	S.W.	Df
Pelham Street	Stepney	E.	He
Pelham Street Bldgs.	Kensington	S.W.	Df
Pelham Terrace	Woolwich		Nk
Pelican House College	Camberwell	S.E.	Gj
Pelican Passage	Stepney	E.	He
Pell Place	Stepney	E.	He
Pell Street	Stepney	E.	He
Pellant Road	Fulham	S.W.	Cg
Pellatt Road	Camberwell	S.E.	Hj
Pellerin Road	Stoke Newington	N.	Hb
Pelling Street	Stepney	E.	Je
Pellipar Place	Woolwich		Mg
Pellipar Road	Woolwich		Mg
Pelly Road	West Ham	E.	Ld
Pelter Street	Bethnal Green	E.	Hd
Pelton Road	Greenwich	S.E.	Kg
Pember Road	Willesden U.D.	N.W.	Cd
Pemberton Avenue	Islington	N.	Fb
Pemberton Gardens	Islington	N.	Fb
Pemberton Place	Hackney	E.	Hc
Pemberton Road	Hornsey & Tottenham U.D.	N.	Ga
Pemberton Row	City of London	E.C.	Ge
Pemberton Terrace	Islington	N.	Fb
Pembridge Crescent	Kensington	W.	De
Pembridge Gardens	Kensington	W.	De
Pembridge Mews	Kensington	W.	De
Pembridge Place	Kensington	W.	De
Pembridge Road	Acton U.D.	W.	Ae
Pembridge Villas	Kensington	W.	De
Pembridge Square	Kensington	W.	De
Pembroke Cottages N.	Kensington	W.	Df
Pembroke Cottages S.	Kensington	W.	Df
Pembroke Gardens	Kensington	W.	Cf
Pembroke Lodge	Richmond		Sk
Pembroke Mews	Kensington	W.	Df
Pembroke Mews	Westminster	S.W.	Ef
Pembroke Mews North	Westminster	S.W.	Ef
Pembroke Place	Kensington	W.	Df
Pembroke Place West.	Kensington	W.	Df
Pembroke Road	Bromley		Lm
Pembroke Road	Kensington	W.	Df
Pembroke Road	Tottenham U.D.	N.	Ha
Pembroke Road	Walthamstow U.D.	E.	Ka
Pembroke Road	Wembley U.D.		Sb
Pembroke Square	Kensington	W.	Df
Pembroke Street	Islington	N.	Fc
Pembroke Terrace	Hackney	E.	Hc
Pembroke Terrace	Marylebone	N.W.	Dc
Pembroke Villas	Kensington	W.	Df
Pembroke Villas	Richmond		Sj
Pembury Grove	Hackney	E.	Hc
Pembury Road	Hackney	E.	Hc
Penell's Place	Camberwell	S.E.	Hh
Pennell's Court	Greenwich	S.E.	Kg
Penarth Street	Camberwell	S.E.	Hg
Penda Road	Hackney	E.	Jb
Pendarves Road	Wimbledon	S.W.	Bm
Pendennis Road	Wandsworth	S.W.	Fl
Pender Street	Deptford and Greenwich	S.E.	Jg
Pendle Road	Wandsworth	S.W.	El
Pendlestone Road	Walthamstow U.D.	E.	Ka
Pendower Road	Wimbledon	S.W.	Bm
Pendrell Road	Deptford	S.E.	Jh

STREET OR PLACE.	BOROUGH.	P.D.	MAP.
Piccadilly Circus Sta.	**Westminster**	W.	Fe
Piccadilly Hotel	Westminster	W.	Fe
Piccadilly Place	Westminster	W.	Fe
Pickard Street	Finsbury	E.C.	Gd
Pickering Mews	Paddington	W.	De
Pickering Place	Paddington	W.	De
Pickering Place	Westminster	S.W.	Ff
Pickering Street	Islington	N.	Gc
Picket Street	Wandsworth	S.W.	Ek
Pickford's Wharf	Southwark	S.E.	Gf
Pickle Buildings	Hackney	E.	Jc
Pickle Herring Stairs	Bermondsey	S.E.	Hf
Pickle Herring Street	Bermondsey	S.E.	Hf
Pickwick Road	Camberwell	S.E.	Gk
Picton Mews	Southwark	S.E.	Gf
Picton Street	Camberwell	S.E.	Gh
Piedmont Road	Woolwich		Ng
Pier Head	Stepney	E.	Hf
Pier Street	Poplar	E.	Kg
Piercefield Street	St. Pancras	N.W.	Ec
Piermont Road	Camberwell	S.E.	Hj
Pierrepoint Road	Acton U.D.	W.	Ae
Pigeon Court	Finsbury	E.C.	Gd
Pigott Street	Stepney	E.	Je
Pigwell Path	Hackney	E.	Hc
Pilgrim Hill	Lambeth	S.E.	Gl
Pilgrim Lane	Hampstead	N.W.	Db
Pilgrim Place	Hampstead	N.W.	Db
Pilgrim Street	City of London	E.C.	Ge
Pilkington Road	Camberwell	S.E.	Hh
Pimlico Pier	Westminster	S.W.	Fg
Pimlico Road	Chelsea and Westminster	S.W.	Eg
Pimlico Walk	Shoreditch	E.C.	Gd
Pinchin Street	Stepney	E.	He
Pincott Road	Merton U.D.	S.W.	Dm
Pine Apple Place	Marylebone	N.W.	Dd
Pine Grove	Islington	N.	Fb
Pine Road	Willesden U.D.	N.W.	Cb
Pine Street	Finsbury	E.C.	Gd
Pines Road	Bromley		Mm
Pines, The	Bromley		Mm
Pinfold Road	Wandsworth	S.W.	Fl
Pinner Road	Wembley U.D.		Sb
Pinner's Cour	City of London	E.C.	Ge
Pinto Place	Camberwell	S.E.	Gh
Piper Road	Kingston		Am
Piquet Road	Beckenham U.D.		Hm
Pirbright Road	Wandsworth	S.W.	Ck
Pitcairn Road	Croydon R.D.		Em
Pitcairn Street	Wandsworth	S.W.	Eh
Pitchford Street	West Ham	E.	Kc
Pitfield Street	Shoreditch	N.	Gd
Pitman Street	Camberwell	S.E.	Gg
Pitney Place	Southwark	S.E.	Gf
Pitt Street	Bethnal Green	E.	Hd
Pitt Street	Camberwell	S.E.	Hh
Pitt Street	Kensington	W.	Df
Pitt Street	St. Pancras	W.	Fe
Pitt Street	Southwark	S.E.	Gf
Pitt Street	West Ham	E.	Le
Pitts Hanger	Ealing	W.	Sd
Pittshanger Road	Ealing	W.	Sd
Pitville Road	Hammersmith	W.	Bf
Pixfield	Bromley		Km
Pixley Street	Stepney	E.	Je
Place, The	Bromley R.D.		Om
Place Green	Chislehurst U.D.		Ol
Placquett Road	Camberwell	S.E.	Hj
Placquett Terrace	Camberwell	S.E.	Hj
Plaistow Cemetery	Bromley		Lm
Plaistow Grove	Bromley		Lm
Plaistow Grove	West Ham	E.	Ld
Plaistow Lane	Bromley		Lm
Plaistow Park Road	West Ham	E.	Ld
Plaistow Road	West Ham	E.	Ld
Plaistow Station	**West Ham**	E.	Ld
Plane Street	Lewisham	S.E.	Hl
Planet Street	Stepney	E.	He
Plasgwyn	Chislehurst U.D.		Nm
Plashet Lane	East Ham	E.	Lc
Plashet Road	West Ham	E.	Lc
Plassey Road	Lewisham	S.E.	Kk
Platina Street	Finsbury	E.C.	Gd
Plato Road	Lambeth	S.W.	Fj
Platt, The	Wandsworth	S.W.	Ch
Platt Lane	Hampstead	N.W.	Db
Platt Street	St. Pancras	N.W.	Fd
Plawsfield Road	Beckenham U.D.		Jm
Playfair Street	Fulham	W.	Cg
Playfield Crescent	Camberwell	S.E.	Hj
Pleasant Court	Stepney	E.	Je
Pleasant Grove	Islington	N.	Fc
Pleasant Place	Lambeth	S.E.	Fg
Pleasant Place	Wandsworth	S.W.	Fj
Pleasant Row	Shoreditch	E.C.	Gd
Pledger Buildings	Lambeth	S.E.	Fg
Plevna Road	Tottenham U.D.	N.	Ha
Plevna Street	East Ham	E.	Me
Plevna Street	Poplar	E.	Kf
Pleydell Court	City of London	E.C.	Ge
Pleydell Street	City of London	E.C.	Ge
Plimsoll Road	Islington	N.	Gb
Plimsoll Street	Poplar	E.	Ke
Plough, The	Wimbledon	S.W.	Dl
Plough Court	City of London	E.C.	Ge
Plough Lane	Camberwell	S.E.	Hk
Plough Lane	Hackney	E.	Jc
Plough Lane	Wandsworth	S.W.	Dk
Plough Lane	Wimbledon	S.W.	Dl
Plough Road	Battersea	S.W.	Dh
Plough Road	Bermondsey and Deptford	S.E.	Jg
Plough Square	Whitechapel	E.	He
Plough Street	Stepney	E.	He
Plough Street Bldgs.	Stepney	E.	He
Plough Terrace	Battersea	S.W.	Dj
Plover Street	Hackney	E.	Jc
Plowden Buildings	City of London	E.C.	Ge
Plum Lane	Woolwich		Nh
Plumber's Place	Finsbury	E.C.	Gd
Plumber's Row	Stepney	E.	He
Plumbridge Street	Greenwich	S.E.	Kh
Plumes Tavern	Willesden U.D.	N.W.	Ad
Plummers Court	Holborn	W.C.	Fe
Plumstead Cemetery	Woolwich		Oh
Plumstead Common	Woolwich		Ng
Plumstead Common Road	Woolwich		Ng
Plumstead Marshes	Woolwich		Of
Plumstead Road	Woolwich		Ng
Plumstead Station	**Woolwich**		Ng
Plumstead Villas Rd.	Woolwich		Ng
Plum Tree Court	City of London	E.C.	Ge
Plymouth Road	Bromley		Lm
Plymouth Road	West Ham	E.	Le
Plympton Avenue	Willesden U.D.	N.W.	Cc
Plympton Road	Willesden U.D.	N.W.	Cc
Pocock Street	Southwark	S.E.	Gf
Podmore Road	Wandsworth	S.W.	Dj
Poets Road	Islington	N.	Gb
Point, The	Greenwich	S.E.	Kh
Point Hill	Greenwich	S.E.	Kh
Point Pleasant	Wandsworth	S.W.	Dj
Poland Street	Westminster	W.	Fe
Pole Street	Stepney	E.	Je
Pollard Road	Hendon U.D.	N.W.	Ba
Pollard Row	Bethnal Green	E.	Hd
Pollard Street	Bethnal Green	E.	Hd
Pollard's Hill, North	Croydon	S.W.	Fm
Pollard's Hill, South	Croydon	S.W.	Fm
Pollardshill Wood	Croydon R.D.	S.W.	Fm
Pollardswood Road	Croydon	S.W.	Fm
Pollen Street	Westminster	W.	Ee
Pollock Road	Southwark	S.E.	Gf
Polworth Road	Wandsworth	S.W.	Fl
Polygon, The	Wandsworth	S.W.	Ej
Polygon Buildings	Wandsworth	S.W.	Ej
Polygon Mews	Paddington	W.	Ee
Polygon Mews, South	Paddington	W.	Ee
Polytechnic	Battersea	S.W.	Eh
Polytechnic	Chelsea	S.W.	Dg
Polytechnic	Islington	N.	Gb
Polytechnic	Southwark	S.E.	Gf
Polytechnic	Woolwich	S.E.	Mg
Polytechnic The	Marylebone	W.	Ee

STREET OR PLACE.	BOROUGH.	P.D.	MAP.
Powis Gardens	Kensington	W.	Ce
Powis Place	Holborn	W.C.	Fe
Powis Road	Poplar	E.	Jg
Powis Road	Poplar	E.	Kd
Powis Square	Kensington	W.	Ce
Powis Street	Woolwich		Mg
Powlett Place	St. Pancras	N.W.	Ec
Pownal Road	Fulham	S.W.	Cg
Pownall Road	Bethnal Green	E.	Hd
Poynders Road	Wandsworth	S.W.	Ej
Poynings Road	Islington	N.	Fb
Poynings Yard	Islington	N.	Fb
Poyntz Road	Battersea	S.W.	Eh
Poyser Street	Bethnal Green	E.	Hd
Praed Street	Paddington	W.	De
Praed Street Station	**Paddington**	W.	De
Pragel Street	West Ham	E.	Ld
Prague Street	Lambeth	S.W.	Fj
Prah Road	Islington	N.	Gb
Prairie Street	Battersea	S.W.	Eh
Pratt Street	Lambeth	S.E.	Ff
Pratt Street	St. Pancras	N.W.	Fc
Pratt Street Mews	St. Pancras	N.W.	Fc
Prebend Gardens	Chiswick U.D.	W.	Bf
Prebend Place	St. Pancras	N.W.	Fc
Prebend Street	Islington	N.	Gd
Prebend Street	St. Pancras	N.W.	Fc
Prentis Road	Wandsworth	S.W.	Fl
Presburg Street	Hackney	E.	Jb
Prescot Place	Wandsworth	S.W.	Fh
President Mews	Finsbury	E.C.	Gd
President Place	Finsbury	E.C.	Gd
President Street	Finsbury	E.C.	Gd
Pressland Street	Paddington	W.	Cd
Pressland Yard	Paddington	W.	Cd
Prestage Buildings	Poplar	E.	Ke
Prestage Street	Poplar	E.	Ke
Prested Road	Battersea	S.W.	Ej
Preston Gardens	Willesden U.D.	N.W.	Bc
Preston House	Wembley U.D.		Sa
Preston Place	Richmond		Sj
Preston Road	Croydon	S.E.	Gm
Preston Road	Leyton U.D.	E.	La
Preston Road	West Ham	E.	Kd
Preston Street	Bethnal Green	E.	Jd
Preston Street	St. Pancras	N.W.	Ec
Preston's Road	Poplar	E.	Kf
Pretoria Avenue	Walthamstow U.D.	E.	Ja
Pretoria Road	Leyton U.D.	E.	Ka
Pretoria Road	Wandsworth	S.W.	El
Price Street	Southwark	S.E.	Gf
Price's Court	Southwark	S.E.	Gf
Prices Yard	Islington	N.	Fc
Prideaux Road	Lambeth	S.W.	Fh
Pridlo Road	Woolwich		Lg
Priestbury Road	East Ham	E.	Lc
Priestlands Park Road	Foots Cray U.D.		Nl
Priestland's Wood	Foots Cray U.D.		Nl
Prima Road	Lambeth	S.W.	Fg
Primrose Hall	West Ham	E.	Lf
Primrose Hill	City of London	E.C.	Ge
Primrose Hill	Hampstead and St. Pancras	N.W.	Ec
Primrose Hill Road	Hampstead and St. Pancras	N.W.	Ec
Primrose Park	Hampstead and St. Pancras	N.W.	Ec
Primrose Road	Leyton U.D.	E.	Ka
Primrose Street	Bethnal Green	E.	Hd
Primrose Street	City of London & Shoreditch	E.C.	Ge
Prince Street	Deptford and Greenwich	S.E.	Jg
Prince Terrace	West Ham	E.	Ld
Prince Arthur Bldgs.	Hampstead	N.W.	Db
Prince Arthur Mews	Hampstead	N.W.	Db
Prince Arthur Road	Hampstead	N.W.	Db
Prince Consort Road	Westminster	S.W.	Df
Prince Edward Road	Hackney	E.	Jc
Prince George Road	Stoke Newington	N.	Hb
Prince George's Ave.	Croydon U.D.	S.W.	Cm
Prince Imperial Memorial	Chislehurst U.D.		Mm
Prince Imperial Monument	Woolwich		Mh
Prince of Wales, The	Croydon R.D.	S.W.	Dm
Prince of Wales, The	Lambeth	S.W.	Fj
Prince of Wales, The	Paddington	W.	Cd
Prince of Wales Cres.	St. Pancras	N.W.	Ec
Prince of Wales Gate	Westminster	S.W.	Ef
Prince of Wales Mns.	Battersea	S.W.	Eh
Prince of Wales Road	Battersea	S.W.	Eh
Prince of Wales Road	St Pancras	N.W.	Ec
Prince of Wales Road	West Ham	E.	Le
Prince of Wales Ter.	Kensington	W.	Df
Prince of Wales Theatre	**Westminster**	W.	Ee
Prince Regent's Lane	West Ham	E.	Le
Prince Regent's Wharf	West Ham	E.	Lf
Princelet Street	Stepney	E.	He
Princes Court	Holborn	W.C.	Fe
Prince's Dock	Bermondsey	S.E.	Jf
Princes Gardens	Westminster	S.W.	Df
Princes Gate	Kensington and Westminster	S.W.	Df
Princes Gate Mews	Kensington	S.W.	Df
Prince's Hall	Westminster	S.W.	Fe
Princes Mews	Kensington	S.W.	Df
Princes Mews	Paddington	W.	De
Princes Place	Kensington	W.	Cf
Princes Place	Stepney	E.	Je
Princes Road	Barnes U.D.	S.W.	Ah
Princes Road	Beckenham U.D.	S.E.	Jm
Princes Road	Croydon R.D.	S.W.	Dm
Princes Road	Ealing	W.	Se
Princes Road	Kensington	W.	Ce
Princes Road	Kingston		Am
Princes Road	Lambeth	S.E.	Fg
Princes Road	Lewisham	S.E.	Kh
Princes Road	Richmond		Sh
Princes Road	Wandsworth	S.W.	Ck
Princes Road	Wimbledon	S.W.	Cm
Princes Road	Woolwich		Ng
Princes Row	Westminster	S.W.	Ef
Princes Square	Lambeth	S.E.	Gg
Princes Square	Paddington	W.	De
Princes Square	Shoreditch	E.C.	Ge
Princes Square	Stepney	E.	He
Princes Stairs	Bermondsey	S.E.	Hf
Princes Street	Bermondsey	S.E.	Hf
Princes Street	Bethnal Green	E.	Hd
Princes Street	City of London	E.C.	Ge
Princes Street	Marylebone	W.	Ee
Princes Street	Shoreditch	E.C.	Ge
Princes Street	Southwark	S.E.	Gf
Princes Street	Stepney	E.	He
Princes Street	Stepney	E.	Je
Princes Street	Westminster	S.W.	Ff
Princes Street	Westminster	W.	Ee
Princes Wharf	Lambeth	S.E.	Fe
Princes Yard	Kensington	W.	Ce
Princess Mews	Hampstead	N.W.	Dc
Princess Place	St. Pancras	N.W.	Ec
Princess Road	St. Pancras	N.W.	Ec
Princess Road	Stoke Newington	N.	Gb
Princess Road	Willesden U.D.	N.W.	Dd
Princess Street	Marylebone	N.W.	Dd
Princess Street	Southwark	S.E.	Gf
Princess's Theatre	**Marylebone**	W.	Fe
Princess Walk	Greenwich	S.E.	Jg
Princess Alice, The	West Ham	E.	Lc
Princess Frederica School	Willesden U.D.	N.W.	Cd
Princess Helena Coll.	Ealing	W.	Sd
Princess Louise Home	Kingston		Am
Princess May Road	Stoke Newington	N.	Hb
Princesthorpe Road	Lewisham	S.E.	Jl
Princeton Street	Holborn	W.C.	Fe
Pringall Street	Camberwell	S.E.	Hh
Printer Street	City of London	E.C.	Ge
Printing House Lane	City of London	E.C.	Ge
Printing House Square	City of London	E.C.	Ge
Priolo Road	Greenwich		Lg
Prior Street	Greenwich	S.E.	Kh

STREET OR PLACE.	BOROUGH.	P.D.	MAP.
Rheidol Mews	Islington	N.	Gd
Rheidol Terrace	Islington	N.	Gd
Rhodes Street	Islington	N.	Fc
Rhodesia Road	Lambeth	S.W.	Fg
Rhodesia Road	Leyton U.D.	E.	Kb
Rhodeswell Road	Stepney	E.	Je
Rhyl Street	St. Pancras	N.W.	Ec
Rhyme Road	Lewisham	S.E.	Kj
Ribblesdale Road	Wandsworth	S.W.	El
Ribston Street	Hackney	E.	Jc
Ricards Lodge	Wimbledon	S.W.	Cl
Ricardo Street	Poplar	E.	Ke
Rich Street	Stepney	E.	Je
Rich Terrace	Kensington	S.W.	Dg
Richard Place	Chelsea	S.W.	Ef
Richard Street	Finsbury and Islington	N.	Gc
Richard Street	Lambeth	S.E.	Gf
Richard Street	Stepney	E.	He
Richard Street	West Ham	E.	Le
Richardson Road	West Ham	E.	Kd
Richardson Street	Bermondsey	S.E.	Gf
Richardson Street	Stepney	E.	He
Richborough Road	Hampstead	N.W.	Cb
Richford Road	West Ham	E.	Ld
Richford Street	Hammersmith	W.	Cf
Richmond Avenue	Merton U.D.	S.W.	Cm
Richmond Avenue	Willesden U.D.	N.W.	Cc
Richmond Bridge	Richmond and Twickenham U.D.		Sj
Richmond Buildings	Westminster	W.	Fe
Richmond Cemetery	Richmond		Aj
Richmond Crescent	Islington	N.	Fc
Richmond Cricket Gd.	Richmond		Sh
Richmond Foot Bridge	Richmond		Sj
Richmond Gardens	Hammersmith	W.	Cf
Richmond Green	Richmond		Sj
Richmond Hill	Richmond		Sj
Richmond House	Wandsworth	S.W.	Ck
Richmond Lock	Richmond		Sj
Richmond Mews	Paddington	W.	De
Richmond Mews	Westminster	W.	Fe
Richmond Park	BarnesU.D.,Ham U.D. and Richmond		Ak
Richmond Park Road	Kingston		Sm
Richmond Place	Fulham	S.W.	Dg
Richmond Place	Islington	N.	Fb
Richmond Place	Lambeth	S.W.	Fh
Richmond Place	Woolwich		Ng
Richmond Road	Ealing	W.	Sf
Richmond Road	Fulham and Kensington	S.W.	Dg
Richmond Road	Hackney	E.	Hc
Richmond Road	Hammersmith	W.	Cf
Richmond Road	Islington	N.	Gc
Richmond Road	Islington	N.	Fc
Richmond Road	Kingston		Sm
Richmond Road	Leyton U.D.	E.	Kb
Richmond Road	Paddington	W.	De
Richmond Road	Richmond		Sh
Richmond Road	Tottenham U.D.	N.	Ha
Richmond Road	TwickenhamU.D.		Sj
Richmond Road	Walthamstow U.D.	E.	Ja
Richmond Road	West Ham	E.	Lc
Richmond Road	Wimbledon	S.W.	Bm
Richmond Station	**Richmond**		Sj
Richmond Street	Finsbury	E.C.	Gd
Richmond Street	Lambeth	S.E.	Ff
Richmond Street	Marylebone	N.W.	Dd
Richmond Street	West Ham	E.	Ld
Richmond Terrace	Hammersmith	W.	Cd
Richmond Terrace	Lambeth	S.W.	Fg
Richmond Terrace	Westminster	S.W.	Ff
Richmond Athletic Association Ground	Richmond		Sh
Riddell Street	Camberwell	S.E.	Hh
Ride, The	Brentford U.D.		Sg
Ridgdale Street	Poplar	E.	Kd
Ridge, The	Hendon U.D.	N.W.	Db
Ridge Road	Croydon R.D.		Em
Ridge Road	Hornsey	N.	Fa
Ridge Street	Bermondsey	S.E.	Hg
Ridgelands	Wimbledon	S.W.	Cm
Ridgeley Road	Hammersmith	W.	Cd
Ridgem Road	Wandsworth	S.W.	Dj
Ridgmount Gardens	Holborn	W.C.	Fe
Ridgmount Road	Wandsworth	S.W.	Dj
Ridgmount Street	Holborn	W.C.	Fe
Ridgway	Wimbledon	S.W.	Cl
Ridgway Place	Wimbledon	S.W.	Cl
Ridgway Road	Lambeth	S.W.	Gh
Riding House Street	Marylebone	W.	Ee
Ridings Road	The Maldens & Coombe U.D.	S.W.	Al
Ridley Mews	Hackney	E.	Hc
Ridley Road	Hackney	E.	Hc
Ridley Road	West Ham	E.	Lb
Ridley Road	Wimbledon	S.W.	Dm
Riffel Road	Willesden U.D.	N.W.	Cb
Rifle Street	Poplar	E.	Ke
Rigault Road	Fulham	S.W.	Dg
Rigden Street	Poplar	E.	Ke
Rigeley Road	Hammersmith	N.W.	Bd
Rigg Street	Stepney	E.	Je
Riggindale Road	Wandsworth	S.W.	Fl
Rignold Road	Camberwell	S.E.	Gh
Riles Road	West Ham	E.	Ld
Riley Road	Willesden U.D.	N.W.	Cd
Riley Street	Bermondsey	S.E.	Hf
Riley Street	Chelsea	S.W.	Dg
Rill Street	Camberwell	S.E.	Gh
Rillington Place	Kensington	W.	Ce
Rinaldo Street	Wandsworth	S.W.	Ek
Ring, The	Westminster	W.	Ee
Ringcroft Street	Islington	N.	Gc
Ringers Road	Bromley		Lm
Ringford Road	Wandsworth	S.W.	Cj
Ringmer Avenue	Fulham	S.W.	Ck
Ringstead Road	Lewisham	S.E.	Kk
Ringwood Road	Walthamstow U.D.	E.	Ja
Ripley Road	West Ham	E.	Le
Ripley Street	Southwark	S.E.	Gf
Ripley Street Mews	Southwark	S.E.	Gf
Ripon Road	Woolwich		Mh
Ripple Castle	Barking U.D.		Od
Ripple Lane	Barking U.D.		Od
Ripple Level	Barking U.D.		Od
Ripple Marsh	Barking U.D.		Oe
Ripple Road	Barking U.D.		Nd
Rippleside	Barking U.D.		Od
Rippolson Road	Woolwich		Ng
Rippoth Road	Poplar	E.	Jc
Risborough Street	Southwark	S.E.	Gf
Risdale Road	Penge U.D.	S.E.	Hm
Risdon Street	Bermondsey	S.E.	Hf
Riseholm Street	Hackney	E.	Jc
Riseldine Street	Lewisham	S.E.	Jk
Risinghill Street	Finsbury	N.	Fd
Rita Road	Lambeth	S.W.	Fg
Ritherdon Road	Wandsworth	S.W.	Ek
Ritson Road	Hackney	E.	Hc
Ritson Road	Tottenham U.D.	N.	Ga
Ritter Street	Woolwich		Mg
River Lane	Richmond		Sk
River Road	Barking U.D.		Nd
River Street	Finsbury	E.C.	Gd
River Street	Islington	N.	Gc
River Street	Poplar	E.	Ke
River Street	Wandsworth	S.W.	Ch
River Terrace	Greenwich	S.E.	Lf
River Terrace	Hammersmith	W.	Cg
Rivercourt Road	Hammersmith	W.	Bg
Riverdale	Lewisham	S.E.	Kj
Riverdale Gardens	Twickenham U.D.		Sj
Riverdale Road	Twickenham U.D.		Sj
Riverdale Road	Woolwich		Ng
Riverhall Street	Lambeth	S.W.	Fh
Riversdale Road	Islington & Stoke Newington	N.	Gb
Riverside House	Barnes U.D.	S.W.	Ah

STREET OR PLACE.	BOROUGH.	P.D.	MAP.
Riverside Road	Wimbledon	S.W.	Dl
River View Park	Lewisham	S.E.	Jk
River View Road	Chiswick U.D.	W.	Ag
Rivet Street	Camberwell	S.E.	Hg
Rivett Street	West Ham	E.	Le
Rivington Place	Shoreditch	N.	Gd
Rivington Street	Shoreditch	N.	Gd
Rivulet Road	Hackney	E.	Ha
Roach Road	Poplar	E.	Jc
Roan Place	Greenwich	S.E.	Kg
Roan School	Greenwich	S.E.	Kh
Roan Street	Greenwich	S.E.	Kg
Robert Street	Bethnal Green	E.	Hd
Robert Street	East Ham	E.	Mf
Robert Street	Holborn	W.C.	Fd
Robert Street	Lambeth	S.E.	Ff
Robert Street	St. Pancras	N.W.	Ed
Robert Street	Southwark	S.E.	Gf
Robert Street	Stepney	E.	Je
Robert Street	West Ham	E.	Kd
Robert Street	Westminster	S.W.	Eg
Robert Street	Westminster	W.	Ee
Robert Street	Westminster	W.C.	Fe
Robert Street	Woolwich		Mg
Robert Street	Woolwich		Ng
Robert's Alley	Ealing	W.	Sf
Robertson Street	Battersea and Wandsworth	S.W.	Eh
Robeson Street	Stepney	E.	Je
Robin Hood Court	City of London	E.C.	Ge
Robin Hood Inn	The Maldens & Coombe U.D.	S.W.	Bl
Robin Hood Lane	The Maldens & Coombe U.D.	S.W.	Bl
Robinhood Lane	Poplar	E.	Ke
Robin Hood Road	Wimbledon	S.W.	Bl
Robinson Lane	Croydon R.D.	S.W.	Em
Robinson Road	Bethnal Green	E.	Hd
Robinson Road	Croydon R.D.	S.W.	Dm
Robinson Street	Chelsea	S.W.	Eg
Robsart Street	Lambeth	S.W.	Fh
Robson Road	Lambeth	S.E.	Gl
Roby Street	Finsbury	E.C.	Gd
Rochdale Road	Leyton U.D.	E.	Ja
Rochdale Road	Woolwich		Og
Rochelle Street	Bethnal Green	E.	Hd
Rochester Avenue	West Ham	E.	Ld
Rochester House	Ealing	W.	Sf
Rochester Mews	St. Pancras	N.W.	Fc
Rochester Place	St. Pancras	N.W.	Fc
Rochester Road	St. Pancras	N.W.	Fc
Rochester Row	Westminster	S.W.	Ff
Rochester Square	St. Pancras	N.W.	Fc
Rochester Street	Westminster	S.W.	Ff
Rochester Street	Southwark	S.E.	Gf
Rochester Terrace	St. Pancras	N.W.	Fc
Rochford Street	St. Pancras	N.W.	Eb
Rock Avenue	Barnes U.D.	S.W.	Ah
Rock Avenue	Fulham	S.W.	Dg
Rock Hill	Camberwell	S.E.	Hl
Rock Place	Camberwell	S.E.	Hj
Rock Road	Hackney	E.	Jb
Rock Street	Islington	N.	Gb
Rockbourne Road	Lewisham	S.E.	Jk
Rockfield Street	Greenwich	S.E.	Kg
Rockhill Road	Hackney	E.	Ha
Rockhall Road	Willesden U.D.	N.W.	Cb
Rockingham Buildings	Southwark	S.E.	Gf
Rockingham Street	Southwark	S.E.	Gf
Rockland Road	Wandsworth	S.W.	Cj
Rockley Road	Hammersmith	W.	Cf
Rockmead Road	Hackney	E.	Jc
Rockmount Road	Croydon	S.E.	Gm
Rockmount Road	Woolwich		Mf
Rockpit Wood	Bromley		Mm
Roclif Street	Islington	N.	Gd
Rod Pond	Bexley		Nj
Roden Street	Islington	N.	Fb
Rodenhurst Road	Wandsworth	S.W.	Fj
Roderick Street	St. Pancras and Hampstead	N.W.	Eb
Roding River		E.	Oe
Roding Road	Hackney		Jb
Rodmere Street	Greenwich	S.E.	Kg
Rodney Residences	Finsbury	N.	Fd
Rodney Road	Southwark	S.E.	Gg
Rodney Street	Finsbury	N.	Fd
Rodney Street	Southwark	S.E.	Gf
Rodney Street	Woolwich		Mf
Rodsley Street	Camberwell	S.E.	Hg
Rodway Road	Bromley		Lm
Rodway Road	Wandsworth	S.W.	Bj
Rodwell Road	Camberwell	S.E.	Hj
Roe Bridge	Wandsworth	S.W.	Em
Roehampton House	Wandsworth	S.W.	Bj
Roehampton Lane	Wandsworth	S.W.	Bj
Roehampton Lodge	Wandsworth	S.W.	Bj
Roehampton Street	Westminster	S.W.	Fg
Roehampton Vale	Wandsworth	S.W.	Bk
Roffey Street	Poplar	E.	Kf
Rojack Road	Lewisham	S.E.	Jk
Rokeby	Wimbledon	S.W.	Cm
Rokeby Crescent	Deptford	S.E.	Jh
Rokeby Road	Deptford	S.E.	Jh
Rokesby Street	West Ham	E.	Kd
Rokesley House	Hornsey	N.	Fa
Roland Gardens	Kensington	S.W.	Dg
Roland Houses	Kensington	S.W.	Dg
Roland Mansions	Kensington	S.W.	Dg
Roland Road	Walthamstow U.D.	E.	Ka
Rolfe Road	Greenwich		Mg
Rollins Street	Deptford	S.E.	Hg
Rollo Street	Battersea	S.W.	Eh
Rolls Road	Bermondsey and Camberwell	S.E.	Hg
Rollscourt Avenue	Lambeth	S.E.	Gj
Rolt Street	Deptford	S.E.	Jg
Roman Road	Bethnal Green & Poplar	E.	Jd
Roman Road	Bexley U.D.		Oj
Roman Road	Chiswick U.D.	W.	Bf
Roman Road	East Ham	E.	Me
Roman Road	Hendon U.D.	N.W.	Ba
Roman Road	Islington	N.	Fc
Roman Road	Lambeth	S.E.	Gl
Roman Bath Street	City of London	E.C.	Ge
Romberg Road	Wandsworth	S.W.	El
Romford Road	East Ham and West Ham	E.	Lc
Romford Street	Stepney	E.	He
Romilly Road	Islington	N.	Gb
Rommany Road	Lambeth	S.E.	Gl
Romney Road	Greenwich	S.E.	Kg
Romney Street	Westminster	S.W.	Ff
Romola Road	Lambeth	S.E.	Gk
Rona Road	St. Pancras	N.W.	Eb
Ronald Street	Stepney	E.	Je
Ronalds Road	Islington	N.	Gc
Rondu Road	Hampstead	N.W.	Cb
Roney Street	Bermondsey	S.E.	Hf
Ronver Road	Lewisham	S.E.	Lk
Rood Lane	City of London	E.C.	Ge
Rook Street	Poplar	E.	Ke
Rookery, The	Wandsworth	S.W.	Bj
Rookery, The	Wandsworth	S.W.	Fl
Rookery, The	Woolwich		Mh
Rookstone Road	Wandsworth	S.W.	El
Rookwood Road	Hackney	N.	Ha
Rope Walk	Greenwich	S.E.	Jg
Rope Walk	Poplar	E.	Jf
Rope Walk	Stepney	E.	He
Rope Yard Rails	Woolwich		Mg
Ropemaker Fields	Stepney	E.	Je
Ropemaker Street	City of London & Finsbury	E.C.	Ge
Roper Street	Woolwich		Mj
Ropery Street	Stepney	E.	Je
Ropley Street	Bethnal Green	E.	Hd
Rosaline Road	Fulham	S.W.	Cg
Rosamond Street	Lewisham	S.E.	Hl
Rosary Gardens	Kensington	S.W.	Dg
Rosaville Road	Fulham	S.W.	Cg
Roscoe Street	Finsbury	E.C.	Gd
Roscoe Street	West Ham	E.	Le
Rose Alley	City of London	E.C.	Ge

STREET OR PLACE.	BOROUGH.	P.D.	MAP.
Rose Court . . .	Finsbury . .	E.C.	Gd
Rose Street . . .	City of London .	E.C.	Ge
Rose Street . . .	Westminster .	W.C.	Fe
Roseacre Road . .	Barnes U.D. .	S.W.	Bh
Rose and Crown . .	Wimbledon .	S.W.	Cl
Rosebank Road . .	Leyton U.D. .	E.	Ka
Rosebank Road . .	Poplar . . .	E.	Jd
Roseberry Place . .	Hackney . .	E.	Hc
Rosebery Avenue . .	Croydon . .		Gm
Rosebery Avenue (part)	Finsbury . .	E.C.	Gd
Rosebery Avenue (part)	Holborn . .	E.C.	Fe
Rosebery Gardens .	Hornsey . .	N.	Fa
Rosebery Gardens .	Tottenham U.D.	N.	Ga
Rosebery Road . .	Fulham . .	S.W.	Dh
Rosebery Road . .	Wandsworth .	S.W.	Fj
Rosebery Square . .	Holborn . .	E.C.	Fe
Rosebery Street . .	Bermondsey .	S.E.	Hg
Rosecroft Avenue . .	Hampstead .	N.W.	Db
Rosedale Road . .	East Ham . .	E.	Lc
Rosedale Street . .	Wandsworth .	S.W.	Ch
Roseford Terrace . .	Hammersmith .	W.	Cf
Roseheath Road . .	Wandsworth .	S.W.	Fk
Rosehill Road . .	Wandsworth .	S.W.	Dj
Roseleigh Avenue .	Islington . .	N.	Gb
Rosemary Road . .	Camberwell .	S.E.	Hh
Rosemary Street . .	Islington . .	N.	Gc
Rosemont Road . .	Acton U.D. . .	W.	Ae
Rosemont Road . .	Hampstead . .	N.W.	Dc
Rosemont Road . .	Richmond . .		Sj
Rosenau Street . .	Battersea . .	S.W.	Eh
Rosendale Road . .	Lambeth . .	S.E.	Gk
Roseneath . . .	Wanstead U.D. .	E.	La
Roseneath Road . .	Battersea . .	S.W.	Ej
Rosenthal Road . .	Lewisham . .	S.E.	Kk
Rosenthorpe Road .	Camberwell .	S.E.	Jj
Roserton Street . .	Poplar . . .	E.	Kf
Rose's Wharf . .	Poplar . . .	E.	Jg
Rosetta Street . .	Lambeth . .	S.W.	Fg
Rosevine Road . .	Wimbledon .	S.W.	Cm
Rosher Road . . .	West Ham . .	E.	Kc
Rosina Street . .	Hackney . .	E.	Jc
Roskell Road . .	Wandsworth .	S.W.	Ch
Roslyn Avenue . .	Lambeth . .	S.E.	Gh
Roslyn Road . . .	Acton U.D. . .	W.	Af
Rosmead Road . .	Kensington . .	W.	Ce
Rosoman Street . .	Finsbury . .	E.C.	Gd
Ross Road . . .	Croydon . .	S.E.	Gm
Ross Street . . .	West Ham . .	E.	Kd
Rossdale Street . .	Wandsworth .	S.W.	Ch
Rossetti Gdns. Mans. .	Chelsea . .	S.W.	Eg
Rossetti Mansions .	Chelsea . .	S.W.	Eg
Rossetti Studios . .	Chelsea . .	S.W.	Eg
Rossington Street .	Hackney . .	E.	Hb
Rossiter Mews . .	Wandsworth .	S.W.	Ek
Rossiter Road . .	Wandsworth .	S.W.	Ek
Rosslyn Hill . . .	Hampstead . .	N.W.	Db
Roslyn Road . . .	Tottenham U.D.	N.	Ga
Rosslyn Road . .	Twickenham U.D.		Sj
Rosslyn Road . .	Walthamstow U.D. . .	E.	Ka
Rossmore Road . .	Marylebone .	N.W.	Ed
Rostella Road . .	Wandsworth .	S.W.	Dl
Rostrevor Mansions .	Fulham . .	S.W.	Cn
Rostrevor Mews . .	Fulham . .	S.W.	Ch
Rostrevor Road . .	Fulham . .	S.W.	Ch
Rostrevor Road . .	Wimbledon .	S.W.	Cl
Rothbury Buildings .	Poplar . . .	E.	Jc
Rothbury Road . .	Poplar . . .	E.	Jc
Rotherfield Street .	Islington . .	N.	Gc
Rotherhill Alley . .	Wandsworth .	S.W.	Fm
Rotherhithe Station .	**Bermondsey** .	S.E.	Hf
Rotherhithe Street .	Bermondsey .	S.E.	Jf
Rotherhithe Tunnel .	Bermondsey and Stepney . .		Je
Rotherhithe New Road	Bermondsey and Camberwell .	S.E.	Hg
Rotherwood Road .	Wandsworth .	S.W.	Ch
Rothery Street . .	Islington . .	N.	Gc
Rothesay Avenue .	Merton U.D. .	S.W.	Cm
Rothie Road . . .	Greenwich . .	S.E.	Lg
Rothsay Road . .	East Ham . .	E.	Lc
Rothsay Street . .	Bermondsey .	S.E.	Gf
Rothschild Gardens .	Lambeth . .	S.E.	Gl
Rothschild Road . .	Acton U.D. . .	W.	Af
Rothschild Street .	Lambeth . .	S.E.	Gl
Rothwell Street . .	St. Pancras .	N.W.	Ec
Rotten Row . . .	Battersea . .	S.W.	Eg
Rotten Row . . .	Westminster .	S.W.	Ef
Rotten Row . . .	Greenwich and Lewisham .	S.E.	Lh
Rouel Road . . .	Bermondsey .	S.E.	Hf
Round Hill . . .	Lewisham . .	S.E.	Hl
Round Pond, The .	Westminster .	W.	Df
Roundell Street . .	Battersea . .	S.W.	Eh
Roundwood House .	Willesden U.D. .	N.W.	Bc
Roundwood Park . .	Willesden U.D. .	N.W.	Bc
Roundwood Road .	Willesden U.D. .	N.W.	Bc
Rounton Road . .	Poplar . . .	E.	Kd
Roupell Road . .	Lambeth and Wandsworth .	S.W.	Fk
Roupell Street . .	Lambeth . .	S.E.	Gf
Routh Road . . .	Wandsworth .	S.W.	Dk
Rowallan Road . .	Fulham . .	S.W.	Cg
Rowan Road . . .	Hammersmith .	W.	Cf
Rowden Road . .	Beckenham U.D.		Jm
Rowena Crescent . .	Battersea . .	S.W.	Eh
Rowfant Road . .	Wandsworth .	S.W.	Ek
Rowhill Road . .	Hackney . .	E.	Hb
Rowland Grove . .	Lewisham . .	S.E.	Hl
Rowle Street . . .	Deptford . .	S.E.	Jg
Rowledge Street . .	Greenwich . .	S.E.	Kg
Rowlett Street . .	Poplar . . .	E.	He
Rowley Road . .	Tottenham U.D.	N.	Ga
Rowley Street . .	Deptford . .	S.E.	Jg
Rowley Street . .	Poplar . . .	E.	Jf
Rowsell Street . .	Stepney . .	E.	Je
Rowton Road . .	Woolwich . .		Nh
Roxley Road . . .	Lewisham . .	S.E.	Kj
Roxwell Road . .	Hammersmith .	W.	Bf
Royal Academy . .	**Westminster** .	W.	Fe
Royal Arcade . .	Westminster .	W.	Ee
Royal Arsenal . .	Woolwich . .		Nf
Royal Avenue . .	Chelsea . .	S.W.	Eg
Royal Circus . . .	Lambeth . .	S.E.	Gl
Royal Crescent . .	Kensington . .	W.	Cf
Royal Crescent Mews .	Kensington . .	W.	Cf
Royal Dockyard . .	Woolwich . .		Mf
Royal Dockyard Wharf	Woolwich . .		Mf
Royal Exchange . .	**City of London** .	E.C.	Ge
Royal Exchange Avenue . . .	City of London .	E.C.	Ge
Royal Exchange Bldgs.	City of London .	E.C.	Ge
Royal Hill . . .	Greenwich . .	S.E.	Kh
Royal Hotel . . .	Woolwich . .		Mk
Royal Hospital . .	Chelsea . .	S.W.	Eg
Royal Hospital Road .	Chelsea . .	S.W.	Eg
Royal Institution . .	Westminster .	W.	Ee
Royal Mews . . .	Westminster .	S.W.	Ef
Royal Mint . . .	**Stepney** . .	E.	He
Royal Music Hall . .	**Holborn** . .	W.C.	Fe
Royal Observatory .	Greenwich . .	S.E.	Kh
Royal Parade . .	Chislehurst U.D.		Nm
Royal Parade . .	Lewisham . .	S.E.	Kh
Royal Place . . .	Greenwich . .	S.E.	Kh
Royal Road . . .	Southwark . .	S.E.	Gg
Royal Road . . .	West Ham . .	E.	Le
Royal Society . .	Westminster .	W.	Fe
Royal Street . . .	Lambeth . .	S.E.	Ff
Royal Terrace . .	Southwark . .	S.E.	Gg
Royal Academy of Music	Westminster .	W.	Ee
Royal Albert and Victoria Docks (part)	West Ham . .	E.	Le
Royal Albert and Victoria Docks (part)	East Ham . .	E.	Me
Royal Albert and Victoria Docks Cut.	East Ham and West Ham .	E.	Me
Royal Architectural Museum . . .	Westminster .	S.W.	Ff
Royal Artillery Barracks . . .	Woolwich . .		Mg
Royal Asiatic Society .	Westminster .	W.	Ee
Royal Botanic Society	Marylebone . .	N.W.	Ed
Royal Cambridge Asylum . . .	Kingston . .		Am
Royal College for Blind	Croydon . .	S.E.	Gm
Royal College of Music	Westminster .	S.W.	Df

STREET OR PLACE.	BOROUGH.	P.D.	MAP.
St. George's Market	Southwark	S.E.	Gf
St. George's Place	Westminster	S.W.	Ef
St. George's Road	Beckenham U.D.		Jm
St. George's Road	Camberwell	S.E.	Gg
St. George's Road.	Croydon R. D.		Em
St. George's Road.	Hampstead	N.W.	Dc
St. George's Road.	Heston & Isleworth U. D. and Twickenham U.D.		Sj
St. George's Road.	Kensington	W.	Ce
St. George's Road.	Kingston		Am
St. George's Road.	Leyton U.D.	E.	Kb
St. George's Road.	Richmond		Ah
St. George's Road.	St. Pancras	N.W.	Ec
St. George's Road.	Southwark	S.E.	Gf
St. George's Road.	Tottenham U.D.	N.	Ha
St. George's Road.	West Ham	E.	Lc
St. George's Road.	Westminster	S.W.	Eg
St. George's Road.	Wimbledon	S.W.	Cm
St. George's Square	West Ham	E.	Lc
St. George's Square	Westminster	S.W.	Fg
St. George's Street	Camberwell.	S.E.	Hh
St. George's Terrace	Kensington	S.W.	Df
St. German's Place	Greenwich	S.E.	Lh
St. German's Road	Lewisham	S.E.	Jk
St. Giles Workhouse	Camberwell.	S.E.	Hh
St. Gothard Road.	Lambeth	S.E.	Gl
St. Helens Road.	Bermondsey	S.E.	Hg
St. Helena Street	Finsbury	W.C.	Fd
St. Helena Terrace	Richmond		Sj
St. Helen's Gardens	Kensington	W.	Ce
St. Helens Place	City of London	E.C.	Ge
St. Helens Road	Croydon	S.W.	Fm
St. Helier Road	Leyton U.D.	E.	Ka
St. Hugh's Road	Penge U.D.	S.E.	Hn
St. James' Avenue	Beckenham U.D.		Jm
St. James' Gardens	St. Pancras	N.W.	Ec
St. James' Gardens	St. Pancras	N.W.	Fd
St. James' Grove	Battersea	S.W.	Eh
St. James' Home	Fulham	S.W	Cn
St. James' Market	Westminster	S.W	Fe
St. James' Palace.	**Westminster**	S.W.	Ff
St. James' Park	Westminster	S.W.	Ff
St. James' Park St.	**Westminster**	S.W.	Ff
St. James' Place	City of London	E.C.	He
St. James' Place	City of London	E.C.	Ge
St. James' Place	Kensington	W.	Ce
St. James' Place	Shoreditch	E.C.	Gd
St. James' Place	Westminster	S.W.	Ff
St. James' Place	Woolwich		Mg
St. James' Place	Woolwich		Ng
St. James' Place East	Greenwich	S.E.	Kg
St. James' Place West.	Greenwich	S.E.	Kg
St. James' Road	Battersea & Wandsworth	S.W.	Ek
St. James' Road	Beckenham U.D.		Jm
St. James' Road	Bermondsey	S.E.	Hf
St. James' Road	Camberwell	S.E.	Hf
St. James' Road	Bethnal Green	E.	Jd
St. James' Road	Deptford	S.E.	Jh
St. James' Road	Kingston		Sm
St. James' Road	Lambeth	S.W.	Gh
St. James' Road	Wandsworth	S.W.	Ek
St. James' Road	West Ham	E.	Lc
St. James' Square	Kensington	W.	Ce
St. James' Square	Westminster	S.W.	Ff
St. James' Street	Finsbury	E.C.	Gd
St. James' Street	Hammersmith	W.	Cg
St. James' Street	Islington	N.	Fc
St. James' Street	Islington	N.	Gd
St. James' Street	Kensington	W.	Ce
St. James' Street	Walthamstow U.D.	E.	Ja
St. James' Street	Westminster	S.W.	Ff
St. James' St. Station	**Walthamstow U.D.**	E.	Ja
St. James' Terrace	Kensington	W.	Ce
St. James' Theatre	**Westminster**	S.W.	Ff
St. James' Walk	Finsbury	E.C.	Gd
St. John Gardens	St. Pancras	N.W.	Ec
St. John Street	Bethnal Green	E.	Hd
St. John Street	Finsbury	E.C.	Gd
St. John Street	Islington	N.	Gd
St. John Street	Westminster	S.W.	Ff
St. John's Avenue	Willesden U.D.	N.W.	Bc
St. John's Church Rd.	Hackney	E.	Hc
St. John's Close	Camberwell.	S.E.	Hh
St. John's Gardens	Kensington	W.	Ce
St. John's Green	Richmond		Sh
St. John's Hill	Battersea & Wandsworth	S.W.	Dj
St. John's Hill Grove	Battersea	S.W.	Dj
St. John's Lane	Finsbury	E.C.	Ge
St. John's Park	Greenwich	S.E.	Lh
St. John's Park	Islington	N.	Fb
St. John's Place	Finsbury	E.C.	Ge
St. John's Place	Kensington	W.	Ce
St. John's Road	Barking U.D.		Nd
St. John's Road	Battersea	S.W.	Ej
St. John's Road	Beckenham U.D. & Penge U.D.	S.E.	Hm
St. John's Road	Deptford	S.E.	Jh
St. John's Road	East Ham	E.	Md
St. John's Road	Foots Cray U.D.		Ol
St. John's Park Road	Greenwich	S.E.	Lh
St. John's Road	Hampton Wick U.D.		Sm
St. John's Road	Islington	N.	Fa
St. John's Gardens	Kensington	W.	Ce
St. John's Road	Lambeth	S.W.	Fh
St. John's Road	The Maldens & Coombe U.D.		Am
St. John's Road	Richmond		Sj
St. John's Road	Shoreditch	N.	Gd
St. John's Road	Tottenham U.D.	N.	Ga
St. John's Road	Wandsworth	S.W.	Cj
St. John's Road	Wembley U.D.		Sb
St. John's Road	West Ham	E.	Le
St. John's Road	Wimbledon	S.W	Cm
St. John's Road	Woolwich		Ng
St. John's Square	Finsbury	E.C.	Gd
St. John's Square Place	Finsbury	E.C.	Ge
St. John's Station	**Deptford**	S.E.	Jh
St. John's Terrace	Deptford	S.E.	Jh
St. John's Terrace	Finsbury	E.C.	Ge
St. John's Terrace	West Ham	E.	Lc
St. John's Terrace	Willesden U.D.	N.W.	Dc
St. John's Villas	Camberwell	S.E.	Hj
St. John's Villas	Islington	N.	Fb
St. John's Wharf	Stepney	E.	Hf
St. John's Wood Barracks	Marylebone	N.W.	Dd
St. John's Wood Park.	Hampstead & Marylebone	N.W.	Dc
St. John's Wood Road.	Marylebone	N.W.	Dd
St. John's Wood Rd. Sta.	**Marylebone**	N.W.	Ed
St. John's Wood Ter.	Marylebone	N.W.	Dd
St. Joseph's Convent	Lambeth	S.W.	Fg
St. Joseph's House Convent	Kensington	W.	Ce
St. Joseph's Retreat	Islington	N.	Ea
St. Jude Street	Bethnal Green	E.	Hd
St. Jude Street	Islington	N.	Hc
St. Julian's Farm Road	Lambeth	S.E.	Fl
St. Julian's Road	Lambeth & Wandsworth	S.W.	Fl
St. Julian's Road	Willesden U.D.	N.W.	Dc
St. Katharine Docks	Stepney	E.	Hf
St. Katharine's	St. Pancras	N.W.	Ed
St. Katharine's	Stepney	E.	Hf
St. Katharine's College	St. Pancras	N.W.	Ed
St. Katharine's Lodge	St. Pancras	N.W.	Ed
St. Katharine's Road	Kensington	W.	Ce
St. Katharine's Wharf.	Stepney	E.	Hf
St. Kilda Road	Stoke Newington	N.	Ga
St. Lawrence Road	Kensington	W.	Ce
St. Lawrence Road	Lambeth	S.W.	Gh
St. Leonard's.	Barnes U.D.	S.W.	Ah
St. Leonard's Avenue	Poplar	E.	Ke
St. Leonard's Place	Westminster	S.W.	Fg
St. Leonard's Road	Acton U.D.	N.W	Bd
St. Leonard's Road	Barnes U.D.	S.W.	Ah
St. Leonard's Road	Ealing	W.	Se
St. Leonard's Road	Poplar	E.	Ke
St. Leonard's Road N.	Ealing	W.	Se

STREET OR PLACE.	BOROUGH.	P.D.	MAP.
St. Stephen's Road	Poplar	E.	Jd
St. Sumner Road	Kensington	S.W.	Dg
St. Swithin's Lane	City of London	E.C.	Ge
St. Tuomas Gardens	St. Pancras	N.W.	Ec
St. Thomas Road	Hackney	E.	Jc
St. Thomas Road	Fulham	S.W.	Cg
St. Thomas Road	Islington	N.	Gb
St. Thomas's Road	Stepney	E.	Je
St. Thomas Road	West Ham	E.	Le
St. Ihomas Road	Willesden U.D.	N.W.	Bc
St. Thomas Square	Hackney	E.	Hc
St. Thomas Street	Bermondsey and Southwark	S.E.	Gf
St. Thomas Street	Islington	N.	Gd
St. Vincent Place	Paddington	W.	De
Salamanca Court	Lambeth	S.E.	Fg
Salamanca Street	Lambeth	S.E.	Fg
Salcombe Road	Leyton U.D.	E.	Ja
Salcombe Road	Stoke Newington	N.	Hc
Salcott Road	Battersea	S.W.	Ej
Sale Street	Bethnal Green	E.	Hd
Sale Street	Paddington	W.	Ee
Salehurst Road	Lewisham	S.E.	Jj
Salem Gardens	Paddington	W.	De
Salem Road	Paddington	W.	De
Salford Road	Wandsworth	S.W.	Fk
Salisbury Arms	Fulham	S.W.	Cg
Salisbury Court	City of London	E.C.	Ge
Salisbury Place	Hendon U.D.	N.W.	Ca
Salisbury Road	Ealing	W.	Sf
Salisbury Road	Hackney	E.	Hc
Salisbury Road	Islington	N.	Fa
Salisbury Road	Leyton U.D.	E.	Kb
Salisbury Road	Richmond		Sh
Salisbury Road	Stepney	E.	Je
Salisbury Road	Tottenham U.D.	N.	Ga
Salisbury Road	Walthamstow U.D.	E.	Ka
Salisbury Road	West Ham	E.	Lc
Salisbury Road	Wimbledon	S.W.	Cm
Salisbury Row	Southwark	S.E.	Gg
Salisbury Square	City of London	E.C.	Ge
Salisbury Street	Acton U.D.	W.	Af
Salisbury Street	Bermondsey	S.E.	Hf
Salisbury Street	Marylebone	N.W.	Ed
Salisbury Street	Shoreditch	N.	Gd
Salisbury Street	Westminster	W.C.	Fe
Salmen Road	West Ham	E.	Ld
Salmen Street	Stepney	E.	Jd
Salmón Lane	Stepney	E.	Je
Salmon Street	Kingsbury U.D.		Aa
Salmon Street	Stepney	E.	Je
Salter Street	Stepney	E.	Je
Salterford Road	Wandsworth	S.W.	El
Salter's Hall Court	City of London	E.C.	Ge
Salter's Hill	Lambeth	S.E.	Gl
Salter's Street	Stepney	E.	He
Saltoun Road	Lambeth	S.W.	Fj
Saltram Crescent	Paddington	W.	Cd
Saltwood Grove	Southwark	S.E.	Gg
Salusbury Road	Willesden U.D.	N.W	Cc
Salusbury Road Sta.	**Willesden U.D.**	**N.W.**	**Cc**
Salvador Place	Wandsworth	S.W.	El
Salvador Row	Wandsworth	S.W.	El
Salvin Road	Wandsworth	S.W.	Ch
Salvin Street	Wandsworth	S.W.	Fh
Salway Road	West Ham	E.	Kc
Samford Street	Marylebone	N.W.	Dd
Samos Road	Beckenham U.D. & Penge U.D.	S.E.	Hm
Samson Street	Camberwell	S.E.	Gh
Samson Street	West Ham	E.	Ld
Samuda Street	Poplar	E.	Kf
Samuda Yard	Poplar	E.	Kf
Samuel Street	Stepney	E.	Je
Samuel Street	Woolwich		Mg
Sancroft Street	Lambeth	S.E.	Fg
Sanctuary, The	Westminster	S.W.	Ff
Sand Street	Woolwich		Mg
Saudal Street	West Ham	E.	Kd
Sandall Road	St. Pancras	N.W.	Fc
Sandall Road Mews	St. Pancras	N.W.	Fc
Sandbach Place	Woolwich		Ng

STREET OR PLACE.	BOROUGH.	P.D.	MAP.
Sandbourne Road	Deptford	S.E.	Jh
Sandbrook Road	Stoke Newington	N.	Gb
Sandell Street	Lambeth	S.E.	Ff
Sander Road	Camberwell	S.E.	Gg
Sander Street	Stepney	E.	He
Sanders Road	Hammersmith	W.	Cf
Sandfield Road	Croydon		Sm
Sandford Lane	Hackney	N.	Hb
Sandford Place	Hackney	N.	Hb
Sandford Road	West Ham	E.	Le
Sandford Row	Southwark	S.E.	Gg
Sandford Street	Deptford	S.E.	Jg
Sanford Terrace	Hackney	N.	Hb
Sandgate Street	Camberwell	S.E.	Hg
Sandhurst Road	Foots Cray U.D.		Ol
Sandhurst Road	Lewisham	S.E.	Kk
Sandilands Road	Fulham	S.W.	Dh
Sandison Street	Camberwell	S.E.	Hh
Sandland Street	Holborn	W.C.	Fe
Sandmere Road	Lambeth	S.W.	Fj
Sandover Road	Camberwell	S.E.	Hg
Sandpits Lane	Richmond		Sk
Sandringham Avenue	Merton U.D.	S.W.	Cm
Sandringham Gardens	Hornsey	N.	Fa
Sandringham Road	East Ham	E.	Lc
Sandringham Road	Hackney	E.	Hc
Sandringham Road	Willesden U.D.	N.W.	Bc
Sandrock Road	Lewisham	S.E.	Jh
Sands End	Fulham	S.W.	Dh
Sands End Cottages	Fulham	S.W.	Dh
Sands End Lane	Fulham	S.W.	Dh
Sandwell Crescent	Hampstead	N.W.	Dc
Sandwell House	Hampstead	N.W.	Dc
Sandwich Street	St. Pancras	W.C.	Fd
Sandy Hill Road	Woolwich		Mg
Sandy Lane	Croydon R.D.		Em
Sandy Lane	Ham U.D. and Richmond		Sk
Sandy Lane	Hampton Wick U.D.		Sm
Sandy Road	Hampstead	N.W.	Db
Sandycombe Road	Richmond		Ah
Sandycombe Road	Twickenham U.D.		Sj
Sandycroft Street	Hammersmith	W.	Ce
Sandy's Row	City of London & Stepney	E.	He
Sanford Lane	Hackney	N.	Hc
Sanford Terrace	Hackney	N.	Hc
Sangley Road	Lewisham	S.E.	Kk
Sangora Road	Battersea	S.W.	Dj
Sansom Street	Camberwell	S.E.	Gh
Sanson Road	Leyton U.D.	E.	Lb
Santley Street	Lambeth	S.W.	Fj
Santoft Road	Greenwich	S.E.	Lg
Santos Road	Wandsworth	S.W.	Cj
Sarah Street	Shoreditch	N.	Hd
Sarah Street	Stepney	E.	He
Sarah Street	West Ham	E.	Le
Saratoga Road	Hackney	E.	Jb
Sardinia Place	Holborn	W.C.	Fe
Sardinia Street	Westminster and Holborn	W.C.	Fe
Sarre Road	Hampstead	N.W.	Cb
Sarsfeld Road	Battersea and Wandsworth	S.W.	Ek
Sartor Road	Camberwell	S.E.	Hj
Sartripp Street	Hackney	E.	Jc
Sash Court	Shoreditch	E.C.	Ge
Satchwell Street	Bethnal Green	E.	Hd
Saunder Place	**Lambeth**	**S.E.**	**Fg**
Saunder Street	**Lambeth**	**S.E.**	**Fg**
Saunders Road	**Hammersmith**	**W.**	**Cf**
Saunders Road	**Woolwich**		**Ng**
Savage Gardens	**City of London**	**E.C.**	**He**
Savage Gardens	East Ham	E.	Me
Saveroake Road	St. Pancras	N.W.	Eb
Savile Place	Westminster	W.	Ee
Savile Row	Westminster	W.	Ee
Saville Place	Lambeth	S.E.	Ff
Saville Road	Acton U.D.	W.	Af
Saville Street	Marylebone	W.	Fe
Savona Place	Battersea	S.W.	Eh
Savona Street	Battersea	S.W.	Eh

STREET OR PLACE.	BOROUGH.	P.D.	MAP.
Savoy Hill	Westminster	W.C.	Fe
Savoy Place	Westminster	W.C.	Fe
Savoy Road	Merton U.D.	S.W.	Cm
Savoy Street	Westminster	W.C.	Fe
Savoy Theatre	**Westminster**	**W.C.**	**Fa**
Sawley Road	Hammersmith	W.	Be
Saw Pit Plantation	Barnes U.D.		Ak
Saxilby Road	West Ham	E.	Le
Saxon Road	Poplar	E.	Jd
Sayer Street	Southwark	S.E.	Gf
Sayes Court Grounds	Greenwich	S.E.	Jg
Sayes Court Street	Greenwich	S.E.	Jg
Sayes Street	Deptford	S.E.	Jg
Sayham Place	St. Pancras	N.W.	Fd
Scadbury Park	Chislehurst U.D.		Nm
Scadbury Park Lodge	Chislehurst U.D.		Nm
Scala Theatre	**St. Pancras**	**W.**	**Fe**
Scarborough Road	Hornsey	N.	Ga
Scarborough Road	Leyton U.D.	E.	Ka
Scarborough Street	Stepney	E.	He
Scarsdale Grove	Camberwell	S.E.	Gg
Scarsdale Road	**Camberwell**	**S.E.**	**Gg**
Scarsdale Villas	**Kensington**	**W.**	**Df**
Scarth Road	**Barnes U.D.**	**S.W.**	**Bh**
Scawen Road	**Deptford**	**S.E.**	**Jg**
Scawfell Street	**Shoreditch**	**E.**	**Hd**
Sceptre Street	**Stepney**	**E.**	**He**
Schofield Place	**Fulham**	**S.W.**	**Ch**
Scholars Road	**Wandsworth**	**S.W.**	**Ek**
Scholefield Road	**Islington**	**N.**	**Fa**
School of Art Needlewk.	**Westminster**	**S.W.**	**Df**
School Lane	**Chiswick U.D.**	**W.**	**Ag**
School of Music	**City of London**	**E.C.**	**Ge**
School Road	**Acton U.D.**	**N.W.**	**Bd**
School Road	**Hampton Wick U.D.**		**Sm**
Schoolhouse Lane	**Stepney**	**E.**	**Je**
Schoolhouse Lane	**Teddington U.D.**		**Sm**
Schubert Road	**Wandsworth**	**S.W.**	**Cj**
Scio House	**Wandsworth**	**S.W.**	**Bk**
Scio Pond	**Wandsworth**	**S.W.**	**Bk**
Scipio Street	**Camberwell**	**S.E.**	**Hg**
Sclater Street	**Bethnal Green**	**E.**	**Hd**
Scoresby Street	**Southwark**	**S.E.**	**Gf**
Scotswood Street	**Finsbury**	**E.C.**	**Gd**
Scott Buildings	**Islington**	**N.**	**Fc**
Scott Road	**Hammersmith**	**W.**	**Cf**
Scott Street	**Bethnal Green**	**E.**	**Hd**
Scott Street	**West Ham**	**E.**	**Le**
Scott-Ellis Gardens	**Marylebone**	**N.W.**	**Dd**
Scott's Lane	**Beckenham U.D.**		**Km**
Scott's Road	**Bromley**		**Lm**
Scott's Road	**Hammersmith**	**W.**	**Bf**
Scott's Road	**Leyton U.D.**	**E.**	**Ka**
Scott's Wharf	**Southwark**	**S.E.**	**Ge**
Scouler Street	**Poplar**	**E.**	**Ke**
Scovell Road	**Southwark**	**S.E.**	**Gf**
Scrooby Street	**Lewisham**	**S.E.**	**Kk**
Scrubs Lane	**Hammersmith**	**N.W. & W.**	**Ce**
Scrubs Road	**Hammersmith**	**N.W.**	**Bd**
Scrutton Street	**Shoreditch**	**N.**	**Gd**
Scutari Road	**Camberwell**	**S.E.**	**Hj**
Scylla Road	**Camberwell**	**S.E.**	**Hh**
Seabright Street	**Bethnal Green**	**E.**	**Hd**
Seacoal Lane	**City of London**	**E.C.**	**Ge**
Seaford Road	**Ealing**	**W.**	**Sf**
Seaford Road	**Tottenham U.D.**	**N.**	**Ga**
Seaford Street	**St. Pancras**	**W.C.**	**Fd**
Seagrave Mews	**Fulham**	**S.W.**	**Dg**
Seagrave Road	**Fulham**	**S.W.**	**Dg**
Seal Street	**Hackney**	**E.**	**Hb**
Seamore Place	**Westminster**	**W.**	**Ef**
Searle Street	**Holborn & Westminster**	**W.C.**	**Fe**
Searle's Road	**Southwark**	**S.E.**	**Gg**
Sears Street	**Camberwell**	**S.E.**	**Gg**
Seaton Street	**Chelsea**	**S.W.**	**Dg**
Seaton Street	**St. Pancras**	**N.W.**	**Fd**
Seaton Street	**West Ham**	**E.**	**Le**
Sea Witch Lane	**Greenwich**	**S.E.**	**Kg**
Sebbon Place	**Islington**	**N.**	**Gc**
Sebbon Street	**Islington**	**N.**	**Gc**

STREET OR PLACE.	BOROUGH.	P.D.	MAP.
Sebert Road	West Ham	E.	Lc
Secker Street	Lambeth	S.E.	Ff
Second Avenue	Barnes U.D.	S.W.	Bh
Second Avenue	Paddington	W.	Cd
Second Avenue	Walthamstow U.D.	E.	Ka
Second Avenue	West Ham	E.	Ld
Second Road	Woolwich		Mg
Secretan Road	Camberwell	S.E.	Hg
Sedan Street	Southwark	S.E.	Gg
Sedgeford Road	Hammersmith	W.	Be
Sedgmoor Place	Camberwell	S.E.	Gh
Sedgwick Road	Leyton U.D.	E.	Kb
Sedgwick Street	Hackney	E.	Jc
Sedgwick Villas	Hackney	E.	Jc
Sedleigh Road	Wandsworth	S.W.	Cj
Sedlescombe Road	Fulham	S.W.	Dg
Sedley Place	Westminster	W.	Ea
Seely Road	Croydon R.D.	S.W.	El
Seething Lane	City of London	E.C.	He
Sefton Street	Wandsworth	S.W.	Ch
Sekforde Street	**Finsbury**	**E.C.**	**Gd**
Selah Place	**Chelsea**	**S.W.**	**Dg**
Selborne Avenue	**Walthamstow U.D.**	**E.**	**Ja**
Selborne Road	**Camberwell**	**S.E.**	**Gh**
Selborne Road	**Walthamstow U.D.**	**E.**	**Ja**
Selby Road	**Croydon and Penge**	**S.E.**	**Hm**
Selby Road	**Leyton U.D.**	**E.**	**Kb**
Selby Street	**Bethnal Green & Stepney**	**E.**	**He**
Selcroft Road	**Greenwich**	**S.E.**	**Lg**
Selden Road	**Deptford and Camberwell**	**S.E.**	**Hh**
Seldon Mews	**Lambeth**	**S.E.**	**Fg**
Seldon Street	**Battersea**	**S.W.**	**Eh**
Selina Place	**Woolwich**		**Mg**
Selkirk Road	**Wandsworth**	**S.W.**	**Dl**
Sellincourt Road	**Wandsworth**	**S.W.**	**El**
Sellons Avenue	**Willesden U.D.**	**N.W.**	**Bc**
Selman Street	**Hackney**	**E.**	**Jc**
Selsdon Road	**Lambeth**	**S.E.**	**Gl**
Selsdon Road	**Wanstead U.D.**	**E.**	**La**
Selsdon Road	**West Ham**	**E.**	**Ld**
Selsey Street	**Stepney**	**E.**	**Je**
Selwin Road	**Willesden U.D.**	**N.W.**	**Bc**
Selwood Place	**Kensington**	**S.W.**	**Dg**
Selwood Terrace	**Kensington**	**S.W.**	**Dg**
Selwyn Avenue	**Richmond**		**Sh**
Selwyn Road	**Poplar & Stepney**	**E.**	**Jd**
Selwyn Road	**West Ham**	**E.**	**Ld**
Semley Place	**Westminster**	**S.W.**	**Eg**
Semley Road	**Croydon**	**S.W.**	**Fm**
Senate Street	**Camberwell and Deptford**	**S.E.**	**Hh**
Seneca Road	**Lambeth**	**S.W.**	**Fj**
Senegal Road	**Deptford**	**S.E.**	**Hg**
Senior Street	**Paddington**	**W.**	**De**
Senrab Street	**Stepney**	**E.**	**Je**
Seravia Road	**Lewisham**	**S.E.**	**Kj**
Serjeant's Inn	**City of London**	**E.C.**	**Fe**
Serle Street	**Westminster and Holborn**	**W.C.**	**Fe**
Sermon Lane	**City of London**	**E.C.**	**Ge**
Sermon Lane	**Finsbury**	**N.**	**Gd**
Serpentine, The	**Westminster**	**W.**	**Ef**
Serpentine Road	**St. Pancras**	**N.W.**	**Ed**
Serpentine Road	**Westminster**	**W.**	**Ef**
Seth Street	**Bermondsey**	**S.E.**	**Hf**
Settles Street	**Stepney**	**E.**	**He**
Settrington Road	**Fulham**	**S.W.**	**Dh**
Seven Dials	**Holborn**	**W.C.**	**Fe**
Seven Step Alley	**Bermondsey**	**E.C.**	**Jf**
Seven Sisters Rd. (pt.)	**Hornsey**	**N.**	**Ga**
Seven Sisters Rd. (pt.)	**Islington**	**N.**	**Fb**
Seven Sisters Rd. (pt.)	**Stoke Newington**	**N.**	**Ga**
Seven Sisters Rd. (pt.)	**Tottenham U.D.**	**N.**	**Ga**
Seven Sisters Station	**Tottenham U.D.**	**N.**	**Ha**
Severn Street	**Stepney**	**E.**	**He**
Severndroog Castle	**Woolwich**		**Mh**
Severus Road	**Battersea**	**S.W.**	**Ej**

STREET OR PLACE.	BOROUGH.	P.D.	MAP.
Silwood	Lambeth	S.W.	Gk
Silwood Street (part)	Bermondsey	S.E.	Hg
Silwood Street (part)	Deptford	S.E.	Jg
Simla Street	Deptford	S.E.	Jg
Simonds Road	Leyton U.D.	E.	Jb
Simpson Road	Poplar	E.	Ke
Simpson Street	Battersea	S.W.	Dh
Simpson Street	Lambeth	S.W.	Fh
Sinclair Gardens	Hammersmith	W.	Cf
Sinclair Road	Hammersmith	W.	Cf
Singer Street	Finsbury and Shoreditch	E.C.	Gd
Singer St. Chambers	Shoreditch	E.C.	Gd
Single Street	Stepney	E.	Jd
Singleton Street South	Shoreditch	N.	Gd
Sion College	City of London	E.C.	Ge
Sion Square	Stepney	E.	He
Sirdar Road	Kensington	W.	Ce
Sise Lane	City of London	E.C.	Ge
Sisters Avenue	Battersea	S.W.	Ej
Sistora Road	Wandsworth	S.W.	Ek
Sitka	Chislehurst U.D.		Mm
Siward Road	Wandsworth	S.W.	Dk
Sixth Avenue	Paddington	W.	Cd
Skardu Road	Hampstead	N.W.	Cb
Skeffington Road	East Ham	E.	Md
Skelbrook Street	Wandsworth	S.W.	Dk
Skelgill Road	Wandsworth	S.W.	Cj
Skelton Road	West Ham	E.	Lc
Skelton's Lane	Leyton U.D.	E.	Ka
Skidmore Street	Stepney	E.	Je
Skiers Street	West Ham	E.	Kd
Skinner Cottages	Chelsea	S.W.	Eg
Skinner Street	Chelsea	S.W.	Eg
Skinner Street	Finsbury	E.C.	Gd
Skipton Street	Southwark	S.E.	Gf
Slade, The	Woolwich		Ng
Sladedale Road	Woolwich		Ng
Slaidburn Street	Chelsea	S.W.	Dg
Slaithwaite Road	Lewisham	S.E.	Kj
Slane Road	Greenwich	S.E.	Lj
Slaney Place	Islington	N.	Fb
Slate Cottages	Barking U.D.		Od
Sleaford Street	Battersea	S.W.	Fg
Slipper's Place	Bermondsey	S.E.	Hf
Sloane Avenue	Chelsea	S.W.	Eg
Sloane Court East	Chelsea	S.W.	Eg
Sloane Court West	Chelsea	S.W.	Eg
Sloane Gardens	Chelsea	S.W.	Dg
Sloane Gardens Mns.	Chelsea	S.W.	Dg
Sloane Square	Chelsea	S.W.	Eg
Sloane Sq. Mansions	Chelsea	S.W.	Eg
Sloane Square Station	**Chelsea**	S.W.	Eg
Sloane Street	Chelsea	S.W.	Ef
Sloane Terrace	Chelsea	S.W.	Ef
Sloane Terrace Mns.	Chelsea	S.W.	Ef
Slough Lane	Kingsbury U.D.		Aa
Sly Street	Stepney	E.	He
Small Street	Bermondsey	S.E.	Gf
Smalley Road	Hackney	N.	Hb
Smallwood Road	Wandsworth	S.W.	Dl
Smardale Road	Wandsworth	S.W.	Dj
Smart Street	Bethnal Green	E.	Jd
Smeaton Road	Wandsworth	S.W.	Ck
Smedley Street	Wandsworth	S.W.	Fh
Smeed Road	Poplar	E.	Jc
Smith Hill	Brentford U.D.		Sg
Smith Square	Westminster	S.W.	Ff
Smith Street	Chelsea	S.W.	Eg
Smith Street	Finsbury	E.C.	Gd
Smith Street	Lambeth	S.E.	Gg
Smith Street	Stepney	E.	He
Smith Terrace	Chelsea	S.W.	Eg
Smith's Buildings	Bethnal Green	E.	Hd
Smithfield Market	City of London	E.C.	Ge
Smithies Road	Woolwich		Og
Smyrke Road	Southwark	S.E.	Hg
Smyrna Road	Hampstead	N.W.	Dc
Snaresbrook Road	Leyton U.D. and Wanstead U.D.	E.	La
Snaresbrook Station	**Wanstead U.D.**	E.	La
Snarsgate Street	Kensington	W.	Ce
Snead Street	Deptford	S.E.	Jh
Sneyd Road	Willesden U.D.	N.W.	Cb
Snow Hill	City of London	E.C.	Ge
Snowhill Avenue	Greenwich	S.E.	Kh
Snowsfield Street	Bermondsey	S.E.	Gf
Soames Street	Camberwell	S.E.	Hj
Soanes Museum	**Holborn**	W.C.	Fe
Socrates Place	Shoreditch	E.C.	Hd
Soho Square	Westminster	W.	Fe
Soho Street	Westminster	W.	Fe
Solander Street	Stepney	E.	He
Solent Road	Hampstead	N.W.	Dc
Solon Road	Lambeth	S.W.	Fj
Solon New Road	Lambeth	S.W.	Fj
Solway Road	Camberwell	S.E.	Hj
Somali Street	Hampstead	N.W.	Cb
Somerfield Road	Stoke Newington	N.	Gb
Somerford Grove	Hackney	N.	Hb
Somerford Street	Bethnal Green	E.	Hd
Somerleyton Road	Lambeth	S.W.	Gj
Somers Place	Paddington	W.	Ea
Somers Road	Lambeth	S.W.	Fj
Somers Road	Walthamstow U.D.	E.	Ja
Somerset House	**Westminster**	W.C.	Fe
Somerset Road	Acton U.D.	W.	Af
Somerset Road	Brentford U.D.		Sg
Somerset Road	Ealing	W.	Se
Somerset Road	Walthamstow U.D.	E.	Ja
Somerset Road	Wimbledon	S.W.	Cl
Somerset Street	Battersea	S.W.	Dh
Somerset Street	City of London	E.	He
Somerset Street	Marylebone	W.	Ee
Somerset Street	Westminster	W.C.	Fe
Somerton Road	Camberwell	S.E.	Hj
Somerville Road	Beckenham U.D.	S.E.	Jm
Somerville Road	Deptford	S.E.	Jh
Sonderburg Road	Islington	N.	Fb
Sondes Street	Southwark	S.E.	Gg
Sonning Street	Islington	N.	Fc
Sophia Road	Leyton U.D.	E.	Ka
Sophia Road	West Ham	E.	Le
Sophia Street	Poplar	E.	Ke
Sotheby Road	Islington	N.	Gb
Sotheron Road	Fulham	S.W.	Dh
Soudan Road	Battersea	S.W.	Eh
Souldern Road	Hammersmith	W.	Cf
Sound Lane	Hammersmith	W.	Cg
South Crescent	Greenwich	S.E.	Kh
South Crescent	Holborn	W.C.	Fe
South Dock	Bermondsey	S.E.	Jf
South Dock	Poplar	E.	Kf
South Drive	Battersea	S.W.	Eh
South End	Kensington	W.	Df
South Grove	Camberwell	S.E.	Hh
South Grove	St. Pancras	N.	Ea
South Grove	Stepney	E.	Jd
South Grove	Tottenham U.D.	N.	Ga
South Grove	Walthamstow U.D.	E.	Ja
South Lane	Kingston		Sm
South Lock	Bermondsey	S.E.	Jf
South Parade	Acton U.D.	W.	Bf
South Parade	Chelsea	S.W.	Dg
South Park	Fulham	S.W.	Dh
South Place	City of London, Finsbury and Shoreditch	E.C.	Ge
South Place	Southwark	S.E.	Gg
South Place	Wandsworth	S.W.	Ch
South Place	Wimbledon	S.W.	Cm
South Quay	Poplar	E.	Ke
South Quay	Stepney	E.	Hf
South Road	Ealing	W.	Sf
South Road	Hackney	E.	Hb
South Road	Lewisham	S.E.	Jk
South Road	Wimbledon	S.W.	Dm
South Row	Greenwich and Lewisham	S.E.	Lh
South Row	Kensington	W.	Cd
South Row	Westminster	W.C.	Fe
South Side	Wandsworth	S.W.	Ej
South Side	Wandsworth	S.W.	Fm

STREET OR PLACE.	BOROUGH.	P.D.	MAP.
Tenter Street South	Stepney	E.	He
Tenter Street West	Stepney	E.	He
Tenterden Street	Westminster	W.	Ee
Terminus Place	Westminster	S.W.	Ef
Terrace, The	Barnes U.D.	S.W.	Bh
Terrace, The	Bermondsey	S.E.	Gf
Terrace, The	Camberwell	S.E.	Gh
Terrace, The	City of London	E.C.	Ge
Terrace, The	Richmond		Sj
Terrace Gardens	Richmond		Sj
Terrace Place	Stepney	E.	Je
Terrace Road	Hackney	E.	Jc
Terrace Road	West Ham	E.	Ld
Terrace Walk, The	Westminster	W.C.	Fe
Terrapin Road	Wandsworth	S.W.	Ek
Terriswood Road	Camberwell	S.E.	Hk
Terront Road	Tottenham U.D.	N.	Ga
Testerton Street	Kensington	W.	Ce
Tetcott Road	Chelsea	S.W.	Dg
Tetley Street	Poplar	E.	Ke
Tetuan Road	Hammersmith	W.	Ce
Teviot Street	Poplar	E.	Ke
Tewkesbury Lodge	Lewisham	S.E.	Hk
Tewkesbury Road	Tottenham U.D.	N.	Ga
Tewsou Road	Woolwich		Ng
Thackeray Road	East Ham	E.	Md
Thackeray Street	Battersea	S.W.	Eh
Thackeray Street	Kensington	W.	Df
Thackeray's Alms-houses	Lewisham	S.E.	Kk
Thames Bank	Chiswick U.D.		Ag
Thames Bank	Fulham	S.W.	Ch
Thames Bank Distillery	Westminster	S.W.	Eg
Thames Conservancy	City of London	E.C.	Ge
Thames Ironworks	West Ham	E.	Ke
Thames Place	Poplar	E.	Jf
Thames Row	Brentford U.D.		Ag
Thames Side	Kingston		Sm
Thames Soap Works	Greenwich	S.E.	Kg
Thames Street	Greenwich	S.E.	Kg
Thames Street	Kingston		Sm
Thames Sugar Refinery	West Ham	E.	Mf
Thames Tunnel	Stepney	E.	Hf
Thane Villas	Islington	N.	Fb
Thanet Place	Westminster	W.C.	Fe
Thanet Street	St. Pancras	W.C.	Fd
Thatchedhouse Lodge	The Maldens and Coombe U.D.		Al
Thatched House Court	Westminster	W.C.	Fe
Thavies Inn	City of London	E.C.	Ga
Thayer Street	Marylebone	W.	Ee
Thayers Farm Road	Beckenham U.D.		Jm
Thaxted Street	Woolwich		Nl
Theatre Street	Battersea	S.W.	Eh
Theban Road	Croydon	S.E.	Gm
Theberton Street	Islington	N.	Gc
Theberton Street West	Islington	N.	Gc
Theed Street	Lambeth	S.E.	Gf
Theobald's Road	Holborn	W.C.	Fe
Theobald Road	Leyton U.D.	E.	Ja
Theobald Street	Southwark	S.E.	Gf
Theodore Road	Lewisham	S.E.	Kj
Theresa Mews	Hammersmith	W.	Bg
Therapia Road	Camberwell	S.E.	Hj
Thesiger Road	Beckenham U.D.	S.E.	Jm
Thessaly Square	Battersea	S.W.	Fh
Thexted Road	Woolwich		Nl
Thetford Place	Bermondsey	S.E.	Hf
Theydon Road	Bethnal Green	E.	Jd
Theydon Road	Hackney	E.	Hb
Thibet Street	Battersea	S.W.	Dh
Thicket Grove	Penge U.D.	S.E.	Hm
Thicket Road	Penge U.D.	S.E.	Hm
Third Avenue	Paddington	W.	Cd
Third Avenue	Walthamstow U.D.	E.	Ka
Third Avenue	West Ham	E.	Ld
Third Road	Woolwich		Mg
Thirleby Road	Westminster	S.W.	Ff
Thirlmere Road	Wandsworth	S.W.	Fl
Thirsk Road	Battersea	S.W.	Ej
Thirza Street	Stepney	E.	Je
Thistle Grove	Kensington	S.W.	Dg
Thistlewaite Road	Hackney	E.	Hb
Thomas Lane	Lewisham	S.E.	Jk
Thomas Passage	Bethnal Green and Stepney	E.	He
Thomas Place	Bethnal Green and Stepney	E.	He
Thomas Place	Finsbury	E.C.	Gd
Thomas Place	Kensington	W.	Ce
Thomas Place	Lambeth	S.E.	Gk
Thomas Square	West Ham	E.	Le
Thomas Street	Finsbury	W.C.	Gd
Thomas Street	Islington	N.	Fa
Thomas Street	Poplar	E.	Kd
Thomas Street	Stepney	E.	He
Thomas Street	Westminster	W.	Ee
Thomas Street	Woolwich		Mg
Thomas Street	Woolwich		Ng
Thompson Road	Camberwell	S.E.	Hj
Thorburn Square	Bermondsey	S.E.	Hg
Thornbury	Bromley		Lm
Thornbury	Islington	N.	Fa
Thornbury Road	Wandsworth	S.W.	Fj
Thornby Road	Hackney	E.	Hb
Thorncliff Road	Wandsworth	S.W.	Fj
Thorncombe Road	Camberwell	S.E.	Gj
Thorndean Street	Wandsworth	S.W.	Dk
Thorne Passage	Barnes U.D.	S.W.	Bh
Thorne Road	Lambeth	S.W.	Fh
Thorne Street	Barnes U.D.	S.W.	Bh
Thorne Street	Lambeth	S.W.	Fh
Thorne Wharf	Lambeth	S.E.	Fe
Thorneyhedge Road	Acton U.D., Brentford U.D. & Chiswick U.D	W.	Af
Thornfield Road	Hammersmith	W.	Bf
Thornford Road	Lewisham	S.E.	Kj
Thorngate Road	Paddington	W.	Dd
Thorngrove Road	West Ham	E.	Ld
Thornham Grove	West Ham	E.	Kc
Thornham Street	Greenwich	S.E.	Kg
Thornhaugh Mews	St. Pancras	W.C.	Fd
Thornhill Crescent	Islington	N.	Fc
Thornhill Grove	Islington	N.	Fc
Thornhill Road	Islington	N.	Gc
Thornhill Road	Leyton U.D.	E.	Kb
Thornhill Square	Camberwell	S.E.	Gh
Thornhill Square	Islington	N.	Fc
Thornlaw Road	Lambeth	S.E.	Gl
Thornley Place	Greenwich	S.E.	Kg
Thornsbeach Road	Lewisham	S.E.	Kk
Thornsett Road	Penge U.D.	S.E.	Hm
Thornsett Road	Wandsworth	S.W.	Dk
Thornton Avenue	Chiswick U.D.	W.	Bf
Thornton Avenue	Wandsworth	S.W.	Fk
Thornton Grove	Southwark	S.E.	Gf
Thornton Hill	Wimbledon	S.W.	Cm
Thornton House	Wandsworth	S.W.	Fk
Thornton Place	Marylebone	W.	Ee
Thornton Road	Barnes U.D.	S.W.	Ah
Thornton Road	Leyton U.D.	E.	Kb
Thornton Road	Wandsworth	S.W.	Fk
Thornton Road	Wimbledon	S.W.	Cm
Thornton Street	Lambeth	S.W.	Fh
Thornton Street	Southwark	S.E.	Gg
Thornville Street	Deptford	S.E.	Jh
Thornwood Road	Lewisham	S.E.	Kj
Thorold Street	Bethnal Green	E.	Hd
Thorparch Road	Battersea & Lambeth	S.W.	Fh
Thorpe Lodge	Kensington	W.	Df
Thorpe Mews	Kensington	W.	Ce
Thorpe Road	East Ham	E.	Md
Thorpe Road	Kingston		Sm
Thorpe Road	Leyton U.D.	E.	Lb
Thorpe Road	Tottenham U.D.	N.	Ha
Thorpebank Road	Hammersmith	W.	Be
Thorpedale Road	Islington	N.	Fa
Thorp's Wharf	Stepney	E.	Hf
Thorverton Road	Hendon U.D.	N.W.	Cb
Thrale Road	Wandsworth	S.W.	El
Thrawl Street	Stepney	E.	He
Threadneedle Street	City of London	E.C.	Ge
Three Colt Corner	Bethnal Green	E.	Hd

STREET OR PLACE.	BOROUGH.	P.D.	MAP.
Warwick Road	Walthamstow U.D.	E.	Ka
Warwick Road	West Ham	E.	Lc
Warwick Square	City of London	E.C.	Ge
Warwick Square	Westminster	S.W.	Eg
Warwick Square Mews	Westminster	S.W.	Eg
Warwick Street	Deptford	S.E.	Jh
Warwick Street	Westminster	S.W.	Eg
Warwick Street	Westminster	W.	Fe
Warwick Street	Woolwich		Mg
Warwick Terrace	Lewisham	S.E.	Hl
Warwick Terrace	Woolwich		Ng
Warwickshire Road	Stoke Newington	N.	Hb
Washington	East Ham	E.	Lc
Washington Road	Kingston		Am
Washington Street	Poplar	E.	Kd
Wastdale Road	Lewisham	S.E.	Jk
Waste Lands Almsho.	Fulham	S.W.	Ch
Waste Lands Almsho.	Hammersmith	W.	Bf
Water Lane	City of London	E.C.	Ge
Water Lane	City of London	E.C.	He
Water Lane	Kingston		Sm
Water Lane	Lambeth	S.W.	Fj
Water Lane	Richmond		Sj
Water Lane	West Ham	E.	Kc
Water Side	City of London	E.C.	Ge
Water Side	Greenwich		Mf
Water Side	Wandsworth	S.W.	Dj
Water Side	Woolwich		Lf
Water Street	City of London	E.C.	Ge
Water Street	Westminster	W.C.	Fe
Water Tower	Chislehurst U.D.		Mm
Waterdale Road	Woolwich		Og
Waterfall Cottages	Croydon R.D.	S.W.	Dl
Waterfall Cottages	Wandsworth	S.W.	Dj
Waterfall Road	Wandsworth	S.W.	Dj
Waterfall Road	Croydon R.D.	S.W.	Dl
Waterford Avenue	Croydon		Sm
Waterford Road	Fulham	S.W.	Dg
Watergate Street	Deptford and Greenwich	S.E.	Jg
Waterloo Bridge	Lambeth and Westminster	S.E.& W.C.	Fe
Waterloo Buildings	Bethnal Green	E.	Hd
Waterloo Cottages	Camberwell	S.E.	Gg
Waterloo Junction Sta.	Lambeth	S.E.	Ff
Waterloo Mews	Camberwell	S.E.	Gh
Waterloo Pier	Westminster	W.C.	Fe
Waterloo Place	Bethnal Green	E.	He
Waterloo Place	Richmond		Ag
Waterloo Place	Shoreditch	E.	Hd
Waterloo Place	Westminster	S.W.	Ff
Waterloo Road	Bethnal Green	E.	Hd
Waterloo Road	East Ham	E.	Lc
Waterloo Road	Lambeth and Southwark	S.E.	Ff Gf
Waterloo Road	Leyton U.D.	E.	Ja
Waterloo Square	Camberwell	S.E.	Gh
Waterloo Station	Lambeth	S.E.	Ff
Waterloo Street	Camberwell	S.E.	Gh
Waterloo Street	Finsbury	E.C.	Gd
Waterloo Street	Hammersmith	W.	Bg
Waterloo Street	Shoreditch	E.	Hc
Waterloo Street	Stepney	E.	Je
Waterloo Terrace	Bethnal Green	E.	He
Waterloo Terrace	Islington	N.	Gc
Waterlow Buildings	Bethnal Green	E.	Hd
Waterlow Park	St. Pancras	N.	Ea
Watermen's Asylum	Penge U.D.	S.E.	Hm
Waterworks Canal	Leyton U.D. and Walthamstow U.D.	E.	Ja
Waterworks River	West Ham	E.	Kc
Waterworks Road	Lambeth	S.W.	Fj
Watery Lane	Chislehurst U.D.		Om
Watery Wood	Finchley U.D.	N.	Da
Watford Road	West Ham	E.	Le
Watford Villas	Battersea	S.W.	Eh
Watling Street	City of London	E.C.	Ge
Watlington Grove	Lewisham	S.E.	Jl
Watling Street	Bexley U.D.		Oj
Watling Street	Hendon U.D.	N.W.	Ba
Watney Passage	Stepney	E.	He
Watney Street	Stepney	E.	He
Watson Place	Shoreditch	E.	Gd
Watson Street	Deptford	S.E.	Jh
Watson Street	Stoke Newington	N.	Gb
Watson Street	West Ham	E.	Ld
Watson's Avenue	Deptford	S.E.	Jh
Watson's Mews	Marylebone	W.	Ee
Watson's Wharf	Stepney	E.	Hf
Wattisfield Road	Hackney	E.	Jb
Watts Street	Stepney	E.	Hf
Watts Street Buildings	Stepney	E.	Hf
Wavel Mews	Hampstead	N.W.	Dc
Waveney Avenue	Camberwell	S.E.	Hj
Waverley Place	Marylebone	N.W.	Dd
Waverley Road	Hornsey	N.	Fa
Waverley Road	Paddington	W.	De
Waverley Road	Woolwich		Ng
Waverley Terrace	Paddington	W.	De
Waverton Street	Westminster	W.	Ee
Wavertree Road	Wandsworth	S.W.	Fk
Waxlow Road	Willesdon U.D.	N.W.	Ad
Waxwell Terrace	Lambeth	S.E.	Ff
Wayford Street	Battersea	S.W.	Eh
Wayland Avenue	Hackney	E.	Hc
Wayland Road	Battersea	S.W.	Dh
Waynfleet Street	Wandsworth	S.W.	Dk
Wealdstone Brook	Wembley U.D.		Sa
Weardale Road	Lewisham	S.E.	Kj
Wearside Road	Lewisham	S.E.	Kj
Weaver Street	Bethnal Green & Stepney	E.	He
Weaver's Alley	Shoreditch	E.C.	Hd
Weaver's Lane	Bermondsey	S.E.	Gf
Weaver's Almshouses	Shoreditch	E.C.	Gd
Weaver's Almshouses	Wanstead U.D.	E.	La
Weaver's Arms	Hackney	E.	Hb
Webb Street	Bermondsey	S.E.	Gf
Webb Street	West Ham	E.	Ld
Webber Row	Southwark	S.E.	Gf
Webber Street	Lambeth and Southwark	S.E.	Gf
Webbs Road	Battersea	S.W.	Ej
Webster Gardens	Ealing	W.	Se
Webster Road	Bermondsey	S.E.	Hf
Webster Road	Leyton U.D.	E.	Kb
Wedderburn Road	Hampstead	N.W.	Dc
Wedlake Street	Kensington and Paddington	W.	Cd
Wedmore Gardens	Islington	N.	Fb
Wedmore Mews	Islington	N.	Fb
Wedmore Street	Islington	N.	Fb
Weech Road	Hampstead	N.W.	Db
Weedington Road	St. Pancras	N.W.	Ec
Weetman Street	Greenwich	S.E.	Kg
Weigall Road	Greenwich and Woolwich	S.E.	Lj
Weighton Road	Penge U.D.	S.E.	Hm
Weimar Street	Wandsworth	S.W.	Ch
Weir Bank	Teddington U.D.		Sl
Weir Street	Wandsworth	S.W.	Em
Weirleigh Road	Wandsworth	S.W.	Dk
Weirs Passage	St. Pancras	N.W.	Fd
Weiss Road	Wandsworth	S.W.	Ch
Welbeck Mansions	Chelsea	S.W.	Eg
Welbeck Mansions	Marylebone	W.	Ee
Welbeck Street	Marylebone	W.	Ee
Welbury Street	Hackney	E.	Hc
Welby Street	Lambeth	S.E.	Gh
Welch's Yard	St. Pancras	N.W.	Fc
Welldale Road	Wandsworth	S.W.	Fm
Weldon Street	Bethnal Green	E.	Hd
Weldrick Road	Fulham	S.W.	Dh
Welham Road	Wandsworth	S.W.	El
Well Lane	Barnes U.D.	S.W.	Aj
Well Court	City of London	E.C.	He
Well Hall	Woolwich		Mj
Well Hall Road	Woolwich		Mj
Well Hall Station	Woolwich		Mj
Well Hall Villas	Woolwich		Mj
Well Road	Hampstead	N.W.	Db
Well Street	City of London	E.C.	Ge
Well Street	Hackney	E.	Hc
Well Street	Stepney	E.	He

STREET OR PLACE.	BOROUGH.	P.D.	MAP.
William Grove	Southwark	S.E.	Gg
William Road	Wimbledon	S.W.	Cm
William Street	Battersea	S.W.	Fh
William Street	Chelsea and Westminster	S.W.	Ef
William Street	Chiswick U.D.	W.	Bg
William Street	City of London	E.C.	Ge
William Street	Finsbury	E.C.	Fd
William Street	Fulham	W.	Cf
William Street	Islington	N.	Fd
William Street	Islington	N.	Gd
William Street	Kensington	W.	De
William Street	Lambeth	S.W.	Fg
William Street	Leyton U.D.	E.	Ka
William Street	Marylebone	N.W.	Ed
William Street	Marylebone	W.	Ee
William Street	St. Pancras	N.W	Ed
William Street	Stepney	E.	He
William Street	West Ham	E.	Kc
William Street	Woolwich		Mg
William the Fourth, The	Wandsworth	S.W.	Fm
William's Lane	Barnes U.D.	S.W.	Ah
William's Mews	Marylebone	W.	Ee
William's Place	Kensington	W.	Df
William's Place	Marylebone	N.W.	Ee
Williamson Street	Islington	N.	Fb
Willifield Way	Hendon U.D.	N.W.	Da
Willingham Terrace	St. Pancras	N.W	Fb
Willington Road	Lambeth	S.W.	Fh
Willis Road	West Ham	E.	Ld
Willis Street	Poplar	E.	Ke
Willis's Court	Wandsworth	S.W.	Fj
Willoughby Road	Hampstead	N.W	Db
Willoughby Road	Kingston		Sm
Willoughby Road	TwickenhamU.D.		Sj
Willow Buildings	Hampstead	N.W	Db
Willow Chambers	Shoreditch	E.C.	Ga
Willow Court	Shoreditch	N.	Ga
Willow Court	Westminster	S.W.	Ff
Willow Grove	Bermondsey	S.E.	Hf
Willow Grove	Chislehurst U.D.		Mm
Willow Grove	West Ham	E.	Ld
Willow Place	Westminster	S.W.	Ff
Willow Road	Hampstead	N.W	Db
Willow Row	Stepney	E.	Je
Willow Street	Bermondsey	S.E.	Hf
Willow Street	Shoreditch	N.	Ga
Willow Street	Stepney	E.	Je
Willow Street	Westminster	S.W.	Ff
Willow Vale	Hammersmith	W.	Bf
Willow Walk	Bermondsey	S.E.	Hg
Willow Walk	Lewisham	S.E.	Hl
Willow Walk	Lewisham	S.E.	Jk
Willow Walk	St. Pancras	N.W	Fb
Willow Walk	Walthamstow U.D.	E.	Ja
Willowbridge Road	Islington	N.	Gc
Willowbrook Grove	Camberwell	S.E.	Hg
Willowbrook Road	Camberwell	S.E.	Hg
Wilman Grove	Hackney	E.	Hc
Wilmer Gardens	Shoreditch	N.	Gd
Wilmer House	Ham U.D.		Sl
Wilmington Avenue	Chiswick U.D.	W.	Ag
Wilmington Place	Finsbury	W.C.	Fd
Wilmington Square	Finsbury	W.C.	Fd
Wilmington Street	Finsbury	W.C.	Fd
Wilmore Place	Lewisham	S.E.	Kh
Wilmot Mews	Bethnal Green	E.	Hd
Wilmot Place	St. Pancras	N.W.	Fc
Wilmot Road	Leyton U.D.	E.	Kb
Wilmot Street	Bethnal Green	E.	Hd
Wilmount Street	Woolwich		Mg
Wilna Road	Wandsworth	S.W.	Dk
Wilson Place	Stepney	E.	Je
Wilson Road	Camberwell	S.E.	Gh
Wilson Road	East Ham	E.	Ld
Wilson Street	Battersea	S.W.	Dj
Wilson Street	Deptford	S.E.	Jh
Wilson Street	Finsbury and Shoreditch	E.C.	Ge
Wilson Street	Hammersmith	W.	Cg
Wilson Street	Poplar	E.	He
Wilson Street	St. Pancras	E.	Fd
Wilson Street	Stepney	W.C.	Je
Wilson Street	Walthamstow U.D.	E.	Ka
Wilson Street	West Ham	E.	Ld
Wilson Street	Westminster	W.C.	Fe
Wilson's Place	St. Pancras	W.C.	Fd
Wilson's Wharf	Bermondsey	S.E.	Gf
Wilstead Court	St. Pancras	N.W.	Fd
Wilton Avenue	Chiswick U.D.	W.	Bg
Wilton Crescent	Merton U.D. and Wimbledon	S.W.	Cm
Wilton Crescent	Westminster	S.W.	Ef
Wilton Crescent Mews	Westminster	S.W.	Ef
Wilton Mews	Westminster	S.W.	Ef
Wilton Place	Westminster	S.W.	Ef
Wilton Road	Croydon R.D.	S.W.	Dm
Wilton Road	Hackney	E.	Hc
Wilton Road	Merton U.D. and Wimbledon	S.W.	Cm
Wilton Road	Westminster	S.W.	Ef
Wilton Road West	Hammersmith	W.	Bf
Wilton Square	Islington	N.	Gc
Wilton Street	Islington	N.	Gc
Wilton Street	West Ham	E.	Mf
Wilton Street	Westminster	S.W.	Ef
Wilton Terrace	Westminster	S.W.	Ef
Wiltshire Road	Croydon		Fm
Wiltshire Road	Lambeth	S.W.	Gh
Wiltshire Row	Shoreditch	N.	Gd
Wimbart Road	Lambeth	S.W.	Fk
Wimbledon Cemetery	Wimbledon	S.W.	Dl
Wimbledon College	Wimbledon	S.W.	Cm
Wimbledon Common	Wandsworth and Wimbledon	S.W.	Bl
Wimbledon Hall Road	Wimbledon	S.W.	Cl
Wimbledon Lodge	Wimbledon	S.W.	Cl
Wimbledon Pk. House	Wimbledon	S.W.	Cl
Wimbledon Park Road	Wandsworth and Wimbledon	S.W.	Ck
Wimbledon Park Side	Wandsworth	S.W.	Bk
Wimbledon Park Sta.	**Wimbledon.**	S.W.	Cl
Wimbledon Road	Wandsworth	S.W.	Dl
Wimbledon Sewage Works	Wimbledon	S.W.	Dl
Wimbledon Station	**Wimbledon.**	S.W.	Cl
Wimbledon Wood	Wimbledon	S.W.	Bm
Wimbolt Street	Bethnal Green	E.	Hd
Wimborne Gardens	Ealing	W.	Se
Wimbourn Mews	Shoreditch	N.	Gd
Wimbourn Street	Shoreditch	N.	Gd
Wimpole Mews	Marylebone	W.	Ee
Wimpole Street	Marylebone	W.	Ee
Wincanton Road	Wandsworth	S.W.	Ck
Wincham Bridge	Bexley U.D.		Nk
Winchendon Road	Fulham	S.W.	Ch
Winchelsea Road	Leyton U.D.	E.	Lb
Winchelsea Road	Willesden U.D.	N.W.	Bc
Winchester Avenue	City of London	E.C.	Ge
Winchester Avenue	Willesden U.D.	N.W.	Cc
Winchester Crescent	Bethnal Green	E.	Hd
Winchester Mews	Hampstead	N.W	Dc
Winchester Place	Bethnal Green	E.	Hd
Winchester Place	Hornsey	N.	Ea
Winchester Road	Hampstead	N.W.	Dc
Winchester Street	Acton U.D.	W.	Af
Winchester Street	Bethnal Green	E.	Hd
Winchester Street	Finsbury and Islington	N.	Fd
Winchester Street	Southwark	S.E.	Gf
Winchester Street	West Ham	E.	Mf
Winchester Street	Westminster	S.W.	Eg
Winchester Terrace	Wandsworth	S.W.	Ch
Winchester Terrace	Westminster	S.W.	Fg
Winchester Wharf	Southwark	S.E.	Gf
Winckfield Road	Lewisham	S.E.	Jl
Wincott Street	Lambeth	S.E.	Gg
Windermere Avenue	Willesden U.D.	N.W.	Cc
Windermere Road	Islington	N.	Fb
Winders Road	Battersea	S.W.	Dh
Windham Road	Kingston		Sm
Windmill Court	City of London	E.C.	Ge
Windmill Court	Lambeth	S.E.	Gf

STREET OR PLACE.	BOROUGH.	P.D.	MAP.
Woodford Road	Wanstead U.D. & West Ham	E.	Lb
Woodford Road	Wanstead U.D.	E.	La
Woodgate Street	Battersea	S.W.	Fg
Woodgrange Hall	West Ham	E.	Lc
Woodgrange Road	West Ham	E.	Lc
Woodhays	Wimbledon	S.W.	Cm
Woodhays Road	Wimbledon	S.W.	Cm
Woodhouse Road	Leyton U.D.	E.	Lb
Woodhurst Road	Acton U.D.	W.	Ae
Woodhurst Road	Woolwich		Og
Woodison Street	Stepney	E.	Jd
Woodland Gardens	Hornsey	N.	Ea
Woodland Grove	Greenwich	S.E.	Kg
Woodland Hill	Lambeth	S.E.	Gl
Woodland House	Lewisham	S.E.	Kk
Woodland Park Road	Greenwich	S.E.	Kg
Woodland Place	Greenwich	S.E.	Kg
Woodland Rise	Hornsey	N.	Ea
Woodland Road	Camberwell and Lambeth	S.E.	Gl
Woodland Road	Stoke Newington	N.	Gb
Woodland Street	Greenwich	S.E.	Kg
Woodland Street	Hackney	E.	Hc
Woodland Terrace	Greenwich		Mg
Woodlands	Bexley U.D.		Oj
Woodlands	Bromley		Mm
Woodlands	Croydon	S.E.	Gm
Woodlands	Croydon R.D.		Em
Woodlands	Greenwich	S.E.	Lg
Woodlands	Hornsey	N.	Fa
Woodlands	Lewisham		Jl
Woodlands	The Maldens and Coombe U.D.		Al
Woodlands	Walthamstow U.D.	E.	Ka
Woodlands	Wandsworth	S.W.	Fl
Woodlands Park Road	Tottenham U.D.	N.	Ga
Woodlands Place	Bermondsey	S.E.	Hf
Woodlands Road	Barnes U.D.	S.W.	Bl
Woodlands Road	Bromley&Chislehurst U.D.		Mm
Woodlands Road	Greenwich	S.E.	Lh
Woodlands Road	Leyton U.D.	E.	Kb
Woodlands Road	Wandsworth	S.W.	Cj
Wooknook Road	Wandsworth	S.W.	El
Woodlands Street	Lewisham	S.E.	Kk
Woodlands Terrace	Camberwell	S.E.	Hk
Woodlawn Road	Fulham	S.W.	Cg
Woodlea Road	Stoke Newington	N.	Gb
Woodman, The	Bexley		Oj
Woodman, The	Wimbledon	S.W.	Dl
Woodman, The	Woolwich		Ng
Woodmans Place	Southwark	S.E.	Gf
Woodneys Road	Willesden U.D.	N.W.	Bb
Woodnook Road	Wandsworth	S.W.	El
Woodpecker Road	Deptford	S.E.	Jg
Woodquest Avenue	Lambeth	S.E.	Gj
Woodrift Road	Leyton U.D.	E.	Ka
Wood's Mews	Westminster	W.	Ee
Wood's Place	Westminster	S.W.	Ff
Wood's Road	Camberwell	S.E.	Hh
Woodside	Ealing	W.	Se
Woodside	Kingston		Sm
Woodside	Wimbledon	S.W.	Cl
Woodside Lane	Bexley U.D.		Ok
Woodside Nursery	Woolwich		Nk
Woodside Road	Kingston		Sm
Woodsome Road	St. Pancras	N.W.	Eb
Woodstock House	Hendon U.D.	N.W.	Ca
Woodstock Road	Acton U.D. and Chiswick	W.	Bf
Woodstock Road	East Ham	E.	Lc
Woodstock Road	Hammersmith	W.	Cf
Woodstock Road	Hornsey	N.	Ga
Woodstock Road	Lewisham	S.E.	Lk
Woodstock Road	Poplar	E.	Ke
Woodstock Street	West Ham	E.	Ke
Woodstock Street	Westminster	W.	Ee
Woodthorpe	Lewisham	S.E.	Hk
Woodthorpe Road	Wandsworth	S.W.	Bj
Woodvale	Camberwell and Lewisham	S.E.	Hk
Woodvale Avenue	Croydon	S.E.	Hm
Woodville Grove	Islington	N.	Gc
Woodville Place	Bexley U.D.		Nh
Woodville Road	Croydon		Sm
Woodville Road	Ealing	W.	Se
Woodville Road	Greenwich	S.E.	Lh
Woodville Road	Hammersmith	W.	Cf
Woodville Road	Islington	N.	Gc
Woodville Road	Leyton U.D.	E.	La
Woodville Road	Tottenham U.D.	N.	Ga
Woodville Road	Walthamstow U.D.	E.	Ja
Woodville Road	Willesden U.D.	N.W.	Cd
Woodville Street	Woolwich		Mg
Woodwarde Road	Camberwell	S.E.	Gj
Woodwell Road	Wandsworth	S.W.	Dj
Wooler Street	Southwark	S.E.	Gg
Woolf Street	Bermondsey	S.E.	Hf
Woolgar Mews	Hackney	N.	Hc
Woollaston Road	Hornsey	N.	Ga
Woollett Street	Poplar	E.	Ke
Woolmore Street	Poplar	E.	Ke
Woolneigh Street	Fulham	S.W.	Dh
Woolpack Place	Hackney	E.	Jc
Wool Quays	City of London	E.C.	Ge
Woolsey Street	Stepney	E.	He
Woolstone Road	Lewisham	S.E.	Jk
Woolwich Cemetery	Bexley U.D. and Woolwich		Nh
Woolwich Common	Greenwich and Woolwich		Mh
Woolwich Dockyard Station	**Woolwich**		Mg
Woolwich Pier	Woolwich		Mf
Woolwich Reach	Woolwich		Mf
Woolwich Road	Erith U.D., Greenwich and Woolwich		Og Lg Mg
Woolwich Theatre	**Woolwich**	S.E.	Mg
Woolwich UnionInfiry.	Woolwich		Ng
Wootton Place	Lambeth	S.E.	Gf
Wootton Street	Lambeth	S.E.	Gf
Worbeck Road	Penge U.D.	S.E.	Hm
Worcester Court	Stepney	E.	Hf
Worcester Road	Wimbledon	S.W.	Cl
Worcester Street	Southwark	S.E.	Gf
Worcester Street	Stepney	E.	Hf
Worcester Street	Westminster	S.W.	Fg
Wordsworth Path	Stoke Newington	N.	Hb
Wordsworth Road	Penge U.D.	S.E.	Hm
Wordsworth Road	Stoke Newington	N.	Hb
Worfield Street	Battersea	S.W.	Eh
Worgate Street	Shoreditch	N.	Gd
Workhouse Lane	Wimbledon	S.W.	Bl
Worland Road	West Ham	E.	Kc
Worlidge Street	Hammersmith	W.	Cg
Worlington Road	Camberwell	S.E.	Hj
Wormholt Road	Hammersmith	W.	Be
Wormwood Scrubs	Hammersmith	N.W.	Bd
Wormwood Scrubs Prison	Hammersmith	W.	Be
Wormwood Street	City of London	E.C.	Ge
Wornington Mews	Kensington	W.	Cd
Wornington Road	Kensington	W.	Cd
Woronzow Road	Hampstead and Marylebone	N.W.	Dc
Woronzow Terrace	Marylebone	N.W.	Dc
Worple Lane	Wimbledon	S.W.	Cm
Worple Road	Wimbledon	S.W.	Cm
Worple Street	Barnes U.D.	S.W.	Ah
Worple Way	Richmond		Sj
Worple Way North	Barnes U.D.	S.W.	Ah
Worple Way South	Barnes U.D.	S.W.	Ah
Worship Court	Shoreditch	E.C.	Ge
Worship Street	Finsbury and Shoreditch	E.C.	Ge
Worslade Road	Wandsworth	S.W.	Dl
Worsley Bridge Road	Beckenham U.D. and Lewisham	S.E.	Jl
Worsley Road	Hampstead	N.W.	Db
Worsley Road	Leyton U.D.	E.	Kb
Worth Grove	Southwark	S.E.	Gg
Wotton Road	Deptford	S.E.	Jg

FIRE STATIONS (L.C.C.)

Ingram Content Group UK Ltd.
Milton Keynes UK
UKHW022123060323
418148UK00005B/223